T0281501

Lecture Notes in Computer Science 14148

Founding Editors

Gerhard Goos
Juris Hartmanis

Editorial Board Members

Elisa Bertino, *Purdue University, West Lafayette, IN, USA*
Wen Gao, *Peking University, Beijing, China*
Bernhard Steffen ⓘ, *TU Dortmund University, Dortmund, Germany*
Moti Yung ⓘ, *Columbia University, New York, NY, USA*

The series Lecture Notes in Computer Science (LNCS), including its subseries Lecture Notes in Artificial Intelligence (LNAI) and Lecture Notes in Bioinformatics (LNBI), has established itself as a medium for the publication of new developments in computer science and information technology research, teaching, and education.

LNCS enjoys close cooperation with the computer science R & D community, the series counts many renowned academics among its volume editors and paper authors, and collaborates with prestigious societies. Its mission is to serve this international community by providing an invaluable service, mainly focused on the publication of conference and workshop proceedings and postproceedings. LNCS commenced publication in 1973.

Robert Wrembel · Johann Gamper ·
Gabriele Kotsis · A Min Tjoa · Ismail Khalil
Editors

Big Data Analytics and Knowledge Discovery

25th International Conference, DaWaK 2023
Penang, Malaysia, August 28–30, 2023
Proceedings

 Springer

Editors
Robert Wrembel 🆔
Poznań University of Technology
Poznan, Poland

Johann Gamper 🆔
Free University of Bozen-Bolzano
Bozen-Bolzano, Italy

Gabriele Kotsis
Johannes Kepler University Linz
Linz, Austria

A Min Tjoa 🆔
Vienna University of Technology
Vienna, Austria

Ismail Khalil
Johannes Kepler University Linz
Linz, Austria

ISSN 0302-9743   ISSN 1611-3349 (electronic)
Lecture Notes in Computer Science
ISBN 978-3-031-39830-8   ISBN 978-3-031-39831-5 (eBook)
https://doi.org/10.1007/978-3-031-39831-5

This Springer imprint is published by the registered company Springer Nature Switzerland AG
The registered company address is: Gewerbestrasse 11, 6330 Cham, Switzerland

Preface

It is with great pleasure that we introduce the proceedings of the 25th International Conference on Big Data Analytics and Knowledge Discovery (DAWAK 2023). This conference brought together leading experts, researchers, and practitioners from around the world to exchange ideas, share insights, and push the boundaries of knowledge in the fields of big data, data analytics, artificial intelligence, and machine learning. The compilation of papers presented in this volume represents a diverse range of topics and showcases the cutting-edge research that was discussed during the conference.

In today's data-driven world, the proliferation of vast and complex datasets has necessitated the development of advanced techniques and methods to extract meaningful insights and to discover valuable knowledge. The advent of big data analytics has transformed the way we process, analyze, and interpret data, empowering organizations and individuals to make informed decisions and gain a competitive edge. Coupled with the power of artificial intelligence and machine learning, these technologies have opened up new frontiers in data exploration, predictive modeling, and automated decision-making.

DAWAK 2023 served as a hub for interdisciplinary collaboration, fostering the exchange of ideas between researchers, practitioners, and industry experts. The papers included in this volume represent the culmination of countless hours of dedicated work, pushing the boundaries of what is possible in the realms of data analytics, artificial intelligence, and machine learning.

The breadth of topics covered in this collection is truly remarkable, encompassing areas such as data mining, pattern recognition, natural language processing, recommendation systems, deep learning, and reinforcement learning, among others. Each paper has undergone an average of 3 single-blind reviews in a rigorous peer-review process, ensuring the highest quality and relevance to the conference's theme.

We are proud to report authors from more than 22 different countries submitted papers to DAWAK this year. Our program committees have conducted close to 300 reviews. From 83 submitted papers the program committee decided to accept 18 full papers with an acceptance rate of 20%, a rate lower than previous DAWAK conferences.

This year again the top DAWAK 2023 papers will be invited to a special issue of the Data & Knowledge Engineering journal. Here we express our gratitude to Carson Woo, the Editor in-Chief of the journal, for hosting these DAWAK 2023 papers in the special issue.

We would like to extend our gratitude to all the authors who have contributed their valuable research to this volume. Their passion, expertise, and dedication to their respective fields have enriched the scientific community and paved the way for groundbreaking discoveries.

We also want to express our appreciation to the program committee members, whose expertise and thorough evaluation played an instrumental role in selecting and refining the papers for inclusion.

Furthermore, we would like to extend our thanks to the conference organizers, keynote speakers, and attendees for their unwavering support and enthusiasm. Their commitment to advancing knowledge and fostering collaboration has been instrumental in creating an environment where innovation flourishes and ideas thrive.

As you peruse the pages of these proceedings, we hope that you find inspiration in the remarkable research and advancements showcased within. Whether you are a seasoned researcher, a student, or an industry professional, we believe that the insights shared here will not only broaden your understanding but also inspire you to embark on your own journey of discovery.

Finally, we would like to express our heartfelt appreciation to the entire conference community. It is your collective efforts and unwavering dedication that have made this event a resounding success, enabling us to push the boundaries of big data analytics, knowledge discovery, artificial intelligence, and machine learning.

August 2023

Robert Wrembel
Johann Gamper
Gabriele Kotsis
A Min Tjoa
Ismail Khalil

Organization

Program Committee Chairs

Robert Wrembel   Poznań University of Technology, Poland
Johann Gamper   Free University of Bozen-Bolzano, Italy

Steering Committee

Gabriele Kotsis   Johannes Kepler University Linz, Austria
A Min Tjoa   Vienna University of Technology, Austria
Robert Wille   Software Competence Center Hagenberg, Austria
Bernhard Moser   Software Competence Center Hagenberg, Austria
Ismail Khalil   Johannes Kepler University Linz, Austria

Program Committee Members

Alberto Abello   Universitat Politècnica de Catalunya, Spain
Cristina Aguiar   Universidade de São Paulo, Brazil
Lars Ailo Bongo   University of Tromsø, Norway
Syed Muhammad Fawad Ali   Accenture, Germany
Witold Andrzejewski   Poznań University of Technology, Poland
Faten Atigui   CNAM, France
Sylvio Barbon Junior   University of Trieste, Italy
Ladjel Bellatreche   LIAS/ENSMA, France
Soumia Benkrid   Ecole Nationale Supérieure d'Informatique, Algeria
Fadila Bentayeb   Université Lumiére Lyon 2, France
Jorge Bernardino   Polytechnic Institute of Coimbra, Portugal
Vasudha Bhatnagar   University of Delhi, India
Besim Bilalli   Universitat Politècnica de Catalunya, Spain
Sandro Bimonte   INRAE, France
Pawel Boinski   Poznań University of Technology, Poland
Kamel Boukhalfa   USTHB, Algeria
Omar Boussaid   Université Lumiére Lyon 2, France
Stephane Bressan   National University of Singapore, Singapore
Tiziana Catarci   Sapienza University of Rome, Italy

Sharma Chakravarthy	University of Texas at Arlington, USA
Silvia Chiusano	Politecnico di Torino, Italy
Frans Coenen	University of Liverpool, UK
Philippe Cudré-Mauroux	Universität Freiburg, Switzerland
Laurent d'Orazio	Université de Rennes 1, CNRS IRISA, France
Jérôme Darmont	Université Lumiére Lyon 2, France
Soumyava Das	Teradata Lab, USA
Karen Davis	Miami University, USA
Anton Dignös	Free University of Bozen-Bolzano, Italy
Christos Doulkeridis	University of Piraeus, Greece
Markus Endres	University of Passau, Germany
Leonidas Fegaras	University of Texas at Arlington, USA
Matteo Francia	University of Bologna, Italy
Filippo Furfaro	University of Calabria, Italy
Pedro Furtado	University of Coimbra, Portugal
Luca Gagliardelli	University of Modena and Reggio Emilia, Italy
Enrico Gallinucci	University of Bologna, Italy
Kazuo Goda	University of Tokyo, Japan
Matteo Golfarelli	University of Bologna, Italy
Marcin Gorawski	Silesian University of Technology, Poland
Sven Groppe	University of Lübeck, Germany
Frank Höppner	Ostfalia University of Applied Sciences, Germany
Mirjana Ivanovic	University of Novi Sad, Serbia
Stéphane Jean	LIAS/ISAE-ENSMA and University of Poitiers, France
Selma Khouri	Ecole Nationale Supérieure d'Informatique, Algeria
Kyoung-Sook Kim	National Institute of Advanced Industrial Science and Technology (AIST), Japan
Young-Kuk Kim	Chungnam National University, South Korea
Nicolas Labroche	University of Tours, France
Jens Lechtenbörger	University of Münster, Germany
Young-Koo Lee	Kyung Hee University, South Korea
Carson Leung	University of Manitoba, Canada
Sebastian Link	University of Auckland, New Zealand
Woong-Kee Loh	Gachon University, South Korea
Sofian Maabout	LaBRI/University of Bordeaux, France
Patrick Marcel	Université de Tours, France
Alex Mircoli	Polytechnic University of Marche, Italy
Jun Miyazaki	Tokyo Institute of Technology, Japan
Rim Moussa	ENICarthage, Tunisia
Sergi Nadal	Universitat Politècnica de Catalunya, Spain

Kjetil Nørvåg   Norwegian University of Science and Technology,
                                  Norway
Boris Novikov   National Research University Higher School of
                                  Economics, Russia
Hiroaki Ohshima   University of Hyogo, Japan
Makoto Onizuka   Osaka University, Japan
Carlos Ordonez   University of Houston, USA
Veronika Peralta   University of Tours, France
Franck Ravat   IRIT/Université Toulouse 1, France
Ilya Safro   University of Delaware, USA
Mahmoud Sakr   Université Libre de Bruxelles, Belgium
Abhishek Santra   University of Texas at Arlington, USA
Sana Sellami   Aix-Marseille University, France
Alkis Simitsis   HP Labs, USA
Darja Solodovnikova   University of Latvia, Latvia
Nadine Steinmetz   TU Ilmenau, Germany
Emanuele Storti   Università Politecnica delle Marche, Italy
Olivier Teste   IRIT, France
Dimitri Theodoratos   New Jersey Institute of Technology, USA
Maik Thiele   Technische Universität Dresden, Germany
Chrisa Tsinaraki   Technical University of Crete, Greece
Alejandro Vaisman   Instituto Tecnológico de Buenos Aires, Argentina
Luca Zecchini   University of Modena and Reggio Emilia, Italy
Yongjun Zhu   Yonsei University, South Korea

External Reviewers

Simone Agostinelli   Sapienza University of Rome, Italy
Guessoum Meriem Amel   University of Science and Technology Houari
                                  Boumediene, Algeria
Bharti Bharti   University of Delhi, India
Tobias Bleifuss   Hasso-Plattner Institute, Germany
Abdesslam Chalallah   UMR Herbivores, INRAE, France
Renukswamy Chikkamath   Munich University of Applied Sciences, Germany
Gianluca Cima   Sapienza University of Rome, Italy
Daniele Cono D'Elia   Sapienza University of Rome, Italy
Deghmani Faiza   University of Science and Technology Houari
                                  Boumediene, Algeria
Chiara Forresi   University of Bologna, Italy
Nguyen Hoang Hai   University of Manitoba, Canada
Vikas Kumar   University of Delhi, India

Maryam Mozaffari	Free University of Bozen-Bolzano, Italy
Toshpulatov Mukhiddin	Inha University, South Korea
Christian Napoli	Sapienza University of Rome, Italy
Anifat Olawayin	University of Manitoba, Canada
Adam Pazdor	University of Manitoba, Canada
Kassem Sabeh	Free University of Bozen-Bolzano, Italy
Ionel Eduard Stan	Free University of Bozen-Bolzano, Italy
Tung Son Tran	Munich University of Applied Sciences, Germany

Organizers

From an Interpretable Predictive Model to a Model Agnostic Explanation (Abstract of Keynote Talk)

Osmar R. Zaïane

Amii Fellow and Canada CIFAR AI Chair, University of Alberta, Canada

Abstract. Today, the limelight is on Deep Learning. With the massive success of deep learning, other machine learning paradigms have had to take a backseat. Yet other models, particularly rule-based learning methods, are more readable and explainable and can even be competitive when labelled data is not abundant, and therefore could be more suitable for some applications where transparency is a must. One such rule-based method is the less-known Associative Classifier. The power of associative classifiers is to determine patterns from the data and perform classification based on the features most indicative of prediction. Early approaches suffer from cumbersome thresholds requiring prior knowledge. We present a new associative classifier approach that is even more accurate while generating a smaller model. It can also be used in an explainable AI pipeline to explain inferences from other classifiers, irrespective of the predictive model used inside the black box.

Contents

Machine Learning

Deep Learning

Data Management

Data Quality

Using Ontologies as Context for Data Warehouse Quality Assessment

Camila Sanz[(✉)] [iD] and Adriana Marotta[iD]

Facultad de Ingeniería, Universidad de la República, Montevideo, Uruguay
{csanz,amarotta}@fing.edu.uy

Abstract. In Data Warehouse systems, Data Quality has an important role since quality of data is compromised throughout the Data Warehouse lifecycle. Data Quality is context-dependent and this fact should be considered in its management. This work is a step forward to a general mechanism for assessing Data Quality in Data Warehouse considering data context. In addition to presenting our general approach, in this paper we propose two particular data quality rules for accuracy dimension, using OWL domain ontologies as context. With our approach, we obtain data quality metrics that can be adapted to any application domain.

Keywords: Data Warehouse · Data Quality · Context

1 Introduction

Data Warehouses (DW) are a fundamental asset for decision making. DWs are populated with data from heterogeneous sources which are transformed to be analyzed with a multidimensional perspective, allowing aggregations by different criteria. In these systems Data Quality (DQ) is an unavoidable issue since it is compromised throughout the DW lifecycle [6,12]. DQ management, allows DQ improvement when it is possible, and also DQ awareness by the user.

DQ in DW systems not only involves DQ management over single data attributes but also over hierarchies and multidimensional operations results. For example, errors in data of a particular dimension of the DW may cause wrong aggregations results in the multidimensional analysis. Such kind of errors are shown in the example of Fig. 1. The example refers to a Sales DW, which consists of a fact table *Sales*, related to dimension tables with data about books, authors, dates, etc. Figure 1 shows *Book Dimension* conceptual representation and a possible dimension instance. There are two DQ problems in this example: rectangle 1 shows a problem over data values, as the name "A Game of Throns" is incorrect and should say "A Game of Thrones", which is known as *Syntactic Accuracy* DQ dimension; and rectangle 2 shows a problem over hierarchies, as the genre of the book "The Witch, the Lion and the Wardrobe" is not "Thriller" in the real world, which is known as *Semantic Accuracy* DQ dimension.

Our approach to DQ management is strongly based on context. The well-known *fitness for use* concept has been widely adopted, accepting that DQ cannot be evaluated nor improved ignoring the context of data [15]. There had

R. Wrembel et al. (Eds.): DaWaK 2023, LNCS 14148, pp. 3–17, 2023.
https://doi.org/10.1007/978-3-031-39831-5_1

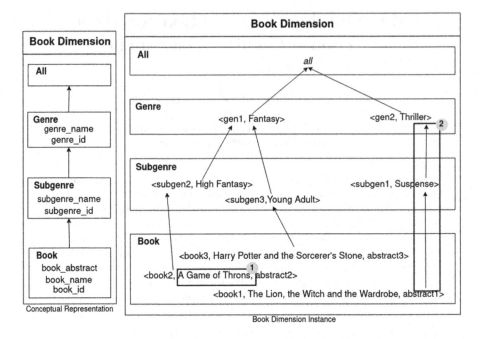

Fig. 1. Data Quality Problems.

been certain progress in research about context-oriented DQ for DW, however, we believe that there is still a gap for arriving to well-formalized integral and robust solutions. There are few works that propose formalizations for these concepts, and they do not address DQ as an integral discipline, considering DQ dimensions and metrics and focusing on DQ evaluation and improvement [12]. As DQ is context dependent, DQ dimensions and metrics are specific for each domain and use case, therefore, in general, existing solutions are dependent on each particular case, as for example in [2,6], where DQ metrics for specific use cases are proposed. It is necessary to provide an abstraction level that gives generality to the DQ metrics, allowing the instantiation of them for each particular case.

In this work, we focus on DQ dimension *accuracy*, the degree up to which data reflects correctly the real world [1], and the context as the application domain.

The main goal of our work is to propose a general mechanism for assessing DQ in DW systems, considering context information [11]. The chosen approach is to represent context through domain ontologies and implement DQ metrics through DQ rules.

Our main contributions are: the specification of a mechanism for assessing DQ in DW, based on DW-ontology mappings, and the proposal of two DQ rules for assessing accuracy that are supported by the proposed mechanism.

The rest of this document is organized as follows. Section 2 presents the related work. Section 3 introduces some previous formalizations: the DW and the context. In Sect. 4 we introduce mappings between the DW and the context. Section 5 contains the definition for our DQ rules and Sect. 6 experimentation with a use case. Finally in Sect. 7 we conclude.

2   Related Work

DQ is modeled through quality dimensions [1], such as consistency, accuracy and completeness. DQ metrics are defined in order to measure DQ dimensions. In literature, DQ is often described as "fitness for use" [1,14], which means DQ is context dependent [3,7,15], considering aspects as the application domain or task at hand [13]. There are some works that study context dependant DQ, such as [17], where context is explicitly defined as an ontology that allows detecting concepts that are syntactically different but equivalent in the real world.

Existing work about contexts, DW and DQ in general, is analyzed in [12], where an exhaustive literature review is presented. The authors show that, although there are many works that consider the context for DQ and for DW systems, there are few that address the problem of managing DQ in DW considering the context. They remark some works that present an approach for managing DQ in DW, implicitly considering context [8,16].

Authors in [2,6] study the importance of data quality considering context. Both works have similarities with our approach. In [2] the authors propose DQ metrics for a relational database, using a multidimensional database as part of the context. On the other hand in [6] they evaluate DQ over a multidimensional database considering context. Quality evaluation is specified using logical rules. In both works, the multidimensional model is represented using Hurtado-Meldezon's proposal [5]. The main difference between [2,6] and our work is that their quality rules are specific for a particular example, whereas our proposal gives general DQ rules that can be instantiated in any use case.

3   Preliminaries

This section presents a running example that will be used in the rest of this work, as well as the formalization of DW and context, which are needed for the specification of our proposal.

3.1   Running Example

We consider a bookshop DW that manages Sales information and we choose an existing domain ontology as context for the DW data.

Bookshop Data Warehouse. Information about books, authors, the sale's date, its price range, the store where it took place, its location and the amount of units sold is stored. The dimensions involved are *Author, Price, Geographic Location, Time, Store* and *Book*. Due to space restrictions we are only focusing on *Book Dimension*, which contains all relevant information about books. Its hierarchy goes as `book → subgenre → genre → All`. In Fig. 1 the conceptual representation and a possible instance of *Book Dimension* are shown.

Context. In order to represent the context of the *Book Dimension*, we choose an existing ontology that models the book domain, with the inclusion of a data property to meet our necessities. The Book Vocabulary Language[1] models information about books; authors, publication date, abstract, number of pages, and genre among others. The Book Vocabulary Language is shown in Fig. 2. For our example, we created an ontology over the Book Vocabulary Language. A partial instance with information about books and genres in Fig. 3.

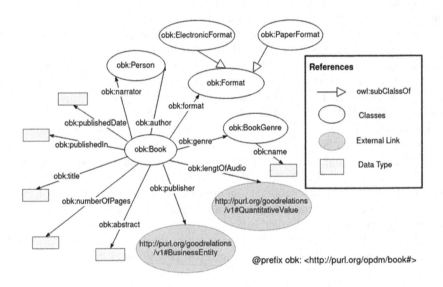

Fig. 2. Book ontology, Book Vocabulary Language.

Other ontologies could have been used, it is up to the domain expert to choose the ones that fit better to the particular necessities. In addition, if no existing ontology meets the needs of the domain, one could decide to create its own ontology to model context.

3.2 Data Warehouse Formal Specification

The specification of the DW is based on Hurtado-Mendelzon's work [5]. Some minor adaptations were made to meet our work necessities. The most important one is the use of tuples instead of atomic elements as members of categories.

---

[1] http://www.ebusiness-unibw.org/ontologies/opdm/book.html.

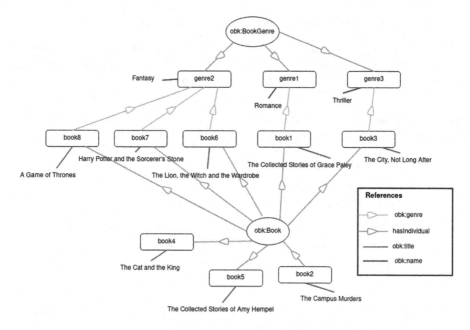

Fig. 3. Ontology instance.

To formally specify a DW it is necessary to define three main aspects: dimensions, multidimensional space, and cubes. Due to space restrictions, we only present the notation for dimensions which is needed to introduce DQ rules presented in Sect. 5. Other definitions and details are available in [2,5].

A *dimension schema*, which provides the structure that allows the definition of dimension instances, is a tuple $S = \langle C, \nearrow \rangle$, where:

- $C = \{c_1, \ldots c_n\}$ is a finite set of categories.
- $c_j = \, < att_{j_1}, \ldots att_{j_a} >$ is a list of attributes, where an attribute is a particular characteristic of the category.
- $\nearrow$: $C \times C$ is a binary relation between elements of C, which represents the hierarchies of dimension S, and $\nearrow^*$ its transitive closure.
- A distinguished element *All* meets the property $(\forall c \in C)(c \nearrow^* All)$.

A *dimension instance* for S, is a tuple $D = \langle U, <, mem \rangle$, where:

- $U = \{t_1, \ldots, t_m\}$ is a non empty finite set of attribute tuples called *members*.
- $t_j = \, < val_1, \ldots, val_p >$, is a tuple of U where val_i is the value for attribute i and p is the number of attributes of the corresponding category of t_j.
- $mem : U \rightarrow C$ is a total function that for each value of U returns the category to which it belongs.
- $<$: $U \times U$ is a binary relation between elements of U, representing hierarchies instances of S. Two elements of U can be related through $<$ if their corresponding categories are related through $\nearrow$. $<^*$ is its transitive closure.

– There is a unique member *all* such that $mem(all) = All$.

For this work, it is not necessary for relation $<$ to be homogeneous and strict (every member of a category has exactly one parent in each of the categories above). Both conditions are required in [5] but not in [2].

Running Example. As mentioned before, we are only focusing on dimension Book, we can define its dimension schema as follows.

$$\mathcal{S}_{Book} = \langle \mathcal{C}_{Book}, \nearrow_{Book} \rangle$$
$$\mathcal{C}_{Book} = \{book, \; subgenre, \; genre, \; All\}$$
$$book =< book\_id, \; book\_name, book\_abstract >$$
$$subgenre =< book\_subgenre\_id, \; book\_subgenre\_name >$$
$$genre =< book\_genre\_id, \; book\_genre\_name >$$
$$\nearrow_{book}= \{(book, \; subgenre), \; (subgenre, \; genre), \; (genre, \; All)\}$$

Considering part of the *Book Dimension* instance shown in Fig. 1, the corresponding instance formalization is presented below.

$$\mathcal{D}_{Book} = \langle \mathcal{U}_{Book}, <_{Book}, mem_{Book} \rangle$$
$$\mathcal{U}_{Book} = \{t_1 =< book1, \; ``TheLion, theWitchandtheWardrobe'', abstract1 >,$$
$$t_2 =< book2, \; ``AGameofThrons'', abstract2 >,$$
$$t_3 =< subgen1, \; ``Suspense'' >,$$
$$t_4 =< subgen2, \; ``Highfantasy'' >,$$
$$t_5 =< gen1, \; ``Fantasy'' >,$$
$$t_6 =< gen2, \; ``Thriller'' >,$$
$$t_7 = all\}$$
$$<_{Book}= \{(t_1, t_3), (t_2, t_4), (t_3, t_6), (t_4, t_5), (t_5, t_7), (t_6, t_7)\}$$
$$mem_{Book}(t_1) = mem_{Book}(t_2) = book$$
$$mem_{Book}(t_3) = mem_{Book}(t_4) = subgenre$$
$$mem_{Book}(t_5) = mem_{Book}(t_6) = genre$$

3.3   Context Formal Specification

We use `OWL` domain ontologies to represent context, as explained in 3.1. In the related works we found a lack of generality, meaning that for each specific domain a new solution is needed. In order to solve this problem, we propose a representation that is based exclusively on the ontology components.

Given an `OWL` ontology schema O, we consider its classes $CL = \{CL_1, \ldots CL_c\}$; its object properties $OP = \{OP_1 \ldots OP_{op}\}$, where $dom(OP_i)$

and $range(OP_i)$ are its domain and range respectively; and its data properties $DP = \{DP_1, \ldots DP_{dp}\}$, where $dom(DP_i)$ is a class and $range(DP_i)$ is a data type dt_i.

An ontology instance o of O is defined as follows:

- $cl_i = \{cl_{i_1}, \ldots, cl_{i_t}\}$ is an instance of CL_i. Therefore, $cl = cl_1 \cup \ldots \cup cl_c$ is an instance of CL.
- $dp_i \subseteq cl_j \times dt_j$, where cl_j is an instance of class $dom(DP_i)$ and dt_j is the data type of $range(DP_i)$. Therefore, $dp \subseteq dp_1 \cup \ldots \cup dp_{dp}$ is an instance of DP.
- $op_i \subseteq cl_j \times cl_k$ is an instance of OP_i, which is a relation between instances of $dom(OP_i) = CL_j$ and $range(OP_i) = CL_k$. Therefore, $op \subseteq op_1 \cup \ldots \cup op_{op}$ is an instance of OP.

Running Example. Considering the Book ontology schema presented in Fig. 2, we can identify its classes, data properties and object properties as follows.
$CL = \{$ obk:Book, obk:Person, obj:Format, obk:ElectronicFormat,
        obk:PaperFormat, obk:BookGenre,
        http://purl.org/goodrelations/v1#BusinessEntity,
        http://purl.org/goodrelations/v1#QuantitativeValue $\}$
$OP = \{$ obk:narrator, obk;author, obj:format, obk:genre,
        obk:lengthOfAudio, obk:publisher $\}$
$DP = \{$ obk:abstract, obk:numberOfPages, obk:title,obk:publishedIn,
        obk:publishedDate $\}$

An example of part of the instance presented in Fig. 3, only considering books *book1* and *book2* and their corresponding genres is shown below:
$cl \supseteq \{obk : book1, obk : book2,$
        $obk : genre1\}$
$op \supseteq \{(obk : book1, obk : genre1)\}$
$dp \supseteq \{(obk : book1, "TheCollectedStoriesofGracePaley"),$
        $(obk : book2, "TheCampusMurders"),$
        $(obk : genre1, "Romance")\}$

4   Data Warehouse to Ontology Mapping

In this section we present DW to ontology mappings definition. Examples on how to apply mappings to our running example are shown in Sect. 5.

Consider a data warehouse DW with n dimensions. The dimension schemas are noted as $S_1, \ldots S_n$ and $D_1, \ldots, D_n$ are their corresponding dimension instances. An OWL ontology schema O and a corresponding ontology o give context to DW.

Mappings are defined as binary relations, where the first argument is the DW element and the second argument is the ontology element. Additionally, preconditions for the definition of each mapping are presented as rules.

Mappings were defined for each of the DW elements. However, in this paper, we are only presenting mappings for the DW elements listed in Sect. 3.2.

Dimensions. $MapDim \subseteq \{\mathcal{S}_1, \ldots, \mathcal{S}_n\} \times CL$ maps DW *dimensions* to ontology schema classes.

Categories. $MapCat \subseteq (\mathcal{C}_1 \cup \ldots \cup \mathcal{C}_n) \times (CL \cup DP)$ maps DW *categories* either to ontology schema classes or to ontology schema data properties. Note that each $\mathcal{C}_i$ is the set of categories of dimension $\mathcal{S}_i$. A category c_j can be mapped to a data property DP_j if: a) its dimension is mapped to the class $dom(DP_j)$ or b) a category c_k such that $c_k \nearrow^* c_j$ is mapped to the class $dom(DP_j)$. In the first case, the category is represented as a property of the class mapped to the dimension, while in the second one, the category is represented as a property of the class mapped to a related category. If a category is mapped to a data property DP_j, then there must exist a mapping between either the dimension to which the category belongs or another related category of the same dimension, and the class $dom(DP_j)$. This condition is formalized in Eq. 1.

$$(\forall i \in \{1, \ldots, n\})(\forall k \in \{1, \ldots, |\mathcal{C}_i|\})(\forall j \in \{1, \ldots, dp\})$$
$$(MapCat(c_{i_k}, DP_j) \rightarrow$$
$$(\exists k' \in \{1, \ldots, |\mathcal{C}_i|\})(MapCat(c_{i_{k'}}, dom(DP_j)) \wedge c_{i_{k'}} \nearrow^* c_{i_k}) \vee \tag{1}$$
$$MapDim(\mathcal{S}_i, dom(DP_j)))$$

At schema level, $MapAtt \subseteq (att_{1_1} \ldots \cup att_{1_{|\mathcal{C}_1|}} \ldots \cup att_{n_1} \ldots \cup att_{n_{|\mathcal{C}_n|}}) \times DP$, maps an *attribute* of a DW category to a data property schema, where $att_{i_j} = \{att_{i_j}[1]\} \cup \ldots \cup \{att_{i_j}[r]\}$ and r is the number of attributes of category j of dimension i.

If an attribute is mapped to a data property DP_d, then there must be a mapping between the category to which the attribute belongs and the class $dom(DP_d)$. This condition is formalized in Eq. 2.

$$(\forall i \in \{1, \ldots, n\})(\forall j \in \{1, \ldots, |\mathcal{C}_i|\})(\forall k \in \{1, \ldots, K\})(\forall d \in \{1, \ldots, dp\}) \tag{2}$$
$$(MapAtt(att_{i_j}[k], DP_d) \rightarrow MapCat(c_{i_j}, dom(DP_d)))$$

At instance level, $MapVal \subseteq (t_{1_1} \ldots \cup t_{1_{|\mathcal{U}_1|}} \ldots \cup t_{n_1} \ldots \cup t_{n_{|\mathcal{U}_n|}}) \times dp$, maps a *value of an attribute* of a DW tuple to an instance of data property, where $t_{i_j} = \{val_{i_j}[1]\} \cup \ldots \cup \{val_{i_j}[r]\}$ and r is the number of attributes of category $mem_i(t_{i_j})$.

For this mapping to be possible, there must exist a mapping at schema level between the corresponding attribute and the data property schema. This condition is formalized in Eq. 3.

$$(\forall i \in \{1, \ldots, n\})(\forall j \in \{1, \ldots, T\})(\forall k \in \{1, \ldots, K\})(\forall r \in \{1, \ldots, dp\})$$
$$(\forall d \in \{1, \ldots, D\})$$
$$(MapVal(t_{i_j}[k], dp_{r_d}) \rightarrow (\exists a \in \{1, \ldots, |\mathcal{C}_i|\})(\tag{3}$$
$$mem_i(t_{i_j}) = c_{i_a} \wedge MapAtt(att_{i_a}[k], DP_r)))$$

For Eqs. 2 and 3 we consider that T is the number of tuples of $\mathcal{U}_i$, the universe of dimension i; K is the number of attributes of category $mem_i(t_{i_j})$; D is the number of data properties instances of DP_r.

Hierarchies. $MapHier \subseteq ((\mathcal{C}_1 \times \mathcal{C}_1) \cup \ldots \cup (\mathcal{C}_n \times \mathcal{C}_n)) \times OP$ maps DW *hierarchies* to ontology schema object properties. If a pair of categories (c_{i_a}, c_{i_b}) is mapped to an object property OP_j then there must exists a mapping between the category c_{i_a} and the class $dom(OP_j)$, and another mapping between the category c_{i_b} and the class $range(OP_j)$. Furthermore, $c_{i_a} \nearrow^* c_{i_b}$ must hold. Both conditions are formalized in Eq. 4

$$
\begin{aligned}
(\forall i \in &\{1, \ldots, n\})(\forall a, b \in \{1, \ldots, |\mathcal{C}_i|\})(\forall j \in \{1, \ldots, op\}) \\
&(MapHier((c_{i_a}, c_{i_b}), OP_j) \rightarrow \\
&\quad (MapCat(c_{i_a}, dom(OP_j)) \wedge \\
&\qquad MapCat(c_{i_b}, range(OP_j)) \wedge c_{i_a} \nearrow^* c_{i_b}))
\end{aligned}
\tag{4}
$$

Table 1 presents a summary of all the proposed mappings between DW and ontology elements, with the needed preconditions.

Table 1. Mappings and pre-conditions.

DW element	Ontology element	Pre-condition
Dimension	Class	None
Category	Class	None
Category	Data Property	Corresponding dimension or related category must be mapped to class
Attribute	Data Property	Category must be mapped to class
Hierarchy (related categories)	Object Property	Related categories must be mapped to classes
Value	Data Property Instance	The attribute of the value must be mapped to the data property schema

5   Context-Based Data Quality Rules

DQ Rules are defined considering the DW, the context and their mappings.

Consider a data warehouse DW with n dimension schemas $\mathcal{S}_1, \ldots, \mathcal{S}_n$ and their corresponding dimension instances $\mathcal{C}_1, \ldots, \mathcal{C}_n$, as presented in Sect. 3.2. An OWL ontology schema O with classes $Cl = \{Cl_1, \ldots Cl_c\}$; object properties $OP = \{OP_1 \ldots OP_{op}\}$; data properties $DP = \{DP_1, \ldots DP_{dp}\}$; and a corresponding ontology o, give context to DW.

Syntactic accuracy Syntactic accuracy is used to know whether or not a value is well written according to a referential. In this case, we check if an attribute value is correctly written according to the context. In order to do so, we check if the value of an attribute of a category of DW belongs to the set of possible values of the data property to which the attribute is mapped.

Considering a data property DP_j and its corresponding instance dp_j, we define V_j the set of possible values for any instance of DP_j as:

$$
V_j = \{v \in dt_j | (\exists k \in \{1, \ldots t\})(cl_{j_k}, v) \in dp_j \wedge Dom(DP_j) = CL_j\}
$$

To define a syntactic accuracy rule there must exist a mapping between an attribute $att_{ik}[p]$ and a data property DP_j: $MapAtt(att_{ik}[p], DP_j)$. Note that $att_{ik}[p]$ is the attribute in position p of category k of dimension i. Therefore, syntactic accuracy rule for a value in position p of a tuple t such that $mem_i(t) = k$, is defined as:

$$t[p] \in V_j \rightarrow SyntAcc(t[p], 1)$$
$$t[p] \notin V_j \rightarrow SyntAcc(t[p], 0)$$

(5)

The result of the measure SyntAcc is boolean, if the value $t[p]$ is well written with respect to V_j, it will return 1, otherwise 0.

Semantic accuracy Semantic accuracy is used to know whether or not values have a correspondence with real world entities. In this case, we check if two related categories in the DW are somehow related in the context. To do so we find two mapped values and a corresponding object property between their classes, then we check if there is a correspondence in the DW and in the context.

To define this semantic accuracy rule some conditions must hold:

- There must exist a mapping between two categories of a hierarchy and an object property $MapHier((c_{i_k}, c_{i_{k'}}), OP_j)$.
- There must exist a mapping between an attribute value of a tuple t of category c_{i_k} and a data property instance dp_r, which domain, cl, must be an instance of class $dom(OP_j)$: $MapVal(t[m], dp_r)$.
- There must exist a mapping between an attribute value of a tuple t' of category $c_{i_{k'}}$ and a data property instance $dp_{r'}$, which domain, cl', must be an instance of class $range(OP_j)$: $MapVal(t'[m'], dp_{r'})$.
- Tuples t and t' must be related through $<^*$: $t <^* t'$

These conditions are shown in Fig. 4. The Figure shows a mapping between the pair of categories $(Category\ 1(c_{i_k}), Category\ 2(c_{i_{k'}}))$ and the object property OP_j. In addition, attribute value $val4$ of tuple $< val4,\ val5,\ val6 >$, which belongs to category c_{i_k}, is mapped to instance dp_r of data property DP_r, whose domain is the class cl. Finally, attribute value $val10$ of tuple $< val9,\ val10 >$, which belongs to category $c_{i_{k'}}$, is mapped to instance $dp_{r'}$ of data property $DP_{r'}$, whose domain is the class cl'.

A semantic accuracy DQ rule for the values $t[m]$ and $t'[m']$ can be defined as

$$(cl, cl') \in op_j \rightarrow SemAcc((t[m], t'[m']), 1))$$
$$(cl, cl') \notin op_j \rightarrow SemAcc((t[m], t'[m']), 0))$$

(6)

In the following, we return to the **running example** in order to illustrate the application of the DQ rules. The proposed DQ rules can be instantiated to the *Book Dimension* presented in Fig. 1 and the context shown in Fig. 3.

Syntactic Accuracy. In order to measure whether a book name is well written or not, we use the instance values of `dct:title` data property as a referential.

As shown in Sect. 5, we need to define $V_{\texttt{dct:title}}$, in this case $V_{\texttt{dct:title}}$ would be the set of all possible book titles, in this case $V_{\texttt{dct:title}} \supseteq \{$ *"A Game of*

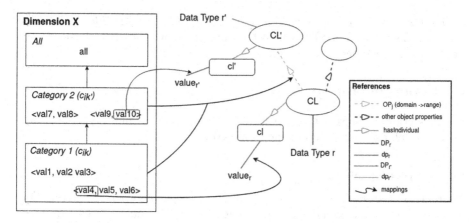

Fig. 4. Semantic Accuracy rule pre-conditions.

Thrones", *"Harry Potter and the Sorcerer's Stone"*}. Then we need to check if the preconditions for this DQ rule hold, in this case a mapping between an attribute of a category and the data property `dct:title`. We can determine that the mapping $MapAtt(title, dct : title)$ holds, which makes it possible to instantiate DQ rule presented in Eq. 5.

$$\text{``}AGameofThrons\text{''} \notin V_{\texttt{dct:title}} \rightarrow SyntAcc(\text{``}AGameofThrons\text{''}, 0) \qquad (7)$$

For the book titled *"A Game of Throns"* the value for the syntactic accuracy DQ rule is 0 because the name does not belong to the set $V_{\texttt{dct:title}}$.

Semantic Accuracy. We want to measure whether a book is related to a correct genre in the DW. To instantiate the semantic accuracy rule, we need to check if the preconditions hold. As the mapping $MapHier((book, genre), obk : genre)$ exists in our reality, it is possible to use an instance of `obk:genere` object property as the referential for this DQ rule.

As mentioned in Sect. 5 besides the mapping $MapHier((book, genre), obk : genre)$, a mapping between some value and a data property instance of an instance of the domain class of `obk:genre` and between some value and an instance of the range class of `obk:genre` need to hold in order to apply DQ rules. In our example, the following mappings hold:

- $MapVal(\text{``}Thriller\text{''}, (obk : genre3, \text{``}Thriller\text{''}))$
- $MapVal(\text{``}TheLion, theWitchandtheWardrobe\text{''},$
    $(obk : book1, \text{``}TheLion, theWitchandtheWardrobe\text{''}))$

Furthermore, the tuples involved in the mappings, are related through $<^*$. These conditions allows the instantiation of semantic accuracy DQ rule presented in Sect. 5 for the pair presented above is shown in Eq. 8.

$$(obk : book1, obk : genre3) \notin obk : genre \rightarrow$$
$$SemAcc((\text{``}TheLion, theWitchandtheWardrobe\text{''}, \text{``}Thriller\text{''}), 0)) \qquad (8)$$

For the pair (*"The Lion, the Witch and the Wardrobe"*, *"Thriller"*) the value for the semantic accuracy DQ rule is 0 because the corresponding classes do not belong to the instance of `obk:genre`.

6   Experimentation

In this section we present the implementation for the DW and the DQ management, as well as a validation of the proposed rules.

6.1   Implementation

In order to have enough expressivity to allow the representation of different DW elements, we use an extended Datalog language that includes aggregation functions. For the sake of clarity, we use standard Datalog syntax with the addition of aggregation functions to present our solution, although we used PyDatalog [4] in our implementation.

Data Warehouse Implementation. The Datalog definition of a DW is based on the formalization presented in Sect. 3.2. As a consequence we are presenting the elements introduced in Sect. 3.2.

Dimensions. Given a fixed number, n, of dimension schemas, $S_i = \langle C_i, \nearrow_i \rangle$, a relational predicate $Dim_i(X_{c_1}, \ldots, X_{c_p})$ is added for each of them. Every predicate Dim_i must follow that $|C_i| = p$ and $(\forall j \in \{1, \ldots, p\})(c_j \in C_i)$.

For the *Book Dimension* presented in Sect. 3.1, the Datalog predicate is:

$$Dim_{Book}(X_{book}, \ X_{sub\_genre}, \ X_{genre})$$

Hierarchies. The rule $Hier_i(X_{c_j}, X_{c_k}) \leftarrow Dim_i(X_{c_1}, \ldots, X_{c_p})$ is added to represent the hierarchy. $Hier_i$ must follow that $c_j \nearrow^* c_k$

For the *Book Dimension* presented in Sect. 3.1 the Datalog rule to represent the hierarchies is the following:

$$HierBook(X_{book}, \ X_{sub\_genre}) \leftarrow$$
$$Dim_{Book}(X_{book}, \ X_{sub\_genre}, \ X_{genre})$$
$$HierBook(X_{sub\_genre}, \ X_{genre}) \leftarrow$$
$$Dim_{Book}(X_{book}, \ X_{sub\_genre}, \ X_{genre})$$
$$HierBook(X_{book}, \ X_{genre}) \leftarrow$$
$$Dim_{Book}(X_{book}, \ X_{sub\_genre}, \ X_{genre})$$

DQ Management Implementation. Even though mappings are formalized as predicates (Sect. 4), we didn't explicitly added them as Datalog predicates. The aim of the mappings is to connect the DW and the ontologies models.

The set of ontologies is managed using a Python library, owlready2 [9] and the DW is represented using PyDatalog. The major problem is how to combine those two different representations in order to use them in DQ rules. The connection between ontologies and Datalog rules was done using simple Python data structures to relate elements in de DW to elements in the ontology. These structures were then used in Datalog DQ rules. Translated rules implementation and used datasets can be found at [10].

Datalog DQ rules for semantic and syntactic accuracy, introduced in Sect. 5, are translated into pyDatalog, completing our implementation.

6.2   Validation

A first validation of our DQ rules was made by comparing the results obtained by the implementation against human observation of a small data set. To this end, syntactic and semantic accuracy problems were detected manually and compared with the automatic DQ assessment. We used the instance of *Book Dimension* presented in Fig. 1 and the book ontology presented in Fig. 3.

Syntactic Accuracy. For the book titled *"A Game of Throns"* shown in Fig. 1, the value for the syntactic accuracy DQ rule is 0 because is not well written with relation to the books presented in the context. For the other two books, as they are well written with relation to the context, the value for the syntactic accuracy DQ rule is 1. The result for the execution of the DQ rule is shown in Fig. 5, and is the same as expected.

```
X1                                           | Y                                         | Z
---------------------------------------------|-------------------------------------------|---
<__main__.Dimbooks object at 0x7fd6c9eaac70> | Harry Potter and the Sorcerer's Stone     | 1
<__main__.Dimbooks object at 0x7fd6c9eaab80> | The Lion, the Witch and the Wardrobe      | 1
<__main__.Dimbooks object at 0x7fd6c9eaac10> | A Game of Throns                          | 0
```

Fig. 5. Syntactic accuracy rule result.

Semantic Accuracy. As we need to use the book title as attribute that represent the book in the ontology, the instance presented in Fig. 1 needs to be corrected and instead of "A Game of Throns", the correct title "A Game of Thrones" is used. The genre of book *"The Lion, the Witch and the Wardrobe"* according to the context is *"Fantasy"* and not *"Thriller"* as is stored in the DW, therefore for the pair (*"The Lion, the Witch and the Wardrobe"*, *"Thriller"*) the value for semantic accuracy DQ rule is 0. For the other two books as the genre stored in the DW is the same as the corresponding in the context, the value for semantic accuracy DQ rule is 1. The result for the execution of the DQ rule is shown in Fig. 6, and is the same as expected.

```
Y1                                           | Y3                                        | Y4       | Z
---------------------------------------------|-------------------------------------------|----------|---
<__main__.Dimbooks object at 0x7f5f47a64160> | A Game of Thrones                         | Fantasy  | 1
<__main__.Dimbooks object at 0x7f5f47a64790> | Harry Potter and the Sorcerer's Stone     | Fantasy  | 1
<__main__.Dimbooks object at 0x7f5f47a64190> | The Lion, the Witch and the Wardrobe      | Thriller | 0
```

Fig. 6. Semantic Accuracy rule result.

7   Conclusions and Future Work

In this work we proposed a general mechanism for DQ evaluation over DW, based on the context, which is represented by domain ontologies. The proposed mechanism is based on mappings between DW and ontology elements, and DQ rules that exploit these mappings for assessing DW quality.

The main focus of our work is to reach a level of abstraction in the formalization of the DW, the context and their interactions, that makes it possible to evaluate certain DQ dimensions for any DW in any context.

In this paper we present the formal specifications for the DW, the context, the mappings and the DQ rules, as well as the implementation as a proof of concept and the experimentation over a small study case. In the implementation, Datalog is used for the DW and for the DQ rules, and is combined with existing OWL domain ontologies. In addition, a running example is presented throughout the paper, in order to illustrate each part of the proposal.

The presented DQ rules evaluate syntactic accuracy and semantic accuracy. Ongoing work is focused on defining DQ rules for other DQ dimensions, as consistency and completeness.

As future work we are planning to apply the proposal to real world cases, evaluating its performance and generality.

References

1. Batini, C., Scannapieco, M.: Data and Information Quality. DSA, Springer, Cham (2016). https://doi.org/10.1007/978-3-319-24106-7
2. Bertossi, L., Milani, M.: Ontological multidimensional data models and contextual data quality. J. Data Inf. Q. **9**(3), 14:1–14:36 (2018)
3. Bertossi, L., Rizzolo, F., Jiang, L.: Data quality is context dependent. In: Castellanos, M., Dayal, U., Markl, V. (eds.) BIRTE 2010. LNBIP, vol. 84, pp. 52–67. Springer, Heidelberg (2011). https://doi.org/10.1007/978-3-642-22970-1_5
4. Carbonnelle, P.: pyDatalog: A pure-python implementation of Datalog, a truly declarative language derived from Prolog., https://sites.google.com/site/pydatalog/
5. Hurtado, C.A., Mendelzon, A.O.: OLAP dimension constraints. In: Proceedings of the Twenty-first ACM SIGMOD-SIGACT-SIGART Symposium on Principles of Database Systems, pp. 169–179. ACM, New York, NY, USA (2002)
6. Marotta, A., Vaisman, A.: Rule-based multidimensional data quality assessment using contexts. In: Madria, S., Hara, T. (eds.) DaWaK 2016. LNCS, vol. 9829, pp. 299–313. Springer, Cham (2016). https://doi.org/10.1007/978-3-319-43946-4_20
7. McNab, A.L., Ladd, D.A.: Information quality: the importance of context and trade-offs. In: 2014 47th Hawaii International Conference on System Sciences, pp. 3525–3532, January 2014
8. Munawar, S.N., Ibrahim, R.: Towards data quality into the data warehouse development. In: 2011 IEEE Ninth International Conference on Dependable, Autonomic and Secure Computing, pp. 1199–1206 (2011)
9. owlready2: owlready2 library. https://owlready2.readthedocs.io/en/v0.37/, Access July 2021

10. Sanz, C.: Sales dw, https://github.com/camila-sanz/sales-datawarehouse
11. Sanz, C.: Context based data quality rules for multidimensional data. In: Bao, Z., Sellis, T.K. (eds.) Proceedings of the VLDB 2022 PhD Workshop co-, Sydney, Australia. CEUR Workshop Proceedings, vol. 3186. CEUR-WS.org (2022)
12. Serra, F., Marotta, A.: Context-based data quality metrics in data warehouse systems. CLEI Electron. J. **20**(2), 3:1–3:23 (2017). number: 2
13. Serra, F., Peralta, V., Marotta, A., Marcel, P.: Modeling context for data quality management. In: Ralyté, J., Chakravarthy, S., Mohania, M., Jeusfeld, M.A., Karlapalem, K. (eds.) Conceptual Modeling. ER 2022, LNCS, vol. 13607, pp. 325–335. Springer, Cham (2022). https://doi.org/10.1007/978-3-031-17995-2_23
14. Shankaranarayanan, G., Blake, R.: From content to context: the evolution and growth of data quality research. J. Data Inf. Q. **8**(2), 9:1–9:28 (2017)
15. Strong, D.M., Lee, Y.W., Wang, R.Y.: Data quality in context. Commun. ACM **40**(5), 103–110 (1997)
16. Sundararaman, A.: A framework for linking data quality to business objectives in decision support systems. In: 3rd International Conference on Trendz in Information Sciences & Computing (TISC2011), pp. 177–181 (2011)
17. Zheng, Z., et al.: Contextual data cleaning with ontology functional dependencies. J. Data Inf. Q. **14**(3), 1–26 (2022)

Preventing Technical Errors in Data Lake Analyses with Type Theory

Alexis Guyot$^{(\boxtimes)}$, Éric Leclercq , Annabelle Gillet , and Nadine Cullot

LIB, Université de Bourgogne, Dijon, France
{alexis.guyot,eric.leclercq,annabelle.gillet,nadine.cullot}@u-bourgogne.fr

Abstract. Data analysts compose various operators provided by data lakes to conduct their analyses on big data through complex analytical workflows. In this article, we present a formal framework based on type theory to prevent technical errors in such compositions of operators. This framework uses restrictions on type definitions to transform technical errors into type errors. We show how to use this framework to prevent errors related to schema or model transformations in analytical workflows. We provide an open-source implementation in Scala which can be used to detect errors at compile time.

Keywords: Data Lakes · Type Theory · Big Data Analytics

1 Introduction

Data analysts use data lakes to conduct their analyses on big data. These platforms provide features to store, manage and analyse big data with a schema-on-read paradigm [2]. In other words, heterogeneous data are ingested and stored as is and only cleansed, structured and integrated when needed. Data lakes associate datasets with insightful metadata to describe and control their content and relationships. Data lakes provide various operators to search, transform, enrich and analyse data at different levels of abstraction (data, schema, model and metadata). For example, they may provide data preparation operators to cleanse data and integrate schemas. They may also provide analytical operators like classifiers or graph miners whose results can be reused as new metadata.

Data analysts compose these various operators to build complex analytical workflows. Composed operators may act on different data models (relational, semi-structured, graph, etc.) and levels of abstraction. They may use different theoretical foundations (relational or linear algebra, graph theory, statistics, etc.) and execution paradigms (stream processing, map-reduce, GPU, etc.). Nevertheless, an invalid composition of operators can be a source of technical errors in analyses.

Therefore, a formal framework based on solid theoretical foundations is required to prevent technical errors in the compositions of operators of data lake analytical workflows. In this article, we propose to use type theory as the

R. Wrembel et al. (Eds.): DaWaK 2023, LNCS 14148, pp. 18–24, 2023.
https://doi.org/10.1007/978-3-031-39831-5_2

foundation of this formal framework. We aim to transform technical errors into type errors through type definitions. Thus, the absence of errors in a composition can be proven using type theory and verified by the compiler. We show how to use types to specify operators acting on multiple data models and levels of abstraction. We also show how our framework prevents errors in a composition of such operators.

2   Related Works

Introducing type safety in operators and languages to prevent errors and misuse is a recurrent solution. Functional languages like OCaml, Haskell or F# use strong static typing based on different type theories to prevent misuse through type inference. Proof assistants like Agda and Coq implement type theories, *resp.* the Unified Theory of dependent Types and the Calculus of Constructions. Recent analytical frameworks also include type safety. For example, type systems for linear algebra are proposed by Griffioen [6] and by Muranushi *et al.* [12]. The Spark framework [15] provides advanced structures like *Datasets* that add type safety guarantees over *DataFrames*. The Tensor Data Model [5] adds types in tensors to provide type safety and schema inference on tensor operators.

Other approaches to prevent errors and misuse of operators focus more on the data they are dealing with. Firstly, metadata can be used to understand the data better. They provide descriptions and constraints on the datasets and their relationships. Metadata models can be based on various formalisms such as logic [7], graphs [13] or UML [16]. Each formalism provides a different trade-off between expressiveness and restrictiveness. Therefore, only some metadata models can formally prevent errors and misuse through mathematical, structural or reasoning properties. Furthermore, existing metadata models are either tailored for specific use cases or not generic enough to be used in different contexts [13]. Secondly, data can be structured using pivot models with well-established algebras. Several pivot models have been proposed for heterogeneous platforms, such as the nested relational model in [1], the JSON semi-structured model in [8], typed tensors in [5] or the RDF graph model in [4]. However, pivot models may hinder the flexibility required in data lakes by restricting the set of available operators. For example, languages based on the nested relational model cannot express transitive closure and, therefore, several graph operators [1]. Finally, foundations of multi-model languages [10,11,14] use category theory to formally represent and control the effects of operators on data. However, this approach mainly focuses on representing different data models with a single formalism. It does not focus on the errors that can occur by composing different operators.

In contrast to the existing approaches, we tackle the problem of preventing errors in analytical workflows by controlling the heterogeneity of data models and the navigation between different levels of abstraction (data, schemas, models, metadata) in a unified way using type theory. This approach allows us to formally ensure the consistency of data manipulations and transformations throughout the data lake workflows.

3   Type-Theoretical Framework

Type theory is a constructive formalism. It can be used to define type systems through construction rules [3]. It defines the concepts of well-formed types and sub-types. A **well-formed type T** (written $T : Type$) is a type having at least one rule allowing its construction. A well-formed type S can be a **sub-type** of another well-formed type T (written $S <: T$). In this case, type S can be used in every situation in which type T is required.

Type theory also defines type constructors. These allow the construction of various composite types such as function, generic, dependent and product types. A **generic type** $A[B]$ is a type that is parameterised by another type like $List[T]$ or $Matrix[T]$. A **dependent type** $\Pi_{(x:B)}A(x)$ is a type that is parameterised by values of another type. For example, a dependent type $\Pi_{(size:Int)}Vec(size)$ represent vectors of a certain size. A dependent type $\Pi_{(name:String)}T(name)$ represent attributes of type T with a certain name. For the sake of readability, we will refer to the latter dependent type as $n \twoheadrightarrow T$ in the following. A **product type** is a composite type (written $A \times B \times ... \times Z$) whose elements are heterogeneous tuples $(a, b, ..., z)$ with $a : A$, $b : B$ and $z : Z$.

We show how to use type theory to prevent technical errors related to schema or model transformations in a composition of operators. Analytical workflows often require such transformations to connect analytical operators acting on different data models. However, transforming the schema or the model of data may be error-prone due to the differences between models. For example, the relational model does not allow multi-valued or nested attributes, whereas semi-structured models such as JSON do. Inconsistencies between data schemas and models due to transformations may result in a technical error. Therefore, we want to verify that a composition of operators does not contain any erroneous schema or model transformation.

Preventing such errors requires representing with types two levels of abstraction (schema, model) and at least two different data models. We propose to represent: 1) one data model allowing multi-valued and nested attributes like JSON, and; 2) one data model not allowing such attributes like the relational model.

$$\frac{\Gamma \vdash S <: Schema}{\Gamma \vdash RelationSchema[S] : Type}$$
$$\frac{}{\Gamma \vdash Relation[S] <: Model[S]}$$

(a) Relational model.

$$\frac{\Gamma \vdash S <: Schema}{\Gamma \vdash JsonSchema[S] : Type}$$
$$\frac{}{\Gamma \vdash JSON[S] <: Model[S]}$$

(b) JSON model.

Fig. 1. Type definitions for data models.

We use products of dependent types to represent data schemas. More precisely, we use products of $n \twoheadrightarrow T$. In schema types, using a dependent type to combine an attribute name with its type can be useful to prevent other errors.

$$\Gamma \vdash RelationValue[Base] : Type$$

(a) Type definition for *RelationValue*.

$$\frac{\begin{array}{c}\Gamma \vdash F <: n \twoheadrightarrow T \\ \Gamma \vdash S <: Schema \\ \Gamma \vdash F \times S <: Schema \\ \Gamma \vdash RelationValue[T] : Type \\ \Gamma \vdash RelationSchema[S] : Type\end{array}}{\Gamma \vdash RelationSchema[F \times S] : Type} \qquad \frac{\begin{array}{c}\Gamma \vdash F <: n \twoheadrightarrow T \\ \Gamma \vdash RelationValue[T] : Type\end{array}}{\Gamma \vdash RelationSchema[F \times SNil] : Type}$$

(b) Type definition for *RelationSchema*.

Fig. 2. Restrictions of the relational model on schemas.

$$\frac{}{\Gamma \vdash JsonValue[Base] : Type} \qquad \frac{\begin{array}{c}\Gamma \vdash T : Type \\ \Gamma \vdash JsonValue[T] : Type \\ \Gamma \vdash List[T] : Type\end{array}}{\Gamma \vdash JsonValue[List[T]] : Type} \qquad \frac{\begin{array}{c}\Gamma \vdash S <: Schema \\ \Gamma \vdash JsonSchema[S] : Type\end{array}}{\Gamma \vdash JsonValue[S] : Type}$$

(a) Type definition for *JsonValue*.

$$\frac{\begin{array}{c}\Gamma \vdash F <: n \twoheadrightarrow T \\ \Gamma \vdash S <: Schema \\ \Gamma \vdash F \times S <: Schema \\ \Gamma \vdash JsonValue[T] : Type \\ \Gamma \vdash JsonSchema[S] : Type\end{array}}{\Gamma \vdash JsonSchema[F \times S] : Type} \qquad \frac{}{\Gamma \vdash JsonSchema[SNil] : Type}$$

(b) Type definition for *JsonSchema*.

Fig. 3. Restrictions of the JSON model on schemas.

For example, it may allow the inference of a schema for the data returned by operators. Analysts can use this inferred schema to prevent inconsistencies or misuses in the analytical workflow. We define a super-type *Schema* and two sub-types *SNil* and $F \times S$, with $F <: n \twoheadrightarrow T$ and $S <: Schema$. This definition allows the construction of schema types with any number of attributes: *SNil* is the type of empty schemas, $F \times SNil$ is the type of schemas with one attribute of type F, $F \times S$ is the type of schemas with at least one attribute of type F. In the following, and for the sake of readability, we assume the existence of a super-type *Base* for the main data types like numbers, strings, booleans, dates, etc.

We use generic types parameterised with schema types to represent data models. We define a super-type *Model[S]* and two sub-types *Relation[S]* and *JSON[S]*, with $S <: Schema$. Figure 1 presents type definitions for these three types. For a given schema type S, type *Relation[S]* (*resp.*, *JSON[S]*) is well-formed if type *RelationSchema[S]* (*resp.*, *JsonSchema[S]*) is well-formed.

We define generic types $RelationSchema[S]$ and $RelationValue[T]$ (*resp.*, $JsonSchema[S]$ and $JsonValue[T]$), with $S <: Schema$ and $T : Type$ (Figs. 2 and 3). These types link the two levels of abstraction by specifying the restrictions that the models apply to schemas. $RelationSchema[S]$ (*resp.*, $JsonSchema[S]$) recursively ensures that all the attributes of the schema conform to the model. $RelationValue[T]$ (*resp.*, $JsonValue[T]$) defines value types that the model allows. The relational model only allows attributes with scalar values (Fig. 2a). The JSON model allows attributes with scalar, multi-valued and nested values (Fig. 3a).

$$\frac{\Gamma \vdash S1 <: Schema \quad \Gamma \vdash S2 <: Schema \quad \Gamma \vdash M1[S1] <: Model[S1] \quad \Gamma \vdash M2[S2] <: Model[S2]}{\Gamma \vdash M1[S1] \Rightarrow M2[S2] : Type}$$

(a) Operators.

$$\frac{\Gamma \vdash M1[S1] <: Model[S1] \quad \Gamma \vdash M2[S2] <: Model[S2] \quad \Gamma \vdash M3[S3] <: Model[S3] \quad \Gamma \vdash f : M1[S1] \Rightarrow M2[S2] \quad \Gamma \vdash g : M2[S2] \Rightarrow M3[S3]}{\Gamma \vdash safeCompo(f, g) : M1[S1] \Rightarrow M3[S3]}$$

(b) Composition of operators.

Fig. 4. Type definitions for operators and compositions.

We represent operators with function types acting on models. Therefore, a composition of operators is a function type taking as inputs two operator types and returning a new operator type if their composition is not erroneous. Figure 4 presents type definitions for operators and compositions. The type of a composition is well-formed if its inputs are well-formed, *i.e.* if a series of construction rules leads to the composition. Any error, for example an erroneous schema or model transformation, breaks the series of rules and results in a type error. According to type theory, a complete series of construction rules leading to a well-formed type is proof of the absence of error.

Compilers can detect type errors. Therefore, we propose an open-source implementation in Scala of the types presented in this article[1]. The implementation uses the rich typing features provided by Scala and the Shapeless library[2]. It can be used to encapsulate operators in analytical workflows and verify the absence of errors related to schema or model transformations at compile time.

4   Conclusion

In this article, we have presented a formal framework to prevent technical errors in data lake analytical workflows by finely controlling the compositions of operators. This framework is based on type theory. It uses types to represent data in different models and levels of abstraction and to express restrictions on operator

[1] https://github.com/AlexisGuyot/type_safe_compo.
[2] https://github.com/milessabin/shapeless.

inputs and outputs. We have proposed an open-source implementation in Scala using the compiler to prevent errors. The current limits of our approach are the lack of representation for constraints that apply to data (*e.g.*, ensuring the uniqueness of values for an attribute) and the lack of representation for metadata. Therefore, as perspectives for future works, we plan to extend our formal framework to support more constraints on data, schemas, models and metadata. As implied by the Curry-Howard isomorphism [9], there is a correspondence between formulae in intuitionistic first-order logic and types (*propositions-as-types*). Therefore, we should be able to handle more logical constraints with new types.

References

1. Alotaibi, R.B.M.: Semantic Optimizations in Modern Hybrid Stores. Ph.D. thesis, University of California, San Diego (2022)
2. Dixon, J.: Pentaho, hadoop, and data lakes - james dixon's blog (2010)
3. Dybjer, P., Palmgren, E.: Intuitionistic type theory. In: Zalta, E.N., Nodelman, U. (eds.) The Stanford Encyclopedia of Philosophy. Metaphysics Research Lab, Stanford University, Spring 2023 edn. (2023)
4. Farid, M., Roatis, A., Ilyas, I.F., Hoffmann, H.F., Chu, X.: Clams: bringing quality to data lakes. In: International Conference on Management of Data, SIGMOD 2016, pp. 2089–2092 (2016)
5. Gillet, A., Leclercq, E., Savonnet, M., Cullot, N.: Empowering big data analytics with polystore and strongly typed functional queries. In: International Database Engineering & Applications Symposium, IDEAS 2020, pp. 1–10 (2020)
6. Griffioen, P.: Type inference for array programming with dimensioned vector spaces. In: Symposium on the Implementation and Application of Functional Programming Languages, IFL 2015, pp. 1–12 (2015)
7. Hai, R., Quix, C.: Rewriting of plain so tgds into nested tgds. Proc. VLDB Endowment **12**(11), 1526–1538 (2019)
8. Hai, R., Quix, C., Zhou, C.: Query rewriting for heterogeneous data lakes. In: Benczúr, A., Thalheim, B., Horváth, T. (eds.) ADBIS 2018. LNCS, vol. 11019, pp. 35–49. Springer, Cham (2018). https://doi.org/10.1007/978-3-319-98398-1_3
9. Howard, W.A.: The formulae-as-types notion of construction. To HB Curry: Essays Combinatory Logic, Lambda Calculus Formalism **44**, 479–490 (1980)
10. Koupil, P., Holubová, I.: A unified representation and transformation of multi-model data using category theory. J. Big Data **9**(1), 61 (2022)
11. Koupil, P., Hricko, S., Holubová, I.: A universal approach for multi-model schema inference. J. Big Data **9**(1), 1–46 (2022)
12. Muranushi, T., Eisenberg, R.A.: Experience report: type-checking polymorphic units for astrophysics research in haskell. ACM SIGPLAN Not. **49**(12), 31–38 (2014)
13. Scholly, E., et al.: Coining goldmedal: a new contribution to data lake generic metadata modeling. In: 23rd International Workshop on Design, Optimization, Languages and Analytical Processing of Big Data (DOLAP@ EDBT/ICDT 2021), vol. 2840, pp. 31–40 (2021)
14. Uotila, V., Lu, J., Gawlick, D., Liu, Z.H., Das, S., Pogossiants, G.: Multicategory: multi-model query processing meets category theory and functional programming. Proc. VLDB Endowment **14**(12), 2663–2666 (2021)

15. Zaharia, M., Chambers, B.: Spark: The Definitive Guide. O'Reilly Media Sebastopol, CA (2018)
16. Zhao, Y., Megdiche, I., Ravat, F., Dang, V.N.: A zone-based data lake architecture for IoT, small and big data. In: International Database Engineering & Applications Symposium, IDEAS 2021, pp. 94–102 (2021)

EXOS: Explaining Outliers in Data Streams

Egawati Panjei[(✉)] and Le Gruenwald

School of Computer Science, The University of Oklahoma, Norman, OK, USA
{egawati.panjei,ggruenwald}@ou.edu

Abstract. Real-time outlier detection is important in many data stream applications. To help analysts understand the detected outliers better, the outliers should be presented with their explanations. One type of explanations for an outlier is its set of outlying attributes which is a subset of features responsible for the outlier's abnormality. There exist techniques that generate outlying attributes in data streams; however, none simultaneously considers the cross-correlation among data streams, the unbounded volume of data, and concept drift. To fill this gap, we propose EXOS, a framework that generates outlying attributes in multi-dimensional data streams. For each outlier, it incrementally finds a local context to determine the decision boundary that separates the outlier from the normal data while handling both the unbounded volume of data and concept drift. It considers the potential data correlation within a data stream and across data streams to estimate the local context. The experiments using three real and two synthetic datasets show that, on average, EXOS achieves up to 49% higher F1 score and 29.6 times lower explanation time than existing algorithms.

Keywords: Outlier Explanation · Outlying Attributes · Data Streams

1 Introduction

An outlier in a dataset is a data point that has a significantly different value compared to other data points in the dataset [1]. The increasing demand for real-time analytics has created a need to apply real-time outlier detection over data streams [2, 3]. Unfortunately, these techniques do not provide explanations of why some data points are deemed to be anomalous, leaving analysts with no guidance to decide whether those objects require further actions. for critical applications like structural health monitoring (SHM) and intrusion detection, the investigation of whether the outliers are subjects of interest should be done fast. However, analysts' effort to investigate an outlier roughly corresponds to the number of attributes associated with the outlier [4]. The investigation takes time, but it can be done faster if each detected outlier is presented with a subset of attributes responsible for its abnormality. This type of explanation is known as *outlying attributes* [5].

A data stream is an infinite sequence of data points with explicit or implicit timestamps [2, 5]. Monitoring and processing data streams in real-time are bound to time and memory constraints as data arrive continuously. They require that every data point be processed online and incrementally [5]. Furthermore, data streams are known for

R. Wrembel et al. (Eds.): DaWaK 2023, LNCS 14148, pp. 25–41, 2023.
https://doi.org/10.1007/978-3-031-39831-5_3

concept drift, where the data distribution changes over time [6]. In data streams, data attributes may be correlated not only within a data stream but also across data streams (cross-correlation [5]) which can then be used to improve outlier explanations. The existing outlying attribute algorithms do not simultaneously consider the unbounded volume of data, concept drift, and cross-correlation in data streams [5]. Thus, to fill this gap, we propose EXOS, an algorithm to generate outlying attributes of each point outlier in data streams in real time that addresses all the three characteristics.

EXOS provides outlying attributes for each point outlier detected by any outlier detectors in data streams. We assume that EXOS and outlier detectors are independent processes that communicate through queues. To deal with the unbounded data volume, outlying attributes are generated based on a time-based tumbling window where a window is used to store a sequence of data in the main memory, and when a specified time period expires, i.e., when the window slides, all the data stored in the window are replaced with the new arriving sequence of data [11]. For each stream, when its window slides, EXOS will read the queue that stores the stream's outliers detected by an outlier detector and apply single-pass incremental computation on the data in the window to generate the outliers' outlying attributes.

EXOS is a local neighborhood-based outlier explanation technique. For each outlier, it defines a local context, which is a set of inlier neighbors of the outlier and uses the local context to find the outlier's outlying attributes. The local context is formed by considering the cross-correlations among data streams. The eigenvectors, used in forming the local context, are initially generated using offline data, yet since data streams are subject to concept drift, the eigenvectors are updated whenever the window slides using a single-pass eigendecomposition.

Our contributions are as follows: 1) we develop an algorithm that generates outlier explanations in terms of outlying attributes that considers the possible data attribute correlation within a data stream and across data streams, while simultaneously addressing the unbounded data volume and concept drift in data streams; and 2) we perform a comprehensive experimental analysis comparing EXOS with existing algorithms in terms of average precision, recall, F1 score, and execution time using real and synthetic datasets.

2 Related Work

In the recent survey of outlier explanations [5], there exist techniques that generate outlying attributes. For example, SFE [4] Applies subspace search with heuristics to find a subset of attributes where a detected outlier has the highest outlier score. Micenková Et Al.[7] and COIN [8] utilize inlier neighbors of an outlier to find its outlying attributes. They use a linear classifier to find a decision boundary between inlier and outlier classes. Attributes whose corresponding weights in the hyperplanes are higher than a threshold are considered outlying attributes. However, all these techniques are designed for static or non-stream data where they assume a finite amount of data and require multiple passes on the dataset; thus, they are unsuitable for dealing with the unbounded volume of data streams in real-time.

There are outlying attribute techniques proposed for data streams, for example, EXAD [9], MICOES [10], MacroBase [15, 16], and EXstream [16, 17]. EXAD employs a decision tree estimated from a neural network trained offline to generate the outlying attributes of each detected outlier in an online manner. However, the decision tree is never updated to address concept drift. MICOES absorbs arriving data points into micro-clusters and uses them to form an inlier class and an outlier class associated with each detected outlier. The inlier and outlier classes are used to generate outlying attributes. Since micro-clusters adapt to the changes in data distribution, the explanations also adapt. MacroBase produces outlier explanations using a prefix tree of frequent itemsets updated periodically, allowing it to handle concept drifts. EXstream generates outlier explanations based on the temporal context of the outliers, which is updated over time, making it hardly unaffected by concept drifts. Still, none of these algorithms consider the relationships among data streams. Thus, in this work, we develop a local context-based outlier explanation for data streams that consider the potential cross-correlations among data streams to improve outlying attribute generation while dealing with the unbounded volume and concept drift.

3 Preliminaries

This section formally defines the problem of finding outlying attributes of an outlier in data streams.

Definition 3.1 (Data Stream). A data stream S is an infinite sequence of data points $\{\mathbb{X}_i | i \geq 0\}$. Each $\mathbb{X}_i$ is a tuple of length $d + 1$ denoted as $\mathbb{X}_i = a_1, a_2, \ldots, a_d, t$ where d is the number of attributes, a_k is the value of the k-th attribute, and t is the associated timestamp when the tuple is recorded or collected.

We consider a set of m concurrent data streams $\mathbb{S} = \{S_1, \ldots, S_m\}$ where $m \geq 1$. The i-th data point in a data stream S_j is denoted as $\mathbb{X}_i^j$. The attributes in S_j can differ from those in S_k. We denote $\mathbb{X}_i^j$ as $\mathbb{O}_i^j$ when $\mathbb{X}_i^j$ is detected as an outlier. We assume that all data streams have the same arrival rate (synchronous data streams) and each data stream has a corresponding outlier detector.

Data streams arrive continuously and are inherently unbounded; hence, keeping all data in the memory for real-time processing is impossible. EXOS handles this by processing data on a time-based tumbling window residing in the memory as defined in Definition 3.2.

Definition 3.2 (Tumbling Window). Given a time interval $\mathbb{T}$ that consists of start and end timestamps, a tumbling window W is a finite sequence of data points or tuples $(\mathbb{X}_n, \mathbb{X}_{n+1}, \ldots, \mathbb{X}_N)$ where $\mathbb{X}_N.t - \mathbb{X}_n.t \leq \mathbb{T}$. All the data points in the window will expire when $\mathbb{T}$ expires, i.e., when the window slides.

The outlying attributes of an outlier are the subset of attributes responsible for the abnormality of the outlier. They are formally defined in Definition 3.3.

Definition 3.3 (Outlying Attributes). [5]. Given an outlier $\mathbb{O}$, a set of d dimensions $\mathcal{D} = \{A_1, A_2, \ldots, A_d\}$ where $\mathbb{O} \in A_1 \times A_2 \times \cdots \times A_d$, an outlier attribute contribution score function $h : \mathcal{D} \to \mathbb{R}$ that generates a real-value quantifying the contribution of each attribute to the abnormality of $\mathbb{O}$, and an outlying contribution score threshold $\gamma \geq 0$, the outlying attributes of $\mathbb{O}$ is a subspace $F \subseteq \mathcal{D}$ such that $\forall A_i \in F, h(A_i) > \gamma$.

Problem Definition: Given a set of m synchronous multi-dimensional data streams $\mathbb{S} = \{S_1, S_2, \ldots, S_m\}$, an outlier $\mathbb{O}_i^j$ identified by an arbitrary outlier detector, and an outlying contribution score threshold $\gamma \geq 0$, the problem is to find all the outlying attributes of $\mathbb{O}_i^j$ accurately and efficiently.

4 The Proposed Algorithm: EXOS

Algorithm 1: EXOS ($\mathbb{S}$, k, D, γ, $\mathbb{A}$, $\mathbb{T}$, $init\_data$, $n\_clusters$, $outlier\_queues$)

Input: $\mathbb{S}$ the set of m concurrent data streams, k the number of eigenvectors, D the total number of attributes, γ the attribute contribution threshold, $\mathbb{A}$ the list of the attribute names in the streams, $\mathbb{T}$ the window size represented as a time interval, $init\_data$ the offline data, $n\_clusters$ the list of the IDs of the clusters for each data stream, $outlier\_queues$ the list of m queues storing the outliers of the m streams.

Output: $\mathbb{F}_j$ the set of outlying attributes of each outlier in each stream j

```
1    Q := initialize_eigen_vectors (init_data)
2    Cₐ := initialize_set_of_clusters(init_data, n_clusters)
3    end_ts := current_time ()
4    start_ts := end_ts – 𝕋
5    while (true): ## repeat indefinitely
6        L₀ := get_outliers(outlier_queues, start_ts, end_ts)
7        N := the number of data points in each window
8        do in parallel
9            Q, est_list := Estimator (𝕊, L₀, k, D, Q, start_ts, end_ts, N) ##Component 1
10           parallel for j in [1: m]
11               Wⱼ := Sⱼ[start_ts, end_ts] # get data in the current window
12               ℂⱼ := TemporalNeighborClustering(Wⱼ, Cₐ[j]) ## Component 2
13               Cₐ[j] := the latest centroids in ℂⱼ
14           end parallel for
15       end do in parallel
16       parallel for j in [1: m]
17           Ô_j := est_list[j] ## get estimated normal values of the outliers in stream j
18           𝕆_j := Wⱼ[L₀[j]] ## get outlier values in stream j
19           𝔽_j := OutlyingAttributesGenerator(Ô_j, 𝕆_j, ℂ_j, γ, A[j]) ) ## Component 3
20           yields(𝔽_j) ## return the outlying attributes of the outliers in the current window of stream j
21       end parallel for
22       while (end_ts + 𝕋 > current_time()) : wait() end while
23       end_ts = current_time (); start_ts = end_ts – 𝕋
24   end while
```

To determine the outlying attributes of the outliers detected in a recent window, EXOS goes through *six steps*: (1) combine the windows from all streams and use the data attribute correlation within each stream and across streams to estimate the normal value of each outlier $\mathbb{O}_i^j$; (2) in parallel with Step (1), incrementally group all the data points in the window into clusters in each data stream S_j; (3) use the estimated normal value of each $\mathbb{O}_i^j$ and its closest cluster to form a local context of the outlier (the inlier class); (4) for each $\mathbb{O}_i^j$, form an outlier class by randomly generating the auxiliary outlier data points that have uncorrelated multivariate normal distribution centered at $\mathbb{O}_i^j$; (5) find a decision boundary that separates the inlier class from the outlier class using a linear classifier; and (6) use the weights of the decision boundary to determine the outlying contribution scores of the attributes of $\mathbb{O}_i^j$. The attributes having the outlying contribution scores greater than the threshold γ are the outlying attributes of $\mathbb{O}_i^j$.

Algorithm 1 shows the overall EXOS algorithm. It consists of the initialization and online phases. Using offline data, EXOS initializes Q, the estimated eigenvectors (Line 1), and C_a the list of the initial clusters' centroids in each stream (Line 2). The online phase indicated in Lines 3–24 has three key components: (i) Estimator *Est* (Line 9), (ii) Temporal Neighbor Clustering $C_1, C_2, \ldots, C_m$ (Line 12) and (iii) Outlying Attribute Generators $\mathcal{G}_1, \mathcal{G}_2, \ldots, \mathcal{G}_m$ (Line 19). The *Est* component corresponding with *Step (1)* uses the eigenvectors to estimate the normal values of outliers. The eigenvectors capture the correlation among attributes from the same data stream and attributes from different streams (cross-correlation). They are incrementally updated whenever the window slides to ensure they deal with concept drift. The C_j component, which handles *Step (2)*, groups the data points in the window of S_j, based on a symmetric distance function. The clusters serve as a temporal context which is the neighborhood that will be used by $\mathcal{G}_j$ to find the outlying attributes of $\mathbb{O}_i^j$. The $\mathcal{G}_j$ component manages *Steps (3)–(6)*.

We now describe the details of the three key components of EXOS.

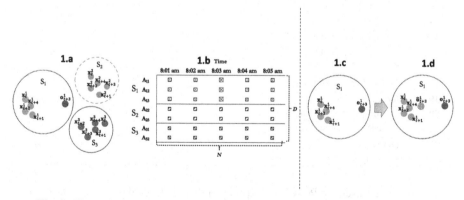

Fig. 1. Illustration of finding a local context of an outlier in concurrent data streams

4.1 Estimator

Estimator, *Est*, approximates the normal value $\hat{\mathbb{O}}_i^j$ of each outlier $\mathbb{O}_i^j$ by using the information from other data streams. Let us consider an example of 3-source data streams depicted in Figs. 1a and b. Between 8:01 am and 8:05 am, the outlier detector at S_1 detects $\mathbb{X}_{t+3}^1$ as an outlier (denoted as $\mathbb{O}_{t+3}^1$). Its normal value estimation is denoted as $\hat{\mathbb{O}}_{t+3}^1$ in Fig. 1d. Suppose S_1, S_2, and S_3 in Fig. 1a have three, two, and two attributes, respectively. The data are observed between 8:01 am and 8:05 am where an outlier is detected in S_1 at 8:03 am. We can represent the data points from those streams as a 7×5 matrix as shown in Fig. 1b.

Algorithm 2: Estimator $(\mathbb{S}, L_\mathbb{O}, k, D, Q, start\_ts, end\_t, N)$

Input: $\mathbb{S}$ the set of m concurrent data streams in the current window, $L_\mathbb{O}$ the list of the indices of the outliers in the window, k the number of eigenvectors, D the total number of attributes, Q the estimated eigenvectors, *start_ts* the starting timestamp of the window, *end_ts* the end timestamp of the window, and N the number of data points in the window,

Output: eigenvector matrix Q, list of estimated outliers $\hat{\mathbb{O}}_j$ *est_list*

1	set O as an empty list
2	create a set of outlier indices from $L_\mathbb{O}$, $I_\mathbb{O} = set(L_\mathbb{O})$
3	Initialize a zero matrix Z of size $D \times k$
4	**for each** t between start_ts and end_ts **do**
5	**if** $|\mathbb{S}| > 1$ **do** ## when m > 1
6	x = combine tuples in $\mathbb{S}$ that share the same t into $x \in \mathbb{R}^D$
7	**else**
8	x = tuple at t
9	**end if**
10	$Z := Z + \frac{1}{N}xx^T Q$
11	**if** t is in $I_\mathbb{O}$ **do**
12	O.append(x)
13	**end if**
14	**end for**
15	$Q := $ QR-decomposition of Z
16	Set O as a matrix $\mathbb{O}$
17	$\hat{\mathbb{O}} = \mathbb{O}QQ^T$
18	Set *est_list* as a list of length $|\mathbb{S}|$
19	**for** $0 \le j < |\mathbb{S}|$ **do**
20	$\hat{\mathbb{O}}_j := $ slice_matrix$(\mathbb{O}, L_\mathbb{O}[j])$
21	*est_list*[j] = $\hat{\mathbb{O}}_j$
22	**end for**
23	**return** Q, *est_list*

To find $\hat{\mathbb{O}}_i^j$, *Est* uses the PCA-based approach that captures the correlations among the observed attributes to derive the k eigenvectors (principal components) and stores them in the matrix Q. In each window, *Est* combines the data points that share the same timestamp from all the data streams. If there is only one data stream, Est will use only the data points in that stream. The combined tuples, after their timestamp is omitted, are used to compute the eigenvectors in the matrix Q. When there are outliers $\mathbb{O}$ in that window, Q is then used to estimate the normal values of those outliers as follows:

$$\widehat{\mathbb{O}} = \mathbb{O}QQ^T \qquad (1)$$

where the matrix $\mathbb{O} \in \mathbb{R}^{|I_{\mathbb{O}}| \times D}$, $|I_{\mathbb{O}}|$ is the number of unique timestamps when the outliers are detected in the current window, D is the total number of attributes in $\mathbb{S}$, and $Q \in \mathbb{R}^{D \times k}$, $k < D$. Referring to the transpose of the matrix in Fig. 1b), $\mathbb{O}$ is a 1×7 matrix because there is one outlier detected between 8:01 am and 8:05 am.

The incremental update of the eigenvectors or principal components allows Est to adapt to concept drift. When the data distribution changes, the eigenvectors adjust to that change. While the naïve PCA algorithm can be used to build new eigenvectors in every window, it ignores the data seen in the previous windows and requires multi-passes on the current window. Hence to update the eigenvector Q, we adopt DBPCA, a single-pass eigendecomposition algorithm described in [25].

Algorithm 2 explains how Est works. In Line 1, the algorithm initializes an empty list O to store the outliers detected in the current window. Line 2 ensures that there are no duplicate outlier timestamps if the outliers are detected in two or more streams at the same timestamp. Continuing our example in Fig. 1, $L_{\mathbb{O}} = [\{8:03\}, \{\}, \{\}]$ and $I_{\mathbb{O}} = \{8:03\}$. Line 3 gets the eigenvector matrix Q generated from the previous window. Lines 4–15, which are the DBPCA approach, are responsible for updating Q of the current window by first absorbing each combined vector x into a Z matrix and then then conduct QR decomposition on Z to get the updated eigenvectors Q. Lines 16–17 estimate the normal values of the outliers as one matrix $\widehat{\mathbb{O}} \in \mathbb{R}^{D \times |I_{\mathbb{O}}|}$. Finally, Lines 18–22 break down $\widehat{\mathbb{O}}$ so that each stream gets the estimated normal values of its outliers. The algorithm returns the updated estimated eigenvector matrix and the list of estimated normal values of outliers.

4.2 Temporal Neighbor Clustering

The temporal neighbor clustering component is a set of m independent functions $C_1, C_2, \ldots, C_m$ that runs in parallel with the estimator component. Each C_j receives data points from the data stream S_j and forms the temporal context that will be used in generating the outlying attributes of each $\mathbb{O}_i^j \in W_j^T$.

As depicted in Fig. 2, the temporal context of a data stream can be further grouped into a set of clusters $\mathbb{C} = \{C_1, \ldots, C_l\}$ such that $C_a \cap C_b = \varnothing$ for all $a \neq b$ and $W_j^T = \bigcup_{a=1}^l C_a$. $\mathbb{C}$ is considered the neighborhood outliers in W_j^T. Each cluster has a center $c_a \in \mathbb{R}^d$ and has the total number of points assigned to the cluster. A point $\mathbb{X} \in W_j^T$ is assigned to the cluster C_a if the distance between $\mathbb{X}$ and c_a is the closest. By excluding the timestamp attribute from $\mathbb{X}$, the distance function denoted as $d(\mathbb{X}, c_a)$ is formulated as $d(\mathbb{X}, c_a) = ||\mathbb{X} - c_a||$.

Recall that the continuous arrival of data points demands a single-pass computation; thus, forming clusters of data points in the window should be done incrementally. We use sequential k-means [12] to form $\mathbb{C}$. Due to the unbounded amount of data points in data streams, we cannot store all the data points in all the clusters in the main memory. To handle this problem, for each cluster $C_a \in \mathbb{C}$ we only keep n_a, the number of points in the cluster, and c_a, the centroid of the cluster.

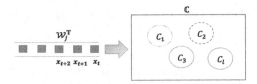

Fig. 2. Grouping Data Points of a Stream Sj

4.3 Outlying Attribute Generators

The outlying attribute generator component is a set of m independent functions $\{\mathcal{G}_1, \mathcal{G}_2, \ldots, \mathcal{G}_m\}$. Each $\mathcal{G}_j$ receives the inputs produced by the estimator component *Est* and the temporal clustering component $\mathcal{C}_j$. For each $\mathbb{O}_i^j \in \mathbb{O}_j^T$, $\mathcal{G}_i$ will form an inlier class and an outlier class. The inlier class is generated by finding a cluster $C_a \in \mathbb{C}_j$ whose centroid c_a is the closest to $\hat{\mathbb{O}}_i^j$. We denote by dc_a the distance between c_a and $\mathbb{O}_i^j$. Recall that the cluster no longer keeps the data points it has absorbed, but it has the information about n_a the number of data points. Initially, the inlier class only has two members: c_a and $\hat{\mathbb{O}}_i^j$. The additional members are then added by generating n_a auxiliary data points whose distances to c_a is less than dc_a. All members of the inlier class are labeled 0. Figure 3a describes how the inlier class is formed.

The outlier class is formed by generating multivariate normal data centered around $\mathbb{O}_i^j$. To ensure that the members of the outlier class do not overlap with those of the inlier class, the standard deviation is set to be less than one-third of the distance between $\mathbb{O}_i^j$ and $\mathbb{O}_i^j$. All objects in the outlier class are labeled 1. Figure 3b illustrates how the outlier class is formed. After forming the inlier and outlier classes, the next step is to find a decision boundary that separates both groups using a linear support vector machine (SVM) [13]. The classifier applies lasso regularization [14] such that the resulting hyperplane has zero weights on any attributes that are not relevant in forming the boundary. The absolute values of the hyperplane's weights, $H_{\mathbb{O}_i^j} \in \mathbb{R}^{d_j}$ are used to calculate the outlying contribution scores stored in the vector $f_{\mathbb{O}_i^j} \in \mathbb{R}^{d_j}$ for the attributes of the outlier $\mathbb{O}_i^j$. This vector is computed using Eq. (2). Each attribute corresponds to a score in f_α. The attributes having a score higher than a given threshold are the outlying attributes of $\mathbb{O}_i^j$.

$$f_{\mathbb{O}_i^j} = \frac{H_{\mathbb{O}_i^j}}{\sum_{b=0}^{d_j-1} H_{\mathbb{O}_i^j}[b]} \tag{2}$$

5 Evaluation

In this section, we discuss our experimental setup and results comparing EXOS with four existing algorithms: COIN [8], MICOES [10], MacroBase [15, 16], and EXstream [9, 16, 17]. Like EXOS, all these algorithms are agnostic to the algorithm used for outlier detection. COIN is one of the algorithms that use inlier neighbors of an outlier as a local

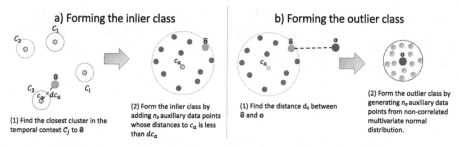

Fig. 3. Forming a) the inlier class and b) the outlier class

context to find the outlier's outlying attributes. It is designed for static data, yet we run it in batches to compare it with EXOS. The other three algorithms are proposed for data streams. EXstream is intended for interval or collective outliers, yet it can also be applied for point outliers. The source code of each algorithm is publicly available.

5.1 Experimental Setup

Dataset. We use three real datasets that reflect multiple-source data streams: Intel, Microtremor, and AMPds2. The Intel dataset [18] is multiple-sensor data where each sensor sent multi-dimensional data with the attributes of humidity, temperature, light, and voltages every 31 s. Since the sensors started sending measurements on different dates, we simulate data gathered between March 1st –31st, 2004 from Sensors 7, 9, 23, 25, 26, 29, 36, 38, and 44. The Microtremor dataset [19] was collected using temporary broadband seismometers over continuous time intervals ranging between 2 and 2.8 h. After examining the metadata, we decide to use the data collected from 4 seismic recording stations, 2030, 2031, 2032, and 2033, because they were collected on the same day. The AMPds dataset [20] contains the measurements of 21 electricity power, 3 water, and 2 natural gas meters at one-minute intervals of a residential house. Each meter sent a total of 1, 051, 200 readings for 2 years (April 2012 to March 2014). Power, water, and gas have 11, 2, and 3 attributes, respectively. These datasets do not have ground truth information for outliers and their outlying attributes. For performance evaluation purposes, we synthetically generate that information. Specifically, for each dataset, we inject outliers into the dataset by randomly selecting around 1% of the original data points to be outliers. For each selected outlier, we randomly select a number of attributes from the dataset to be the outlying attributes and set the value of each of those attributes to be far away from its mean value. For example, suppose in the Intel dataset, the data point #100 is chosen as an outlier and the temperature as the outlying attribute. The temperature value of the data point #100 is set to be either *max (temperature)* + *delta * standard deviation(temperature)* or *min(temperature)* – *delta * standard deviation(temperature)*, where *delta* >3. The summary of the datasets used for the evaluation is provided in Table 1.

In addition, we create synthetic datasets that consist of five and four groups of data whose attributes are correlated. Each group represents a data stream and has 10 attributes. To ensure that the data streams are correlated, we first generate a correlation matrix using a technique described in [21]. This technique allows us to specify whether the data attributes within and across data streams are highly or weakly correlated. The correlation matrix is then used to generate multivariate normal data. We inject outliers into the dataset in the same way as we do for the real datasets.

Table 1. Summary of the Datasets

Datasets	# Streams	Total number of data points	# Outliers injected	# Attributes per stream
Microtremor	4	7 M	70 K (1%)	3
AMPds2	25	26 M	260 K (1%)	2, 3, 11
Intel	9	401 K	4.5 K (1.12%)	4
Synthetic 1	5	50 K	500 (1%)	10
Synthetic 2	4	40 K	400 (1%)	10

Evaluation Metrics. We measure the effectiveness of the algorithms in finding outlying attributes using average precision, average recall, and average F1 score. For each outlier, we compute the numbers of True Positives (TP), False Positives (FP), False Negatives (FN), and True Negatives (TN) of its outlying attributes. For each outlier, we compute the numbers of True Positives (TP), False Positives (FP), False Negatives (FN), and True Negatives (TN) of its outlying attributes. We then use the information to compute Precision $= \frac{TP}{TP+FP}$, Recall $= \frac{TP}{TP+FN}$, F1 Score $= \frac{2*Recall*Precision}{Recall+Precision}$. In addition to measuring the accuracy of the algorithms, we also consider their efficiency. Given that we are dealing with a continuous arrival of data, we need to ensure that the algorithms can keep up with the incoming data in a timely manner. Therefore, we also measure the average explanation time of the algorithms as part of our evaluation process.

Software and Hardware. We implement EXOS in Python 3 and simulate the multi-source data streams in parallel using the multiprocessing package. Our source code is available on GitHub https://github.com/egawati/exos. We use the Python implementations of MacroBase and EXstream that are provided for the Exathlon benchmark [16]. We add some helper functions to the competitive algorithms so that they can run in parallel as we simulate synchronous data streams. The experiments were conducted on a MacBook Pro: macOS Catalina, Processor 2.7 GHz Quad-Core Intel Core i7, Memory 16 GB 1600 MHz DDR3.

5.2 Results and Analysis

We evaluate EXOS by simulating each dataset described in Sect. 5.1 as data streams. Since EXOS generates outlying attributes in the tumbling window where the number of data points inside the window depends on the data point arrival rate; we set the window

Table 2. Algorithm Performance Evaluation

Datasets	Algorithms	Avg precision	Avg recall	Avg F1 Score	Avg Execution Time (second)
Intel	COIN	0.49	**0.99**	0.63	0.042
	MICOES	0.5	0.91	0.61	0.020
	MacroBase	0.49	0.92	0.59	0.016
	EXstream	0.85	0.59	0.66	0.011
	EXOS	**0.95**	0.84	**0.85**	**0.010**
Microtremor	COIN	0.33	**0.99**	0.49	0.045
	MICOES	0.33	**0.99**	0.50	0.089
	MacroBase	0.34	**0.99**	0.51	0.007
	Exstream	**0.87**	0.87	0.87	**0.003**
	EXOS	0.83	**0.99**	**0.89**	**0.003**
AMPds2	COIN	0.43	0.86	0.54	0.045
	MICOES	0.39	0.80	0.48	0.020
	MacroBase	0.56	**0.89**	**0.64**	0.097
	Exstream	0.91	0.42	0.52	**0.007**
	EXOS	**0.94**	0.48	0.58	**0.007**
Synthetic 1	COIN	**1.00**	0.52	0.64	0.072
	MICOES	0.47	0.79	0.54	0.055
	MacroBase	0.3	**1.00**	0.44	0.599
	Exstream	0.98	0.45	0.56	0.050
	EXOS	0.84	0.87	**0.82**	**0.047**
Synthetic 2	COIN	0.42	0.86	0.54	0.076
	MICOES	0.51	0.58	0.51	0.234
	MacroBase	0.34	**1.00**	0.50	0.629
	Exstream	**0.99**	0.44	0.58	0.093
	EXOS	0.88	0.97	**0.91**	**0.058**

size to be 1,440 data points (1 day) for Intel and AMPds2 and 1,000 data points for Microtremor and Synthetic. Table 2 shows that EXOS has the highest average precision for Intel and AMPds2. Exstream get the best average precision for Microtremor and Synthetic 2 while COIN wins on Synthetic 1. When it comes to average recall, MacroBase wins for AMPds2, Synthetic 1 and 2, while COIN dominates for Intel. For Microtremor, the four algorithms achieve an average recall of 0.99, while Exstream 0.87.

Obviously, there is a tradeoff between precision and recall among these algorithms. Precision measures the extent of error by False Positives, while recall deals with the error caused by False Negatives. F1 score balances those scores. EXOS achieves the best F1 score in all the datasets except for AMPds2. The F1 scores of EXOS are 22%, 40%, 4%, 18%, and 37% better than those of COIN, 24%, 49%, 10%, 28%, and 40% better than those of MICOES, 26%, 38%, −6%, 38%, 41% better than those of MacroBase, and 19%, 2%, 6%, 26%, and 33% better than those of Exstream for the Intel, Microtremor, AMPds2, Synthetic 1, and Synthetic 2 datasets, respectively.

Table 2 shows that for all the datasets, EXOS has the fastest average execution time for generating outlying attributes for outliers. EXOS is 4.2, 15, 6.4, 1.5, and 1.3 times faster than COIN, 2, 29.6, 2.8, 1.1, and 4 times faster than MICOES, and 1.6, 2.3, 13.8, 12.7, and 10.8 times faster than MacroBase for the Intel, Microtremor, AMPds2, Synthetic 1, and Synthetic 2 datasets, respectively. Even though EXOS and Exstream have similar average execution times, EXOS outperforms Exstream in the average F1 scores for all the datasets.

The Impact of Concept Drift. We investigate the impact of data distribution changes, known as concept drift, on the performance of the algorithms. We simulate four data streams, each having ten attributes, eleven windows, and 1K data points per window. Our experiments focus on two key questions: (Q1) Does the performance of the algorithms alter when a concept drift occurs in a specific window?, and (Q2) How does the frequency of concept drift occurrences affect the algorithms' performance?

To address Question (Q1), we simulate the synthetic datasets with varying locations of concept drift occurrences, i.e., varying the IDs of the windows where concept drifts occur. Following the similar approaches used to verify that concept drifts indeed occur in a dataset [22, 23], we confirm the presence of concept drift by adding binary classification labels to the datasets and then training a logistic regression classifier using the data in Window 0 before testing it using the data in Windows 1 through 10. It is worth noting that this classification model is designed for static data and thus is expected to perform similarly across all the windows in the absence of concept drifts. However, as shown in Fig. 4, once there is a change in data distribution at Window i, the classifier performance begins to decline, indicating the presence of concept drifts in the datasets.

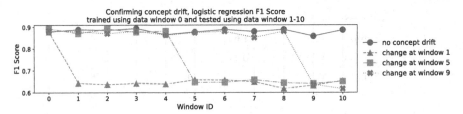

Fig. 4. Using a static data-based classifier to confirm whether a dataset indeed has concept drift

Figure 5 demonstrates the performance of the outlying attribute algorithms when the location of a concept drift occurrence is varied from Window 0 to Window 10. EXOS achieves the highest average F1 scores, followed by Exstream, MacroBase, MICOES, and COIN. However, MacroBase has the highest average explanation time, while other algorithms performed similarly. The consistent performance is maintained by all the algorithms regardless of the concept drift locations as their model and/or temporal context are updated in each window. Thus, we can conclude that the performance of the algorithms is unaffected by the locations of concept drifts.

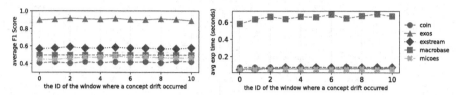

Fig. 5. The impact of the location of concept drift on the algorithms' average F1 Score and explanation time

To answer Question (Q2), we vary the occurrences of concept drifts across eleven windows ranging from 0% to 100%. A 0% concept drift indicates that Windows 1–10 have the same data distribution as Window 0. A 50% concept drift means from Window 1 through Window 10; there are five windows where data distribution changes. A 100% concept drift means the data distribution changes every time the window slides. We run experiments on four synthetic datasets. Sets 1, 2, and 3 are composed of multivariate Gaussian data. In Set 1, a change in the mean vectors occurs during specific windows while the covariance matrices remain unchanged. In contrast, in Set 2, only the covariance matrices change while the mean vectors stay constant. Finally, both the mean vectors and covariance matrices change simultaneously in Set 3.

Set 4 has mixed data distributions consisting of randomly chosen multivariate Gaussian, Beta, Gamma, and Exponential distributions. For instance, when we set the percentage concept drifts to be 50%; Windows 0–2 can have a Gaussian distribution, while Window 3 may use a gamma distribution. Likewise, Windows 4–7 could be exponential, Window 8 beta and finally, Windows 9–10 are gamma distributed.

Figure 6 displays the performance of the algorithms on Sets 1–4, with EXOS having the best average F1 score. All the algorithms have similar average explanation times while MacroBase has the highest one. The frequency of concept drifts hardly has any effect on the performance of EXOS, EXstream, Coin, and MacroBase. With EXOS, the eigenvectors used to estimate the normal values of the outliers in a window are updated when the window slides. The same update applies to its temporal context component, which, together with the estimated normal values of the outliers, forms the inlier and outlier classes. Therefore, even when a window data distribution differs from that in the previous window, the EXOS components are constantly adjusted, allowing it to maintain its performance. A similar window-sliding update on the explanation model also applies to the other algorithms. Even though COIN is intended for static data, we run it in batches such that the subsets of data points used to build its outlier explanation model are continually updated as the window slides. The average F1 score slightly fluctuates on MICOES, indicating that the temporal context formed using denstream-based micro-clusters in a window still mixes with some micro-clusters from the previous window. MICOES's average explanation time also slightly increases in Sets 2–4 when the frequency of concept drifts is getting higher. This explanation time increase relates to the maintenance of its micro-clusters. When the data distribution changes more often, the state of its micro-clusters also changes, requiring more time to maintain/update.

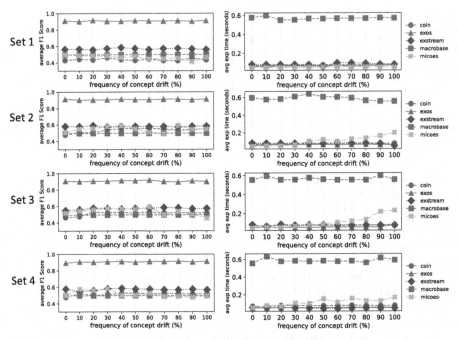

Fig. 6. The Impact of the frequency of concept drifts on average F1 score and explanation time

EXOS' Parameter Study. We also studied the impact of three EXOS user-defined parameters using the synthetic dataset, which we report below.

1) Impact of the number of eigenvectors (k)

In order to generate the estimated normal value of each outlier found in each window, EXOS needs the estimated eigenvectors (principal components) whose size depends on k. k can be varied from 1 to D (the total number of attributes). In our study, we vary k from 1 to 10. Figure 7 tells us that k has an impact on the average precision, recall, and F1 score. As k increases, the average recall also increases but the average precision and F1 score decrease. However, the average F1 score remains almost constant. It is not surprising that as k gets larger, the average explanation time gets longer as the EXOS' estimation component uses more eigenvectors. Therefore, to provide the explanation quickly, a small value of k such as $k = 1$ would be preferred.

2) Impact of the outlier attribute contribution threshold (γ)

The threshold γ tells EXOS which attributes to consider as the outlying attributes of an outlier after finding the decision boundary that separates the inlier class from the outlier class for that outlier. $\gamma = 0.01$ indicates that only attributes that have the outlying contribution scores greater than 1% are considered the outlying attributes. The study on the synthetic dataset reveals that γ has a small impact on the algorithm's performance as shown in Fig. 7.

3) Impact of the window size

In generating outlier explanations, EXOS depends on the time period that determines the size of the tumbling window used by its components. For the synthetic data, we varied the window size from 1,000 to 10,000 data points. Figure 7 shows that when the window size increases, EXOS' average recall increases but average precision and F1 score decrease. Thus, more inliers compared with the outlier does not mean a more accurate explanation of outlying attributes. The larger the window size also means the larger the average running time. The running time using the window size of 10K is about three times of that using the window size of 1K.

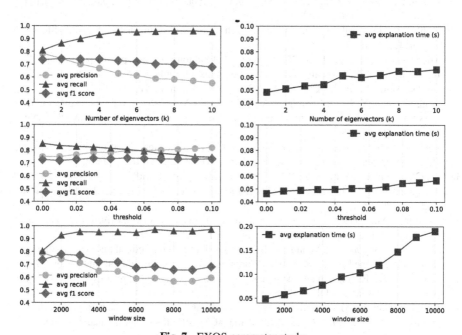

Fig. 7. EXOS parameter study

6 Conclusions

Our experiments on the three real datasets and two synthetic datasets show that EXOS is promising in generating outlying attributes for each outlier in data streams. It addresses the data streams' characteristics of the unbounded data volume, concept drift, and cross-correlation. For all the studied datasets, it achieves the best average F1 score (except for one dataset) and the fastest average running time compared with the existing algorithms. For future work, we plan to extend the estimator component to deal with asynchronous data streams.

References

1. Chandola, V., Banerjee, A., Kumar, V.: Anomaly detection: A survey. ACM Comput. Surv. **41**, 15:1–15:58 (2009)
2. Tran, L., Mun, M.Y., Shahabi, C.: Real-time distance-based outlier detection in data streams. In: Proceedings of the VLDB Endowment (2020)
3. Yoon, S., Lee, J.G., Lee, B.S.: NETS: Extremely fast outlier detection from a data stream via set-based processing. In: Proceedings of the VLDB Endowment (2018)
4. Siddiqui, M.A., Fern, A., Dietterich, T.G., Wong, W.-K.: Sequential feature explanations for anomaly detection. ACM Trans. Knowl. Discov. Data **13**, 1–22 (2019)
5. Panjei, E., Gruenwald, L., Leal, E., Nguyen, C., Silvia, S.: A survey on outlier explanations. VLDB J. **31**, 977–1008 (2022)
6. Sadik, M., Gruenwald, L.: Research issues in outlier detection for data streams. SIGKDD Explor. **15**, 33–40 (2014)
7. Micenková, B., Ng, R.T., Dang, X.-H., Assent, I.: Explaining outliers by subspace separability. In: 2013 IEEE 13th International Conference on Data Mining, pp. 518–527 (2013)
8. Liu, N., Shin, D., Hu, X.: Contextual outlier interpretation. In: IJCAI (2018)
9. Song, F., Diao, Y., Read, J., Stiegler, A., Bifet, A.: EXAD: a system for explainable anomaly detection on big data traces. In: IEEE International Conference on Data Mining Workshops, pp. 1435–1440 (2018)
10. Panjei, E., Gruenwald, L., Leal, E., Nguyen, C.: Micro-clusters-based outlier explanations for data (2021) Streams. https://sites.google.com/view/andea2021/accepted-papers
11. Li, C.L., Lin, H. ten, Lu, C.J.: Rivalry of two families of algorithms for memory-restricted streaming PCA. In: Proceedings of International Conference on Artificial Intelligence and Statistics (2016)
12. Ackerman, M., Dasgupta, S.: Incremental clustering: the case for extra clusters. In: Advances in Neural Information Processing Systems (2014)
13. Boser, B.E., Guyon, I.M., Vapnik, V.N.: Training algorithm for optimal margin classifiers. In: Proceedings of the Fifth Annual ACM Workshop on Computational Learning Theory (1992)
14. Tibshirani, R.: Regression shrinkage and selection via the lasso: a retrospective. J. R. Stat. Soc. Ser. B Stat. Methodol. **73** (2011)
15. Bailis, P., Gan, E., Madden, S., Narayanan, D., Rong, K., Suri, S.: MacroBase: prioritizing attention in fast data. In: Proceedings of the ACM SIGMOD International Conference on Management of Data (2017)
16. Jacob, V., Song, F., Stiegler, A., Rad, B., Diao, Y., Tatbul, N.: Exathlon: A benchmark for explainable anomaly detection over time series. In: Proceedings of the VLDB Endowment (2021)
17. Zhang, H., Diao, Y., Meliou, A.: EXstream: explaining anomalies in event stream monitoring. In: International Conference on Extending Database Technology (2017)
18. Bodik, P., Hong, W., Guestrin, C., Madden, S., Paskin, M., Thibaux, R.: Intel Lab Data (2004). http://db.csail.mit.edu/labdata/labdata.html
19. Buckreis, T., Winders, A., Wang, P., Brandenberg, S., Stewart, J.: Microtremor Data Collected in Sacramento-San Joaquin Delta Region of California (2021). https://doi.org/10.17603/ds2-dk6t-8610

20. Makonin, S.: AMPds2: The Almanac of Minutely Power Dataset (Version 2) (2016). https://doi.org/10.7910/DVN/FIE0S4
21. Hardin, J., Garcia, S.R., Golan, D.: A method for generating realistic correlation matrices. Ann Appl Stat. **7**, 1733–1762 (2013)
22. Gu, F.: Concept Drift Detection for Machine Learning with Stream Data (2019). https://opus.lib.uts.edu.au/bitstream/10453/140165/2/02whole.pdf
23. Das, S.: Best Practices for Dealing with Concept Drift. https://neptune.ai/blog/concept-drift-best-practices. Accessed 03 Apr 2023

Motif Alignment for Time Series Data Augmentation

Omar Bahri[✉][ID], Peiyu Li[ID], Soukaina Filali Boubrahimi[ID], and Shah Muhammad Hamdi[ID]

Computer Science Department, Utah State University, Logan, UT, USA
{omar.bahri,peiyu.li,soukaina.boubrahimi,s.hamdi}@usu.edu

Abstract. In this paper, we propose MotifAug, a parameter-free, pattern mixing-based time series data augmentation method that improves previous approaches in the literature. MotifAug leverages the warping path constructed by MotifDTW, a novel alignment method that uses the Matrix Profile (MP) motif discovery mechanism and Dynamic Time Warping (DTW) to align two time series data instances.

Keywords: Data Augmentation · Time Series Data · Time Series Motifs

1 Introduction

In contrast to other domains such as image recognition, data augmentation techniques for time series datasets are still in their early stages. Iwana et al. [4] categorize the current efforts into four distinct groups: random transformations, generative models, decomposition methods, and pattern mixing. In this paper, we focus on the latter. Pattern mixing time series augmentation methods are based on the idea that by combining two data samples belonging to the same dataset, and particularly to the same dataset class, it is possible to generate new, realistic data samples. The main advantages of these methods is that they neither make assumptions about the validity of specific transformations in certain data domains, nor do they require extensive training to learn models that mimic the dataset distribution. Thus, they are domain agnostic and can be used on-the-fly. Notable examples include the Synthetic Minority Over-sampling Technique (SMOTE) [1], Weighted-Dynamic Barycenter Averaging (wDBA) [2], SuboPtimAl Warped time series geNEratoR (SPAWNER) [5], and guided warping [3].

The Matrix Profile (MP) time series data structure has been introduced recently by Yeh et al. [8] and was met with huge success in the time series mining community. By considerably reducing the time complexity required for the all-pairs similarity search for time series subsequences, the MP has been quickly leveraged to solve a plethora of problems and improve algorithms. However, to the extent of our knowledge, it has not been used yet in the context of data augmentation. In this paper, we introduce (1) MotifDTW, a novel algorithm

R. Wrembel et al. (Eds.): DaWaK 2023, LNCS 14148, pp. 42–48, 2023.
https://doi.org/10.1007/978-3-031-39831-5_4

that leverages the MP to match motifs between two time series instances and constructs a warping path by aligning them using DTW, and (2) MotifAug, a time series data augmentation approach that uses MotifDTW at its core to generate meaningful intra-class data samples from existing datasets.

2   Preliminaries

2.1   Matrix Profile

Let T be a time series vector of length l: $T = (t_1, t_2, ..., t_l)$ and T' a time series vector of length l': $T' = (t'_1, t'_2, ..., t'_{l'})$. If T is the query time series and T' is the reference time series, the aim of the MP is to provide fast access to the nearest neighbor of each subsequence of T of length m defined as $sub_{i,m} = (t_i, t_{i+1}, ..., t_{i+m})$ to every other subsequence $sub'_{j,m}$ of T' of the same length m. This is achieved by computing a distance profile vector $D_i = (d_{i,0}, d_{i,1}, ..., d_{i,l-m+1})$ for each query subsequence $sub_{i,m}$ s.t. $d_{i,j} = dist(z(sub_{i,m}), z(sub'_{j,m}))$, where z is the z-normalization function and $dist$ is the Euclidean distance. Then, the matrix profile vector $MP(T, T') = (mp_1, mp_2, ..., mp_{l-m+1})$, which provides direct access to the distance separating each $sub_{i,m}$ from its nearest neighbor in T', is created such that $mp_i = min(D_i)$. Similarly, the matrix profile index vector $MPI(T, T') = (mpi_1, mpi_2, ..., mpi_{l-m+1})$ with $mpi_i = argmin(D_i)$ indicates the location of each nearest neighbor subsequence.

2.2   Pan-Matrix Profile

Determining the window size parameter m for the MP is a challenging task that usually requires prior knowledge from experts. Therefore, Madrid et al. [6] introduced the Pan Matrix Profile (PMP), a new data structure that combines information from MPs with a range r of window sizes lengths, and proposed the Scalable KInetoscopic Matrix Profile (SKIMP) algorithm to compute it with a time complexity and space complexity of $O(l^2 r)$ and $O(lr)$ respectively. Given a range of window lengths $m = [m_{min}, m_{max}]$ such that $m_{min} > 0$ and $m_{max} < l$, a query time series T, and a reference time series T', $PMP(T, T')$ is simply a matrix where each row consists in $MP(T, T')$ with window length m. Similarly, $PMPI(T, T')$ is the matrix where each row consists in $MPI(T, T')$.

2.3   DTW Alignment for Time Series Data Augmentation

Besides its use as a distance measure, DTW [7] can be adopted to align two time series instances. In the context of time series data augmentation, several methods have taken advantage of this feature to generate new data samples. wDBA [2] uses DBA, which aligns two or more time series samples before computing their medoid, to create new data samples. SPAWNER [5] generates new time series by selecting two intra-class instances, splitting them into two segments each, aligning the first two segments and the last two segments separately using DTW, averaging the result, and adding some noise to ensure that the instance generated is different than the original ones. As to Guided Warping [3], it generates time series instances by selecting two intra-class instances from the original data set and aligning one to the other using either DTW or a modified version of it described below.

One of the main drawbacks of aligning time series instances using DTW for time series data augmentation is that the result might contain sudden, non-smooth changes that disrupt the continuity of the time series. This is most relevant when the warping path maps two adjacent time steps to two distant points in the reference time series. Therefore, Iwana and Uchida [3] introduced shapeDTW, a modified version of DTW that takes into consideration high-level shape descriptors to optimize the warping path. To this end, shapeDTW proceeds similarly to DTW but considers the immediate neighborhood of each time series data point (time steps within $m/2$ on each side) in its distance calculation.

3   Proposed Method

3.1   Motif Mapping

By introducing data points' neighborhoods in distance calculations, shapeDTW aims to avoid abrupt changes in the warping path and preserve high-level shape descriptors. However, shapeDTW suffers from two main problems: **(1) It often fails to avoid abrupt changes.** Consider the case where the reference time series T' contains repeating motifs, i.e. quasi-identical shape descriptors, occurring at different time steps. There is no guarantee under the shapeDTW constraints that two adjacent data points t_i and t_{i+1} in the query time series, whether or not they belong to the same shape descriptor, will not be mapped to two distant data points t'_j and t'_k belonging to two distant occurrences of the same motif in the target time series. See Sect. 4.2 for visual examples. **(2) It is dependent on the window length parameter.** Determining the window length m that defines shape descriptors requires prior expert knowledge of the domain. Moreover, it might be necessary to use several lengths. Therefore, the fact that shapeDTW is highly dependent on m makes it vulnerable to issues resulting from the use of a wrong value. For example, using too large of a value will result in missing important shape descriptors, while setting it too small might emphasize meaningless intervals.

Thus, we propose MotifDTW, a novel time series alignment method that relies on motifs to construct a meaningful warping path. MotifDTW solves the issues described above by (1) being parameter-free and (2) mapping all the time steps of a motif in the query time series to the time steps of one single matching motif in the reference time series instance.

Algorithm 1 describes MotifDTW. Given query time series $T = (t_1, t_2, \ldots, t_l)$ and reference time series $T' = (t'_1, t'_2, \ldots, t'_l)$, MotifDTW starts by initializing a motif alignment path to a vector $MAP(T, T') = (map_1, map_2, \ldots, map_l)$ of length l. Then, it computes $PMP(T, T')$ matrix. Since each row of $PMP(T, T')$ represents $MP(T, T')$ with a different window length m, the higher the length value m the higher the corresponding row values will be. Thus, MotifDTW divides the values in each row by their corresponding m. This allows ranking motifs of different lengths by according equal importance to longer ones. Then, starting with the motif $sub_{i,m}$ with the lowest value in $PMP(T,T')$, the time steps $(map_i, map_{i+1}, ..., map_{i+m})$ in $MAP(T,T')$ are set to the corresponding time steps of its nearest neighbor in T'. This is repeated for each motif, in descending order of importance until all the time steps are mapped. During this process, overlapping motif indices are resolved as follows. If the current overlapping motif extends a previously mapped one (with a shorter length), the non-mapped indices are introduced in $MAP(T,T')$. Otherwise, the overlapping motif is ignored. Then, the final warping path $WP(T,T') = (wp_1, wp_2, ..., wp_l)$ is constructed by aligning each mapped motif $(t_i, t_{i+1}, ..., t_{i+m})$ to its nearest neighbor subsequence $(T'_{map_i}, T'_{map_{i+1}}, ..., T'_{map_{i+m}})$ using DTW.

Algorithm 1. MotifDTW

Inputs: Time series samples T and T'.
Output: Warping Path $WP(T,T')$.
1: Initialize the motif alignment path and the warping path:
2:   $MAP(T,T') \leftarrow (map_1, map_2, ..., map_l)$ **s.t.** $\forall map_i = -1$
3:   $WP(T,T') \leftarrow (wp_1, wp_2, ..., wp_l)$
4: Compute PMP and $PMPI$ matrices:
5:   $PMP(T,T'), PMPI(T,T') \leftarrow SKIMP(T,T')$ ▷ $SKIMP()$ **from [6]**
6: Divide each row (MP) in PMP by its window length m:
7: **for** $MP_i(T,T')$ in $PMP(T,T')$
8:   $MP_i(T,T') \leftarrow MP_i(T,T')/m$
9: Map the motifs with the highest importance in T (smallest value in $PMP(T,T')$) to its nearest neighbor in T' until all T time steps are mapped:
10: **while** $\exists map \in MAP(T,T')$ **s.t.** $map = -1$
11:   $m, i \leftarrow argmin(PMP(T,T'))$
12:   $nn\_i \leftarrow PMPI(T,T')_{m,i}$
13:   **if** $\forall map \in (map_i, map_{i+1}, ..., map_{i+m})$, $map = -1$
14:   **or** $\forall map_{i+j} \in (map_i, map_{i+1}, ..., map_{i+m})$ **s.t.** $map_{i+j} > 0$, $map_{i+j} = nn_i + j$ **then**
15:     $(map_i, map_{i+1}, ..., map_{i+m}) \leftarrow (nn\_i, nn\_i + 1, ..., nn\_i + m)$
16:     $PMP(T,T')_{m,i} \leftarrow \infty$
17: Record the DTW path between each mapped motif $sub_{i,m}$ in $MAP(T,T')$ and its
18: nearest neighbor
19: **for** $sub_{i,m}$ in $MAP(T,T')$
20:   $nn \leftarrow (T'_{map_i}, T'_{map_{i+1}}, ..., T'_{map_{i+m}})$
21:   $(wp_i, wp_{i+1}, ..., wp_{i+m}) \leftarrow DTW(sub_{i,m}, nn)$
22: **return** $WP(T,T')$

3.2   Time Series Augmentation

We introduce MotifAug, a novel parameter-free time series augmentation method that leverages this mechanism for time series data augmentation. Given a time series dataset containing N time series instances $\mathcal{D} = \{T_1, T_2, ..., T_N\}$ with each T_i belonging to a class $C_m \in \{C_1, C_2, ..., C_M\}$, MotifAug generates a new instance T_{aug} with class label C_m by selecting two intra-class instances from D, a query and a reference $T, T' \in C_m$. Then, it uses MotifDTW to construct the warping path $WP(T,T')$ and aligns T to T'. The result is T_{aug}, a new time series that has the time motifs of the query T aligned to the time steps of T'.

Since $WP(T,T') \neq WP(T',T)$, for each dataset class C_m containing N_m data samples, MotifAug can generate a total of $N_m \times (N_m - 1)$ new instances.

4   Experimental Evaluation

Due to space restrictions, we encourage the reader to visit our project website[1] and repository[2] for additional details and to access the source code.

4.1   Setup

We compare the performance of MotifAug on the 110 fixed-length datasets from the UCR archive that contain at least two intra-class instances per class to the following baseline methods (with default parameters): **SMOTE** [1], **wDBA** [2] with the ASD weighting scheme since it had the best results in the original paper, **SPAWNER** [5], Discriminative GW using DTW alignment (**DGW-D**), Random GW using DTW alignment (**RGW-D**), DGW using shapeDTW alignment (**DGW-sD**), and RGW using shapeDTW alignment (**RGW-sD**) [4].

4.2   Aligning Time Series Using MotifDTW

The warping path constructed by MotifDTW between two time series instances allows their alignment in a realistic way, avoiding any abrupt changes by ensuring that all time steps contained in a single motif are aligned to adjacent time steps belonging to a single subsequence. In Fig. 1, we visualize the time series instances generated from two samples from ECG200, a dataset of healthy and unhealthy electrocardiograms, using RGW-sD, DGW-sD, and MotifAug. The gray lines represent the original query time series and the red lines represents the generated instances. In Fig. 1.a and 1.b, the red arrows point to short horizontal subsequences in the data generated by RGW-sD and DGW-sD that are unusual in the electrocardiograms present in the dataset. The segments are a direct result of the abrupt changes in shapeDTW's warping path described in Sect. 3.1. This problem is avoided by MotifAug as shown in Fig. 1.c.

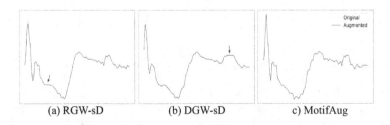

(a) RGW-sD   (b) DGW-sD   c) MotifAug

Fig. 1. Visually inspecting data generated from the ECG200 dataset. The subsequences pointed to by the arrows represent unusual heartbeat segments.

---

[1] https://sites.google.com/view/MotifAug/home.
[2] https://github.com/omarbahri/MotifAug.

4.3   Performance Gain

To evaluate the performance of MotifAug in comparison to the time series data augmentation benchmarks, we adopt the classification performance gained from extending the training set with the augmented data samples to double the original size as the main criteria. We adopt two main state-of-the-art time series classification models: RandOm Convolutional KErnel Transform (ROCKET) and Residual Network (ResNet). For both classifiers, we used the default training and testing splits provided in the UCR archive and combine the original training data with the augmented data samples to form a larger training set. Then, we train the classification model and test its performance on the original test set. Since some datasets in the UCR archive suffer from data imbalance, our classification measure of choice is the f1-score. For ROCKET, we repeat this procedure ten times with different initialization seeds and consider the mean f1-scores of the ten trials. Since training ResNet is considerably more time-consuming, the mean f1-scores are taken from only 3 runs. In addition, we narrow down the number of datasets to 50 randomly selected ones.

In Table 1, we display the average ranks of the mean f1-scores of all methods including the classification results using the original datasets (without any augmentation). The detailed results can be found in our project website. An interesting result we noted using ROCKET is that only half of the augmentation methods —including MotifAug — have led to an overall increase in performance. MotifAug had the first average rank using both classifiers, followed by SMOTE with ROCKET and DGW-sD with ResNet.

Table 1. Classification comparison: average rank of f1-scores.

Average Rank	No Aug	SMOTE	wDBA	SPAWNER	RGW-D	DGW-D	RGW-sD	DGW-sD	MotifAug
ROCKET	4.64	4.13	5.14	6.45	5.21	5.65	4.45	4.89	**4.00**
ResNet	5.96	4.36	4.44	5.94	4.9	4.1	3.56	3.88	**3.28**

5   Conclusion

We introduced MotifDTW, an alignment method for time series data that leverages the fast MP motif discovery process. MotifDTW allows constructing a meaningful warping path by ensuring that all adjacent motif time steps in the query data sample are mapped to adjacent time steps in the reference. We also introduced MotifAug, a time series augmentation algorithm that uses MotifDTW as its core. To our knowledge, this is the first effort to leverage the MP motif discovery for time series data augmentation.

Acknowledgments. This project has been supported in part by funding from GEO Directorate under NSF awards #2204363, #2240022, and #2301397 and the CISE Directorate under NSF award #2305781.

References

1. Chawla, N.V., Bowyer, K.W., Hall, L.O., Kegelmeyer, W.P.: SMOTE: synthetic minority over-sampling technique. J. Artif. Intell. Res. **16**, 321–357 (2002)
2. Forestier, G., Petitjean, F., Dau, H.A., Webb, G.I., Keogh, E.: Generating synthetic time series to augment sparse datasets. In: Proceedings - IEEE International Conference on Data Mining, ICDM, vol. 2017-Novem, pp. 865–870, December 2017
3. Iwana, B.K., Uchida, S.: Time series data augmentation for neural networks by time warping with a discriminative teacher, April 2020
4. Iwana, B.K., Uchida, S.: An empirical survey of data augmentation for time series classification with neural networks. PLOS ONE **16**(7), e0254841 (2021)
5. Kamycki, K., Kapuscinski, T., Oszust, M.: Data augmentation with suboptimal warping for time-series classification. Sensors **20**(1), 98 (2019)
6. Madrid, F., Imani, S., Mercer, R., Zimmerman, Z., Shakibay, N., Keogh, E.: Matrix profile XX: finding and visualizing time series motifs of all lengths using the matrix profile. In: International Conference on Big Knowledge, ICBK 2019, pp. 175–182, November 2019
7. Sakoe, H., Chiba, S.: Dynamic programming algorithm optimization for spoken word recognition. IEEE Trans. Acoust. Speech Sig. Process. **26**(1), 43–49 (1978)
8. Yeh, C.C.M., et al.: Matrix profile I: all pairs similarity joins for time series: a unifying view that includes motifs, discords and Shapelets, pp. 1317–1322, February 2017

State-Transition-Aware Anomaly Detection Under Concept Drifts

Bin Li$^{(\boxtimes)}$ and Emmanuel Müller

TU Dortmund University, Dortmund, Germany
{bin.li,emmanuel.mueller}@tu-dortmund.de

Abstract. Detecting temporal abnormal patterns over streaming data is challenging due to volatile data properties and the lack of real-time labels. The abnormal patterns are usually hidden in the temporal context, which cannot be detected by evaluating single points. Furthermore, the normal state evolves over time due to concept drifts. A single model does not fit all data over time. Autoencoders are recently applied for unsupervised anomaly detection. However, they are trained on a single normal state and usually become invalid after distributional drifts in the data stream. This paper uses an Autoencoder-based approach STAD for anomaly detection under concept drifts. In particular, we propose a state-transition-aware model to map different data distributions in each period of the data stream into states, thereby addressing the model adaptation problem in an interpretable way. Our experiments evaluate the proposed method on synthetic and real-world datasets. While delivering comparable anomaly detection performance as the state-of-the-art approaches, STAD works more efficiently and provides extra interpretability.

Keywords: State transition · Anomaly detection · Concept drift · Autoencoder

1 Introduction

Anomaly detection in streaming data is gaining traction in the current big data research. Despite the high demand in a variety of real-world applications [22] (e.g., health care, device monitoring, and predictive maintenance), rare existing models show convincing performance in real-time deployment. The detection of abnormal patterns in streaming data is challenging. On the one hand, labels are unavailable or expensive to acquire in real-time, such that supervised approaches usually fail. On the other hand, the conventional batch models easily expire, while a single stationary model does not fit the ever-changing data stream.

Recently, Autoencoders have been employed for anomaly detection in an unsupervised manner [14,26]. Autoencoders are trained to reconstruct the normal data[1], such that for any unknown data instance, a high reconstruction error

---

[1] Unless specifically stated, instead of normally distributed data, normal data refers to the opposite of abnormal data in the anomaly detection context.

© The Author(s), under exclusive license to Springer Nature Switzerland AG 2023
R. Wrembel et al. (Eds.): DaWaK 2023, LNCS 14148, pp. 49–63, 2023.
https://doi.org/10.1007/978-3-031-39831-5_5

indicates an anomaly. Specifically, for time series data, the temporal dependencies between data points can be captured by constructing Autoencoders using Recurrent Neural Networks (RNNs) and their variants [14,16]. Although such methods show impressive performance on time series data, they usually ignore the fact that such data is commonly collected in a streaming way and does not allow full access during the training phase. Therefore, an adaptive Autoencoder is desired, which can be initialized with a few normal data and continuously capture the latest knowledge from the real-time data stream. Another major challenge of anomaly detection in streaming data is distinguishing between abnormal patterns and concept drifts. Once the data stream drifts to a novel distribution, a stationary model trained only on outdated data may detect most of the upcoming data undesirably as anomalies.

Given the severe problems, we aim to consider the concept drift detection and anomaly detection holistically, adapt the model to the latest data distribution, and detect anomalies only concerning the temporal context where they are located. Previous concept drift detection researches focus on detecting changes of the joint probability $P(X, y)$ under a supervised setting, namely, the decision boundary changes along with the distributional changes in the input data [13]. However, for anomaly detection, the class distribution between normal and abnormal is extremely unbalanced, and labels are usually missing or delayed, so it is impractical to use traditional supervised approaches [4,11], e.g., detecting drifts based on the changes of real-time prediction error rate. Instead, the adaptation based on changes of the prior $P(X)$ will ensure the Autoencoder learns the normal data pattern from the latest data distribution.

Statistical tests are commonly used for unsupervised drift detection [13]. For instance, the two-sample tests examine whether samples from two collections are generated from the same data distribution. However, many existing methods conduct tests mostly in the original input space, which only works for linearly detectable drifts. Ceci et al. [7] introduce both PCA and Autoencoder to embed features into a latent space for the change detection in power grid data. However, they use a feed-forward Autoencoder, which does not directly capture the temporal information in the data.

In this paper, we propose STAD (State-Transition-aware Anomaly Detection). In STAD, data distribution in a time period is defined as a state. We use state transitions to model the concept drifts between periods. As Autoencoders are well-studied for non-linear time series anomaly detection, we are motivated to extend the state transition paradigm to Autoencoders. We follow the standard usage of Autoencoders for anomaly detection and novelly couple the detection of concept drifts and anomalies with the informative latent representation of Autoencoders. An existing Autoencoder can be reused when a data concept reappears in the stream. A state transition is triggered by the detection of a concept drift, and this will further guide the reuse or adaptation of Autoencoders for the next period. The states raise interpretability in understanding the decisions of Autoencoders and changes in the data stream.

2 Related Works

Online Anomaly Detection. A major category of online anomaly detection methods is based on a prediction model, which employs historical data to predict the near future. Abnormal data may not fit the normal prediction and therefore cause a large prediction error. The widely used ARIMA model in time series analysis is also used in anomaly detection [3]. However, specific adaptation strategies are to be made to use it in online fashion. The Hierarchical Temporal Memory (HTM) model [1] is designed for real-time application, while it can automatically adapt to changing statistics. One issue with models in this category is that they are usually designed for univariate data. Therefore, deep neural networks are also used recently to model higher dimensional and more complex data. [15] use LSTMs as a basic prediction model, which can capture the high-dimensional contextual information between different timestamps. [12] also employs an LSTMs-based prediction model for anomaly detection. However, their semi-supervised approach requires partial labels from the history, which is not always possible in the streaming processing scenario.

Reconstruction-based approaches train models to reconstruct the normal data so that unknown abnormal data in the test phase will cause larger reconstruction errors due to the lack of knowledge. Autoencoders are used as an unsupervised approach for anomaly detection. [26] adopts a Gaussian Mixture Model to detect anomalies from the reconstruction error. However, they use the feed-forward network, which cannot deal with inter-dependent data points as in the data stream. [14] builds the Autoencoder with LSTM units to capture temporal information. Similarly, [17] constructs the Autoencoder with Transformers. These models assume that the sequential data are generated from the same distribution. Therefore they are vulnerable to drifts. In the worst case, every data point that arrives after the drifts will be predicted as an anomaly.

Drift Detection. Recent drift detection approaches are well-summarized in [13]. Common processing paradigms aggregate the historical data, extract data features and conduct statistical tests. Many works contribute to the streaming data classification problem [4,18], where the real-time classification error is used as an indicator of drift detection. Unfortunately, the labels are not always immediately available in real time. On the contrary, unsupervised drift detection methods detect changes in $P(X)$, namely the distributional changes in the streaming data. Statistical tests are usually applied to detect drifts in univariate streaming data [18,20]. For multivariate streaming data, each dimension can be tested individually and aggregated afterward [19].

Finally, the model's trustworthiness and reliability are important for real-time anomaly detection, especially in safety-crucial applications. However, the interpretation of black-box anomaly detection models and complex streaming data is still under-studied. [22] interprets device anomalies by feature responsibility gained from Integrated Gradient [24]. [2] uses a graph-based framework to model recurring concepts in the data stream. None of them has a focus on the drift detection perspective.

3   Problem Definition

3.1   Terminology

Data Stream and Concept Drift. Let $\mathcal{X} = \{X_t\}_{t\in\mathbb{N}^*}^{D}$ be a D-dimensional data stream, where X_t denotes the observation at timestamp t. The data stream contains unlabeled anomalies as well as distributional changes caused by concept drifts. Instead of explicitly categorizing different concept drift types [13], we uniformly consider that a concept drift occurs in the data stream between timestamps t and $t + c$ if the prior probability $P_{<t}(X) \neq P_{>t+c}(X)$, where $P_{<t}$ and $P_{>t+c}$ are respectively the data distribution from the last concept drift to t and from $t + c$ to the next concept drift. The period $[t, t + c]$ is the drift period, defined as the minimum period that covers the whole distributional change. The data distribution other than drift periods is assumed to be stable. Due to the lack of labels under the unsupervised setting, we only consider the prior (virtual) shifts [13] in the data stream.

State Transition. Imitating the automata theory, we formulate concept drifts in streaming data with a state transition model $\mathcal{M} = \langle \mathcal{X}, \mathcal{S}, \delta \rangle$ where $\mathcal{X}$ is a multivariate data stream, $\mathcal{S} = \{S_1, S_2, ..., S_N\}$ is a set of states (N is the user-defined maximum number of states that can be maintained), δ is a set of transition functions $\delta : \{S_i \Rightarrow S_j\}(S_i, S_j \in \mathcal{S}, i \neq j)$. For each state $S_i = \langle P_i, AE_i \rangle (i = 1, ..., N)$, AE_i is the Autoencoder trained on the current concept data, P_i is the empirically estimated distribution in the Autoencoder latent space. In this work, we assume sufficient data after the concept drifts is available to learn P_i and AE_i.

Considering that no information about the upcoming new concept is accessible, despite a potential high error rate, we still keep using the previous model for anomaly detection until the model adaptation is finished. Or in other words, the previous model is used during the upcoming *drift period*. For distributional stationary data streams where no concept drift occurs, there will be only a single state without transition, and the model reduces to a single conventional Autoencoder for stationary data.

Anomaly. An observed data snippet $X_t^w = \{x_{t+1}, ..., x_{t+w}\}(t, w \in \mathbb{N}^*)$ is abnormal if it significantly deviates from its temporal neighbors (data snippets in the same *state*). The significance of the deviation can be determined by thresholding or statistical techniques. Both concept drifts and anomaly snippets are distributionally deviating from their temporal neighbors. In our study, we distinguish them in terms of length. After the concept drifts, we assume that the data distribution stays stationary in the new concept for a significantly longer period. In contrast, the data stream returns to the previous distribution after a short anomaly snippet.

3.2 Problem Statement

Given a D-dimensional data stream $\mathcal{X} = \{X_t\}_{t \in \mathbb{N}^*}^D$, we aim to identify any period $[t+1, t+w]$ where the corresponding data snippet X_t^w is abnormal. The detection process should be unsupervised and in real time. We also detect concept drifts in the data stream and switch to an existing Autoencoder or train a new one on the newly arrived data.

4 State-Transition-Aware Anomaly Detection

In this section, we propose STAD, a state-transition-aware anomaly detection model, which employs Autoencoder as the base model. The latent representations of Autoencoders are used to detect concept drifts, which consequently trigger state transitions. An overview of STAD is shown in Fig. 1.

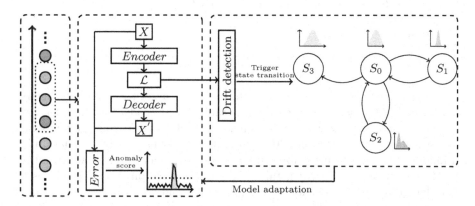

Fig. 1. STAD overview: The left block is a multivariate data stream, where red dots denote abnormal data points and the dashed box is a data snippet. The middle block is an conventional autoencoder-based anomaly detection module, which detects abnormal snippets from the data stream. The right block takes latent representations from the autoencoder and conducts concept drift detection, which consequently triggers state transition and model adaptation. (Color figure online)

4.1 Reconstruction and Latent Representation Learning

Let $f_{Enc}\colon \mathbb{R}^{w \times D} \to \mathbb{R}^H$ and $f_{Dec}\colon \mathbb{R}^H \to \mathbb{R}^{w \times D}$ be the encoder and decoder of an Autoencoder. The encoder maps a snippet X_t^w of the multivariate streaming data into an H dimensional latent representation $L \in \mathbb{R}^H$, while the decoder reconstructs the same format snippet $X_t'^w$ from L, where w is the snippet length and $t, w \in \mathbb{N}^*$. A common assumption for anomaly detection using Autoencoders is that pure normal data are available for the initial model training. The reconstruction error $e_t^w = |X_t^w - X_t'^w|$ indicates the goodness of fit to the normal

---

Algorithm 1. Latent Space Drift Detection

Input: $\mathcal{L}_{hist}$ with maximum size m, $\mathcal{L}_{new}$ with maximum size n, minimum $\mathcal{L}_{hist}$ size m^* trigger test, current state $S = \langle P, AE \rangle$, state transition model $\mathcal{M} = \langle \mathcal{X}, \mathcal{S}, \delta \rangle$

1: **while** stream does not end **do**
2:     $L_t \leftarrow$ ANOMALYDETECTION(AE, X_t^{t+w})   ▷ Get latent representation
3:     $\mathcal{L}_{new} \leftarrow \mathcal{L}_{new} \cup L_t$
4:     **if** $\mathcal{L}_{new}.size > n$ **then**   ▷ Move the oldest element of $\mathcal{L}_{new}$ to $\mathcal{L}_{hist}$
5:         $L_{t-n+1} = \mathcal{L}_{new}.pop()$
6:         $\mathcal{L}_{hist} \leftarrow \mathcal{L}_{hist} \cup L_{t-n+1}$
7:     **end if**
8:     **if** $\mathcal{L}_{hist}.size > m$ **then**
9:         $\mathcal{L}_{hist}.pop()$
10:     **end if**
11:     **if** $\mathcal{L}_{hist}.size \geq m^*$ and $\mathcal{L}_{new}.size = n$ **then**
12:         **if** KSTEST($\mathcal{L}_{hist}^h, \mathcal{L}_{new}^h$) is True **then**   ▷ Equation 1
13:             $S \leftarrow$ STATETRANSITION(S, $\mathcal{L}_{new}$, $\mathcal{S}$, δ)   ▷ Section 4.3
14:             Report concept drift, clear $\mathcal{L}_{hist}$ and $\mathcal{L}_{new}$
15:         **end if**
16:     **end if**
17: **end while**

---

data. In the test phase, abnormal snippets will cause larger reconstruction errors than normal data such that they are separable. The encoder and decoder can be implemented with a variety of deep models [25, 26]. Considering the temporal dependencies in streaming data, RNNs and their variants [14, 16] are naturally suitable for the target. In the following illustration, as an example, we take the LSTM-Autoencoder [14], which takes data snippets as input and produces a single latent representation for each snippet. To map the multivariate reconstruction error to the likelihood of anomalies, a commonly used approach is to estimate a multivariate Gaussian distribution from the reconstruction error of normal data and measure the Mahalanobis distance between the reconstruction error of an unknown data point to the estimated distribution [14]. Moreover, the Gaussian Mixture Model (GMM) [26] and energy-based model [25] can also be used for likelihood estimation. The thresholding over the estimated anomaly likelihood in an unsupervised manner is challenging, especially in the real-time prediction scenario. A possible non-parametric dynamic thresholding technique is proposed in [12]. The unsupervised approach for the adaptive threshold in different periods is not the main focus of this paper and will be addressed in our future work. In the following sections, we focus on adapting Autoencoders based on the state transitions.

4.2   Drift Detection in the Latent Space

In real-time, the latent representations of the Autoencoder are accumulated for concept drift detection. Existing concept drift detection approaches mostly work in the original space, targeting linear separable concept drifts. Considering the complex concept drifts in multivariate streaming data, even non-linear distributional changes can be observed in the Autoencoder latent space. We perform the non-parametric and distribution-free two-sample Kolmogorov-Smirnov Test (KS-Test) [8,9] on each latent space dimension to check whether two latent representations are drawn from the same continuous distribution. Algorithm 1 shows the online concept drift detection process.

Formally, let $\mathcal{L}_{hist} = \{L_{t-\hat{m}-n+1}, L_{t-\hat{m}-n+2}, ..., L_{t-n}\}$ $(m^* \leq \hat{m} \leq m)$ be the accumulated latent representation since the last concept drift and $\mathcal{L}_{new} = \{L_{t-n+1}, L_{t-n+2}, ..., L_t\}$ be the latest latent representations. m and n are the maximum size of $\mathcal{L}_{hist}$ and $\mathcal{L}_{new}$, m^* is the minimum size of $\mathcal{L}_{hist}$ to trigger a statistical test. F_{hist} and F_{new} are the empirical estimated cumulative distribution functions from the two latent representation sets. The null hypothesis (i.e., the observations in $\mathcal{L}_{hist}$ and $\mathcal{L}_{new}$ are from the same distribution) will be rejected if

$$\sup_{L}|F_{hist}(L) - F_{new}(L)| > c(\alpha)\sqrt{\frac{\hat{m} + n}{\hat{m} \cdot n}} \qquad (1)$$

where sup is the supremum function, α is the significance level, $c(\alpha) = \sqrt{-\ln(\frac{\alpha}{2}) \cdot \frac{1}{2}}$. We maintain both $\mathcal{L}_{hist}$ and $\mathcal{L}_{new}$ as queues. m is larger than n such that $\mathcal{L}_{hist}$ contains longer and more stable historical information, while $\mathcal{L}_{new}$ captures the latest data characteristic. The drift detector will only start if $\mathcal{L}_{hist}$ contains at least m^* samples, such that the procedure starts smoothly.

Since the KS-test is designed for univariate data, we conduct parallel tests in each latent dimension and report concept drift if the null hypothesis is rejected on all the dimensions. Once a concept drift is detected, we will conduct the state transition procedure for model adaptation (Sect. 4.3). The historical and latest sample sets are emptied, and we further collect samples from the new data distribution.

4.3   State Transition Model

Modeling reoccurring data distributions (e.g., seasonal changes), coupling Autoencoders with drift detection, and reusing models based on the distributional features can increase the efficiency of updating a deep model in real time. We represent every stable data distribution (concept) and the corresponding Autoencoder as a *state* $S \in \mathcal{S}$. In STAD, for each period between two concept drifts in the data stream, the data distribution, as well as the corresponding Autoencoder, are represented in a queue $\mathcal{S}$ with limited size. The first state $S_0 \in \mathcal{S}$ represents the beginning period of the data stream before the first concept drift. After a concept drift, a new Autoencoder will be trained from scratch with the latest m input data snippets, if no existing element in $\mathcal{S}$ fits the current data distribution; Otherwise, the state will transit to the existing one and reuse the corresponding Autoencoder. In our study, we assume that sufficient data after the concept drifts can be accumulated to initialize a new Autoencoder.

To compare the distributional similarity between the newly arrived latent representations Q and the distributions of existing states $\{P_i|i = 1, ..., N\}$, we employ the symmetrized Kullback-Leibler Divergence. The similarity between Q and an existing state distribution P_i is defined as

$$D_{KL}(P_i, Q) = \sum_{L \in \mathcal{L}} P_i(L) log \frac{P_i(L)}{Q(L)} + Q(L) log \frac{Q(L)}{P_i(L)} \tag{2}$$

The next step is to estimate the corresponding probability distributions from the sequence of latent representations. In [8,9], the probability distribution of categorical data is estimated by the number of object appearances in each category. In our case, the target is to estimate the probability distribution of fixed-length real-valued latent representations. In previous research, one possibility for density estimation of streaming data is to maintain histograms of the raw data stream [21]. In STAD, we take advantage of the fix-sized latent representation of Autoencoders and maintain histograms of each period in the latent space for the density estimation.

Let $\mathcal{L} = \{L_1, L_2, ..., L_t\}$ be a sequence of observed latent representations, where $L_i = \langle h_1^i, h_2^i, ..., h_H^i \rangle$ and H is the latent space size, the histogram of $\mathcal{L}$ is

$$g(k) = \frac{1}{t} \sum_{L_i \in \mathcal{L}} \frac{e^{h_k^i}}{\sum_{j=1}^{H} e^{h_j^i}} \quad (k = 1...H) \tag{3}$$

and the density of a given period is estimated by $P(k) = g(k)$. Hence, Eq. 2 can be converted to

$$D_{KL}(P_i, Q) = \sum_{k=1...H} P_i(k) log \frac{P_i(k)}{Q(k)} + Q(k) log \frac{Q(k)}{P_i(k)} \tag{4}$$

For a newly detected concept with distribution Q, if there exist a state $S_i (i \in [1, N])$ with corresponding probability distribution P_i satisfies $D_{KL}(P_i, Q) \leq \epsilon$, where ϵ is a tolerant factor, and S_i is not the direct last state, the concept drift can be treated as a reoccurrence of the existing concept. Therefore the corresponding Autoencoder can be reused, and the state transfers to the existing state. If no Autoencoder is reusable, a new one will be trained on the latest arrived data after concept drift. To prevent an explosion in the number of states, the state transition model $\mathcal{M} = \langle \mathcal{X}, \mathcal{S}, \delta \rangle$ only maintains the N latest states. Considering that no information about the upcoming new concept is accessible, despite a potentially high error rate, we still keep using the previous model for anomaly detection until the model adaptation is finished. Or in other words, the previous model is used for prediction during the upcoming *drift period*. The state transition procedure is described in Algorithm 2.

5   Experiment

Common time series anomaly detection benchmark datasets are often stationary without concept drift. Although some claim that their datasets contain distributional changes, the drift positions are not explicitly labeled and are hard for us

Algorithm 2. State Transition Procedure

1: **function** STATETRANSITION($S_{hist}, \mathcal{L}_{new}, \mathcal{S}, \delta$)
2:     $P_{new} = $ DENSITYESTIMATION($\mathcal{L}_{new}$)
3:     **if** $\min\limits_{S_i = \langle P_i, AE_i \rangle \in \mathcal{S}} \{D_{KL}(P_{new}, P_i)\} \leq \epsilon$ **then**   ▷ Equation 4
4:         $\delta \leftarrow \delta \cup (S_{hist} \Rightarrow S_{min})$
5:         **return** S_{min}
6:     **end if**
7:     $S_{new} \leftarrow \langle P_{new}, AE_{new} \rangle$ ▷ AE_{new}: Trained on new concept data
8:     $\mathcal{S} \leftarrow \mathcal{S} \cup S_{new}$
9:     $\delta \leftarrow \delta \cup (S_{hist} \Rightarrow S_{new})$
10:    **if** $\mathcal{S}.size > N$ **then**
11:        Remove the oldest state and relevant transitions
12:    **end if**
13:    **return** S_{new}
14: **end function**

to evaluate. To this end, we introduce multiple synthetic datasets with known positions of abnormal events and concept drifts. Furthermore, we concatenate selected real-world datasets to simulate concept drifts. We evaluate the anomaly detection performance and show the effectiveness of model adaptation based on the detected drifts.

5.1   Experiment Setup

Datasets. We first generate multiple synthetic datasets from a sine and a cosine wave with anomalies and concept drifts. For initialization, we generate 5000 in purely normal data points with amplitude 1, period 25 for the two wave dimensions. For real-time testing, we generate 60000 samples containing 300 point anomalies. All synthetic datasets contain reoccurring concepts, such that we can evaluate the state-transition and model reusing of STAD. Following [18], we create the drifts in three fashions, abrupt (A-*), gradual (G-*) and incremental (I-*). For each type of drift, we create a standard version (*-$easy$) and a hard version (*-$hard$) with more frequent drifts leaving the model less time for reaction. The drifts are created by either swapping the feature dimensions (-$Swap$-) or multiplying a factor by the amplitude (-$Ampl$-). The abrupt drifts are created by directly concatenating two concepts. The gradual drifts take place in a 2000 timestamp period with partial instances changing to the new concepts. The incremental drifts also take 2000 timestamps, while the drift features incrementally change at every timestamp. Anomaly points are introduced by swapping the values on the two dimensions.

SMD (Server Machine Dataset) [23] is a real-world multivariate dataset containing anomalies. To simulate concept drifts, we manually compose *SMD-small* and *SMD-large*. Both only contain abrupt drifts. *SMD-small* consists of test data from *machine-1-1* to *machine-1-3*, which are concatenated in the order of *machine-1-1* ⇒ *machine-1-2* ⇒ *machine-1-1* ⇒ *machine-1-3*. We take each machine as a concept and *machine-1-1* appears twice. *SMD-large* consists of data from *machine-1-1* to *machine-1-8* and is composed in the same fashion with *machine-1-1* recurring after each concept. For both datasets, the training set of *machine-1-1* is used for the model initialization.

Forest (Forest CoverType) [5] is another widely used multivariate dataset in drift detection. To examine the performance in a real-world scenario, we do not introduce any artificial drift here, but only consider the forest cover type changes as implicit drifts. As in [10], we consider the smallest class *Cottonwood/Willow* as abnormal.

Evaluation Metrics. We adopt the AUROC (AUC) score to evaluate the anomaly detection performance. An anomaly score $a \in [0, 1]$ is predicted for each timestamp. The larger a, the more likely it is to be abnormal. The labels are either 0 (normal) or 1 (anomaly). We evaluate the AUC score over anomaly scores without applying any threshold [6] so that the performance is not impacted by the quality of the selected threshold technique.

Competitors. We compare our model with two commonly used unsupervised streaming anomaly detectors. The LSTM-AD [15] is a prediction-based approach. Using the near history to predict the near future, the model is less impacted by concept drifts. The prediction deviation to real values of the data stream indicates the likelihood of being abnormal. The HTM [1] model is able to detect anomalies from streaming data with concept drifts. Neither LSTM-AD nor HTM provides an interpretation of the evolving data stream besides anomaly detection.

Experimental Details. We construct the Autoencoders with two single-layer LSTM units. All training processes are configured with a 0.2 dropout rate, $1e-5$ weight decay, $1e-4$ learning rate, and a batch size of 8. All Autoencoders are trained for 20 epochs with early stopping. We detect drifts with the KS-Tests at a significance level of $\alpha = 0.05$. We restrict that $\mathcal{L}_{hist}$ has to contain at least $m^* = 50$ data point to trigger the KS-Tests. We set the input snippet size as the sine curve period 25. For the *SMD*-based datasets, following [23], the snippet size is set to 100. We process the snippets of the data stream as a sliding window without overlap. All experiments are conducted on an NVIDIA Quadro RTX 6000 24GB GPU and are averaged over three runs.

5.2   Performance

Overall Anomaly Detection Performance Comparison. We compare the AUC score in the streaming data anomaly detection task between STAD and the competitors. In STAD, we set the latent representation size $H = 50$, and the sizes of the two buffers during the online prediction phase as $m = 200$ and $n = 50$. The threshold ϵ is set to 0.0005. We evaluate the performance of STAD in each *state* and report the average AUC. The results are shown in Table 1. STAD achieves the best performance on all synthetic datasets with abrupt and gradual drifts. In the two more complicated real-world datasets, STAD outperforms LSTM-AD and stays comparable to HTM, while requiring significantly less processing time (see Sect. 5.2). LSTM-AD shows a dominating performance on the two incremental datasets. Due to the fact that the value at every single timestamp changes in *I-Ampl-easy* and *I-Ampl-hard*, LSTM-AD benefits from its dynamic forecasting at every timestamp, while STAD suffers under the delay between state transitions.

Table 1. Anomaly detection performance (AUC).

	STAD (Ours)	LSTM-AD	HTM
A-Swap-easy	0.986 ± 0.005	$\mathbf{0.994} \pm 0.005$	0.535 ± 0.008
A-Swap-hard	$\mathbf{0.883} \pm 0.016$	0.742 ± 0.076	0.440 ± 0.017
A-Ampl-easy	$\mathbf{0.816} \pm 0.025$	0.717 ± 0.052	0.500 ± 0.006
A-Ampl-hard	$\mathbf{0.810} \pm 0.012$	0.715 ± 0.051	0.499 ± 0.006
G-Swap-easy	$\mathbf{0.948} \pm 0.019$	0.854 ± 0.064	0.506 ± 0.008
G-Swap-hard	$\mathbf{0.926} \pm 0.030$	0.800 ± 0.082	0.502 ± 0.005
I-Ampl-easy	0.911 ± 0.014	$\mathbf{0.975} \pm 0.018$	0.488 ± 0.003
I-Ampl-hard	0.970 ± 0.017	$\mathbf{1.000} \pm 0.000$	0.470 ± 0.003
SMD-small	0.755 ± 0.067	0.562 ± 0.001	$\mathbf{0.813} \pm 0.001$
SMD-large	$\mathbf{0.763} \pm 0.016$	0.578 ± 0.002	0.762 ± 0.003
Forest	0.751 ± 0.022	$\mathbf{0.977} \pm 0.001$	0.211 ± 0.001

Parameter Sensitivity. In this section, we conduct multiple experiments to examine the impact of several parameters to STAD. We maintain two data buffers $\mathcal{L}_{hist}$ and $\mathcal{L}_{new}$ to collect data from the Autoencoder latent space to detect drifts. We set the upper bound of $\mathcal{L}_{hist}$'s size $m = 200$ for all experiments. Depending on the computational resource, larger m will lead to more stable test results. Here we examine the effect of the lower bound m^*. Similarly, we also experiment with different sizes n of $\mathcal{L}_{new}$. Additionally, the latent representation size H of Autoencoders is a parameter depending on the complexity of the input data.

In Fig. 2, we check the impact of the three parameters H, n and m^* on abrupt drifting datasets. We try different values on each parameter while keeping the other two parameters equal to 50. The model is not sensitive to either of the three parameters on abrupt drifting datasets. Specifically for the two buffers, 20 data windows of both the historical (m^*) and the latest (n) latent representations are sufficient for drift detection. Similar results have been shown on the datasets with gradual and incremental drifts. The performance is stably better than the

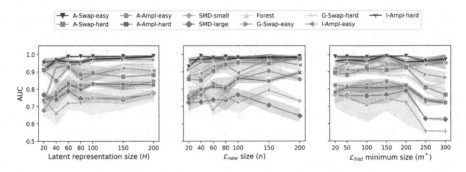

Fig. 2. Parameter sensitivity: AUC scores under different settings of latent representation size H, $\mathcal{L}_{new}$ size n and minimum size m^* of $\mathcal{L}_{hist}$ to trigger KS-Tests.

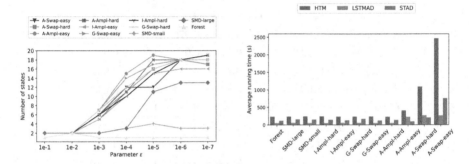

Fig. 3. Number of distinct states under different settings of threshold ϵ.

Fig. 4. Average running time comparison.

abrupt drifting dataset. One reason is that a longer drifting period leaves the model more time for detecting the drifts and conducting the state transition. On the contrary, the model may make mistakes after an abrupt drift until sufficient data is collected and the state transition is triggered.

The other parameter ϵ controls the sensitivity of re-identifying an existing state. The larger ϵ, the more likely for the model to transfer to a similar existing state. We set all H, m, and n to 50 and examine ϵ with a value that varies from 0.1 to $1e-7$, and observe the total number of distinct states created during the online prediction. As shown in Fig. 3, with large ϵ's (0.1 or 0.01), the model only creates two states and transits only between them once a drift is detected. On the contrary, too small ϵ will lead to an explosion of state. The model seldom matches an existing state but creates a new state and trains a new model after each detected drift. Currently, we determine a proper value of ϵ heuristically during the online prediction.

Running Time Analysis. Finally, we compare the running time (including training, prediction, and updating time) of the three models on all datasets in

Fig. 4. It turns out that the efficient reusing of existing models especially benefits large and complex datasets, where the model adaptation is time-consuming. STAD costs a similar processing time as LSTMAD in synthetic datasets and less in real-world datasets. The HTM always takes significantly more processing time.

6 Conclusion

We proposed the state-transition-aware streaming data anomaly detection approach STAD. With a reconstruction-based Autoencoder model, STAD detects abnormal patterns from data streams in an unsupervised manner. Based on the latent representation, STAD maintains states for concepts and detects drifts with a state transition model. With this, STAD can identify recurring concepts and reuse existing Autoencoders efficiently; or train a new Autoencoder when no existing model fits the new data distribution. Our empirical results have shown that STAD achieves comparable performance as the state-of-the-art streaming data anomaly detectors. Beyond that, the states and transitions also shed light on the complex and evolving data stream for more interpretability.

There are still some challenges in the current model. The current selection of parameter ϵ is still heuristic-based. We assume sufficient data is available to train a new Autoencoder if a drift has been detected. And we did not investigate the variety of drift types, especially gradual drifts with different lengths of *drift periods*. We plan to address the challenges above in future work.

Acknowledgement. This work was supported by the Research Center Trustworthy Data Science and Security, an institution of the University Alliance Ruhr.

References

1. Ahmad, S., Lavin, A., Purdy, S., Agha, Z.: Unsupervised real-time anomaly detection for streaming data. Neurocomputing **262**, 134–147 (2017)
2. Ahmadi, Z., Kramer, S.: Modeling recurring concepts in data streams: a graph-based framework. Knowl. Inf. Syst. **55**(1), 15–44 (2018)
3. Bianco, A.M., Garcia Ben, M., Martinez, E., Yohai, V.J.: Outlier detection in regression models with Arima errors using robust estimates. J. Forecast. **20**(8), 565–579 (2001)
4. Bifet, A., Gavalda, R.: Learning from time-changing data with adaptive windowing. In: Proceedings of the 2007 SIAM International Conference on Data Mining, pp. 443–448. SIAM (2007)
5. Blackard, J.A., Dean, D.J.: Comparative accuracies of artificial neural networks and discriminant analysis in predicting forest cover types from cartographic variables. Comput. Electron. Agric. **24**(3), 131–151 (1999)
6. Campos, G.O., et al.: On the evaluation of unsupervised outlier detection: measures, datasets, and an empirical study. Data Min. Knowl. Disc. **30**(4), 891–927 (2016)

7. Ceci, M., Corizzo, R., Japkowicz, N., Mignone, P., Pio, G.: Echad: embedding-based change detection from multivariate time series in smart grids. IEEE Access **8**, 156053–156066 (2020)
8. Chen, C., Wang, Y., Zhang, J., Xiang, Y., Zhou, W., Min, G.: Statistical features-based real-time detection of drifted twitter spam. IEEE Trans. Inf. Forensics Secur. **12**(4), 914–925 (2016)
9. Dasu, T., Krishnan, S., Venkatasubramanian, S., Yi, K.: An information-theoretic approach to detecting changes in multi-dimensional data streams. In: In Proceedings of Symposium on the Interface of Statistics, Computing Science, and Applications. Citeseer (2006)
10. Dong, Y., Japkowicz, N.: Threaded ensembles of autoencoders for stream learning. Comput. Intell. **34**(1), 261–281 (2018)
11. Gama, J., Medas, P., Castillo, G., Rodrigues, P.: Learning with drift detection. In: Bazzan, A.L.C., Labidi, S. (eds.) SBIA 2004. LNCS (LNAI), vol. 3171, pp. 286–295. Springer, Heidelberg (2004). https://doi.org/10.1007/978-3-540-28645-5_29
12. Hundman, K., Constantinou, V., Laporte, C., Colwell, I., Soderstrom, T.: Detecting spacecraft anomalies using LSTMS and nonparametric dynamic thresholding. In: Proceedings of the 24th ACM SIGKDD International Conference on Knowledge Discovery & Data Mining, pp. 387–395 (2018)
13. Lu, J., Liu, A., Dong, F., Gu, F., Gama, J., Zhang, G.: Learning under concept drift: a review. IEEE Trans. Knowl. Data Eng. **31**(12), 2346–2363 (2018)
14. Malhotra, P., Ramakrishnan, A., Anand, G., Vig, L., Agarwal, P., Shroff, G.: Lstm-based encoder-decoder for multi-sensor anomaly detection. arXiv preprint arXiv:1607.00148 (2016)
15. Malhotra, P., Vig, L., Shroff, G., Agarwal, P., et al.: Long short term memory networks for anomaly detection in time series. In: Proceedings, vol. 89, pp. 89–94 (2015)
16. Marchi, E., Vesperini, F., Weninger, F., Eyben, F., Squartini, S., Schuller, B.: Non-linear prediction with lstm recurrent neural networks for acoustic novelty detection. In: 2015 International Joint Conference on Neural Networks (IJCNN), pp. 1–7. IEEE (2015)
17. Meng, H., Zhang, Y., Li, Y., Zhao, H.: Spacecraft anomaly detection via transformer reconstruction error. In: Jing, Z. (ed.) ICASSE 2019. LNEE, vol. 622, pp. 351–362. Springer, Singapore (2020). https://doi.org/10.1007/978-981-15-1773-0_28
18. Pesaranghader, A., Viktor, H.L., Paquet, E.: Mcdiarmid drift detection methods for evolving data streams. In: 2018 International Joint Conference on Neural Networks (IJCNN), pp. 1–9. IEEE (2018)
19. Rabanser, S., Günnemann, S., Lipton, Z.: Failing loudly: An empirical study of methods for detecting dataset shift. Adv. Neural Inf. Process. Syst. **32** (2019)
20. dos Reis, D.M., Flach, P., Matwin, S., Batista, G.: Fast unsupervised online drift detection using incremental kolmogorov-smirnov test. In: Proceedings of the 22nd ACM SIGKDD International Conference on Knowledge Discovery and Data Mining, pp. 1545–1554 (2016)
21. Sebastião, R., Gama, J.: Change detection in learning histograms from data streams. In: Neves, J., Santos, M.F., Machado, J.M. (eds.) EPIA 2007. LNCS (LNAI), vol. 4874, pp. 112–123. Springer, Heidelberg (2007). https://doi.org/10.1007/978-3-540-77002-2_10
22. Sipple, J.: Interpretable, multidimensional, multimodal anomaly detection with negative sampling for detection of device failure. In: International Conference on Machine Learning, pp. 9016–9025. PMLR (2020)

23. Su, Y., Zhao, Y., Niu, C., Liu, R., Sun, W., Pei, D.: Robust anomaly detection for multivariate time series through stochastic recurrent neural network. In: Proceedings of the 25th ACM SIGKDD International Conference on Knowledge Discovery & Data Mining, pp. 2828–2837 (2019)
24. Sundararajan, M., Taly, A., Yan, Q.: Axiomatic attribution for deep networks. In: International Conference on Machine Learning, pp. 3319–3328. PMLR (2017)
25. Zhai, S., Cheng, Y., Lu, W., Zhang, Z.: Deep structured energy based models for anomaly detection. In: International Conference on Machine Learning, pp. 1100–1109. PMLR (2016)
26. Zong, B., et al.: Deep autoencoding gaussian mixture model for unsupervised anomaly detection. In: International Conference on Learning Representations (2018)

Anomaly Detection in Financial Transactions Via Graph-Based Feature Aggregations

Hewen Wang[1]([✉]), Renchi Yang[2], and Jieming Shi[3]

[1] National University of Singapore, Singapore, Singapore
wanghewen@u.nus.edu
[2] Hong Kong Baptist University, Kowloon Tong, Hong Kong
renchi@hkbu.edu.hk
[3] Hong Kong Polytechnic University, Hung Hom, Hong Kong
jieming.shi@polyu.edu.hk

Abstract. Anomaly detection in the financial domain aims to detect abnormal transactions such as fraudulent transactions that can lead to loss of revenues to financial institutions. Existing solutions utilize solely transaction attributes as feature representations without the consideration of direct/indirect interactions between users and transactions, leading to limited accuracy. We formulate anomaly detection in financial transactions as the problem of edge classification in an edge-attributed multigraph, where each transaction is regarded as an edge, and each user is represented by a node. Then, we propose an effective solution `DoubleFA`, which contains two novel schemes: proximal feature aggregation and anomaly feature aggregation. The former is to aggregate features from neighborhoods into edges based on top-k Personalized PageRank (PPR). In anomaly feature aggregation, we employ a predict-and-aggregate strategy to accurately preserve anomaly information, thereby alleviating the over-smoothing issue incurred by proximal feature aggregation. Our experiments comparing `DoubleFA` against 10 baselines on real transaction datasets from PayPal demonstrate that `DoubleFA` consistently outperforms all baselines in terms of anomaly detection accuracy. In particular, on the full PayPal dataset with 160 million users and 470 million transactions, our method achieves a significant improvement of at least 23% in F1 score compared to the best competitors.

Keywords: Anomaly Detection · Financial Transaction Network · Graph Embedding

1 Introduction

Online payment services generate numerous transactions every day and play an important role in web-based services, *e.g.*, e-commerce and e-banking. Anomaly detection in financial transactions is to identify the transactions that may generate potential losses to the service providers, common users, or merchants.

R. Wrembel et al. (Eds.): DaWaK 2023, LNCS 14148, pp. 64–79, 2023.
https://doi.org/10.1007/978-3-031-39831-5_6

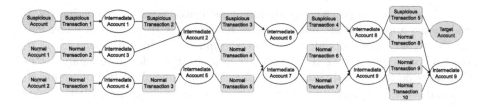

Fig. 1. Layering for disguising the source of funds in money laundering.

For instance, PayPal suffers from anomalous transactions and incurs significant losses. According to the latest statistics [15], PayPal has transaction losses increasing from 1.092 billion USD to 1.135 billion USD from the year 2019 to 2020, which is around a 4% increase per year. Therefore, there is an urgent need to automatically detect abnormal transactions so as to improve the risk management capabilities of online payment platforms and consequently prevent transaction losses from happening.

A transaction represents the interaction between a buyer and a seller, and it usually contains various types of information, such as the transaction amount, merchant and consumer segmentations, seller types, etc. It is challenging to properly analyze massive transactions and build effective anomaly detection models to identify suspicious transactions. There exist traditional ways for anomaly detection on transaction data, including manual variable creation using statistical analysis and machine learning techniques to construct regression and classification models using various classifiers such as neural networks and XGBoost, e.g. [19]. However, such a process can rarely capture the interactions among the parties in the transactions.

Effective graph analytics provides users with a deeper understanding of what is behind the data and thus can benefit a lot of useful applications such as node classification, node recommendation, link prediction, etc. Some recent works use attributed graphs to consider interactions between entities in the transaction network. For instance, in [23], graph neural network models are applied on constructed heterogeneous networks to do financial fraud detection; in [26], it explores graph convolution networks on anti-money laundering on bitcoin networks. However, these existing solutions are devised for node-wise features and cannot be directly applied to edge features, especially when a node pair is linked via multiple edges. Moreover, they often incur expensive computation overheads in searching proximal nodes for large graphs. An example in Fig. 1 shows a common technique called *layering* in the money laundering process. Fraudsters (top left) will disguise themselves across multiple accounts and a long chain of transactions. By doing this, individual transactions often look ordinary and become difficult to trace.

In this paper, we model the transaction graph as an edge-attributed multigraph (i.e., a directed graph with multiple edges between the sender node and receiver node). This multigraph representation captures the fact that transactions between two users may have different attributes and anomaly labels, which necessitates analyzing each transaction independently. While most existing graph embedding approaches are not applicable for graphs with millions of nodes and attributed edges, we intend to develop a graph embedding approach for huge transaction graphs and verify its effectiveness through different anomaly detection tasks on the PayPal transaction dataset. Our anomaly detection approach is based on the following two observations on transaction graphs. Firstly, similar users can often be grouped together on the transaction graph, which can be modeled using graph structure information. Secondly, similar users will have similar edges on the transaction graph, which can be modeled using edge attribute information.

While we can treat these two kinds of information separately, we may lose the node-attribute affinity, i.e., we can't incorporate attributes that can be reached by a node through one or more hops along the edges of the graph. Based on these observations, we propose our own graph embedding methods and conduct experiments on large real-world transaction datasets to evaluate their effectiveness and applicability.

In summary, our contributions are as follows:

- We formulate the financial transaction anomaly detection problem as a supervised edge classification task on the graph.
- We propose an effective edge embedding algorithm that incorporates multihop transaction information, graph structural information, and transaction attribute anomalies.
- Comprehensive experiments on real-world transaction datasets that compare our results with several strong baselines demonstrate the effectiveness and applicability of our proposed method. Ablation studies and model analysis confirm the reasonableness of the proposed model.

2   Related Work

In this section, we review the state-of-the-art techniques in graph representation learning and anomaly detection. We also discuss how graph embeddings are applied to graph anomaly detection and their real-world applications.

2.1   Graph Embedding

Graph embedding (a.k.a. graph representation learning) captures topological and attribute information inside the input graph and can be commonly applied to various graph mining tasks.

One line of research is to build embeddings using unsupervised approaches. The seminal works [6,16] on graph embedding proposed to learn node embeddings using random walks to distinguish positive and negative context nodes. These embeddings can be further fed to downstream machine applications such as node classification, link prediction or graph reconstruction. Many subsequent efforts are devoted to improving the embedding efficiency and effectiveness.

These methods can be broadly categorized into deep-learning-based methods (e.g. [3,5,22]) and matrix-factorization-based methods. (e.g. [1,6,17,30,31]).

Another line of research uses graph neural networks (GNNs) to do supervised or semi-supervised learning to build embeddings. GNNs adopt ideas from convolutional neural networks to generalize to graph-structured data. They have shown promising results in modeling graph-structured data. [10] proposes simple filters known as graph convolutional networks (GCNs) to operate in one-step neighborhood around each node. To make GCN more scalable on large networks, [7] proposes GraphSAGE and uses neighborhood sampling to support inductive learning. [27] directly removes intermediate activation functions and weight parameters to save computation time. [11] introduces PPR to consider multiple propagation steps for GCN. Other GNN works introduce extra parameters to enhance model capacity by sacrificing computation time and memory consumption. [21] adds an attention layer to compute the importance of different neighborhood nodes. [28] introduces transformer layers in heterogeneous graphs to softly select edge types and composite relations for meta-paths.

2.2 Anomaly Detection

Anomaly detection (a.k.a. outlier detection) refers to the process of detecting data instances that significantly deviate from the majority of data instances. Due to the difficulty and cost of collecting large-scale labeled anomaly data, existing work on anomaly detection focuses on using unsupervised learning or weakly supervised learning to obtain anomaly scores: [18] uses the distance to the k-th nearest neighbor as the outlier score; [12] uses the ensemble method to get the outlier score from subsets of features. [20] studies several one-class classification methods to detect anomalies. These anomaly detection methods focus more on attribute information and fail to consider interactions between data instances. To tackle this, there are several graph-based anomaly detection methods that utilize different graph algorithms like PPR or TrustRank to indicate node anomalies. A plethora of subsequent efforts involves using graph neural networks (GNNs) to do supervised or semi-supervised learning to build embeddings for anomaly detection tasks. Specifically, to detect anomalies in financial networks, [13,26,32] propose different GNN models and use graph node embeddings to model graph attribute information. While most existing works focus more on node anomalies and have difficulties considering anomalies on edges, few of these methods scale to large attributed graphs due to the costly training courses of deep neural networks.

3   Problem Formalization

In this section, we first transform the financial transaction data into a financial transaction graph and then formalize anomaly detection in financial transactions as a graph edge classification problem.

Financial Transaction Graph. Given a set of financial transactions, each transaction consists of the user IDs of a sender and a receiver, as well as a variety of attributes describing the transaction, such as the transaction amount, the number of days the user has been registered, and the account balance. Based on such transaction records, we construct an edge-attributed multigraph as follows. First of all, we create a node for each sender/receiver with user IDs in transactions. Afterwards, each transaction is converted into a directed edge connecting the corresponding nodes of the sender and receiver in this transaction. Each directed edge signifies that there is a money flow from sender to receiver. Note that we allow multiple edges between a sender and a receiver since multiple transactions could occur between two users. Next, we apply a pre-processing step (including standardization and one-hot encoding) to transform the attributes of each transaction into the attribute vector $\mathbf{x}_i$ of each edge e_i. More specifically, for each numerical attribute, $e.g.$, transaction amount, registered days, and account balance, we standardize it such that the mean and standard deviation are 0 and 1, respectively. In terms of categorical attributes such as customer region and transaction funding source, we transform the attribute into a set of binary ones through one-hot encoding. Eventually, we obtain an edge-attributed multigraph $G = (V, E, \{\mathbf{x}_i\})$, where V is a node-set with cardinality n, E is an edge set of size m and $\mathbf{x}_i$ denotes the length-d attribute vector of edge $e_i \in E$. It is important to note that transactions between two users in transaction networks may have varying attributes and anomaly labels, which emphasizes the need to analyze each transaction independently. We denote by $E(u)$ the set of edges connecting node u. Definition 1 presents the formal definition of the pre-processed financial transaction graph. For each transaction, there is a sender and a receiver so we model financial transactions as a multi-edge directed attributed network. Each edge signifies that there is a money flow from sender to receiver.

Definition 1. (Financial Transaction Graph). A financial transaction graph is denoted as $G = (V, E, \{\mathbf{x}_i\})$, where V is a node set with cardinality n, E is an edge set of size m, the element $e_i = (u_i, v_i) \in E$ represents a transaction from user u_i to v_i; $\mathbf{x}_i$ denotes the length-d attribute vector of edge e_i.

An anomaly transaction is defined by Definition 2.

Definition 2. (Anomaly transaction). Given the edge label list E_L indicating whether a transaction will incur a financial loss for a subset of edges, an anomaly transaction is defined to be an edge in E which incurs a financial loss.

Graph-based Financial Transaction Anomaly Detection. An edge e_i in a financial transaction graph G is said to be an $anomaly$ if it is associated with an extra label indicating that the corresponding transaction incurs a financial loss, i.e., the loss amount for that transaction is greater than 0. Given a financial transaction graph $G = (V, E, \{\mathbf{x}_i\})$ and a set E_{tr} of edges ($E_{tr} \subseteq E$) with associated labels $\mathbf{y}$ where $\mathbf{y}_i = 1$ if edge $e_i \in E_{tr}$ is an anomaly, otherwise $\mathbf{y}_i = 0$, a $graph\text{-}based$ $financial$ $transaction$ $anomaly$ $detection$ task aims to predict the

---

Algorithm 1: DoubleFA

Input: Financial transaction graph G, decay factor α, sampling ratio τ, residue
threshold r_{max}, number of nodes k
Output: Edge feature vectors $\{\hat{\mathbf{z}}_i | e_i \in E\}$
1 $\{\mathbf{z}_i^\alpha | e_i \in E\} \leftarrow$PFA$(G, \alpha, r_{max}, k)$;
2 $\{\mathbf{z}_i^\tau | e_i \in E\} \leftarrow$AFA$(G, \tau)$;
3 **for** $e_i \in E$ **do** $\hat{\mathbf{z}}_i \leftarrow \mathbf{z}_i^\alpha \parallel \mathbf{z}_i^\tau$

---

labels for edges in $E_{ts} = E \backslash E_{tr}$. We model the objective function with the
maximum likelihood estimation, which can be formulated by minimizing the
following binary cross-entropy loss function:

$$\mathcal{L}(\Theta) = \sum_{e_i \in E_{tr}} \mathbf{y}_i \cdot \log(g(f(e_i))) + (1 - \mathbf{y}_i) \cdot \log(1 - g(f(e_i))) + \lambda \cdot \|\Theta\|_2^2, \quad (1)$$

where Θ are the model parameters to be optimized, λ is the regularization
parameter, $f(\cdot)$ is a feature generation function which maps each edge e_i to a
feature vector $f(e_i)$, and $g(\cdot)$ is a label prediction function (*i.e.*, classifier such
as *Multilayer perceptron*, *Logistic regression* and *XGBoost* [2]) which maps the
edge feature vector to the labels. Intuitively, an effective feature vector $f(e_i)$
can capture the graph structure and attribute information surrounding edge e_i,
thereby achieving high classification accuracy. In this paper, we mainly focus on
designing an effective feature generation method $f(\cdot)$.

4   Proposed Method

In this section, we propose DoubleFA as the methodology for generating informative feature vectors for edges. In DoubleFA, we obtain edge feature vectors
through a widely-adopted technique called *feature aggregation*, which refers to
aggregating feature information from selected objects into a given object based
on graph structure. At a high level, DoubleFA consists of two feature aggregation phases: (i) feature aggregation from proximal nodes (*i.e.*, PFA in Sect. 4.1),
and (ii) feature aggregation from anomaly neighbors (*i.e.*, AFA in Sect. 4.2).
Algorithm 1 outlines the proposed DoubleFA, which initially takes a financial
transaction graph G and parameters α, τ, r_{max} and k as inputs, and generates
feature vectors $\mathbf{z}_i^\alpha$ and $\mathbf{z}_i^\tau$ for each edge e_i by invoking PFA and AFA respectively
(Lines 1–2). The final feature vector $\hat{\mathbf{z}}_i$ for edge e_i is obtained by concatenating
the aggregated feature vectors returned by PFA and AFA, *i.e.*, $\mathbf{z}_i^\alpha \parallel \mathbf{z}_i^\tau$ (Line 3).

4.1   PFA: Proximal Feature Aggregation

Recently, several decoupling graph neural networks (GNN) models [11,27]
demonstrate that the superior performance of GNNs in node classification tasks
is achieved via linearly aggregating features from proximal nodes (including

---

Algorithm 2: PFA

Input: Financial transaction graph G, decay factor α, residue threshold r_{max} and number of nodes k.

Output: Edge feature vectors for all directed edges $\{\mathbf{z}_i^\alpha | \forall e_i \in E\}$.

1 Obtain top-k PPR values $\mathcal{T} \leftarrow \texttt{ForwardPush}(G, \alpha, r_{max}, k)$;

2 **for** $u \in V$ **do** $\hat{\mathbf{x}}_u \leftarrow \sum_{e_i \in E(u)} \mathbf{x}_i$;

3 **for** $e_i \in E$ **do**

4   **for** $(w, \hat{\pi}_{u_i}(w)) \in \mathcal{T}(u_i) \setminus (u_i, \hat{\pi}_{u_i}(u_i))$ **do** $\hat{\mathbf{z}}_{u_i} \leftarrow \hat{\mathbf{z}}_{u_i} + \hat{\pi}_{u_i}(w) \cdot \hat{\mathbf{x}}_w$

5   **for** $(w, \hat{\pi}_{v_i}(w)) \in \mathcal{T}(v_i) \setminus (v_i, \hat{\pi}_{v_i}(v_i))$ **do** $\hat{\mathbf{z}}_{v_i} \leftarrow \hat{\mathbf{z}}_{v_i} + \hat{\pi}_{v_i}(w) \cdot \hat{\mathbf{x}}_w$

    $\mathbf{z}_{u_i} \leftarrow \hat{\mathbf{z}}_{u_i} + \hat{\pi}_{u_i}(u_i) \cdot \mathbf{x}_i;$

6      $\mathbf{z}_{v_i} \leftarrow \hat{\mathbf{z}}_{v_i} + \hat{\pi}_{v_i}(v_i) \cdot \mathbf{x}_i;$

7      $\mathbf{z}_i^\alpha \leftarrow \mathbf{z}_{u_i} \parallel \mathbf{z}_{v_i}$

---

direct and indirect neighbors) of a given node v before feeding v's aggregated feature vector to classifiers. Thus, a promising idea is to utilize these models for edge feature generation. However, these feature aggregation methods are devised for node-wise features and, therefore, cannot be directly applied to obtain edge features, especially in our case where a node pair is linked via multiple edges.

Basic Idea. To address the above challenges, we propose PFA, a *personalized PageRank*-based feature aggregation scheme for edge feature generation. For a given edge $e_i = (u_i, v_i) \in E$, the basic idea of PFA is to represent each endpoint (i.e., u_i and v_i) of e_i by a feature vector, and then combine the node features into a feature vector for edge e_i. Thus, the problem turns to be constructing a feature vector $\mathbf{z}_{u_i}$ for u_i w.r.t. edge e_i. A simple and straightforward way to obtain feature representations of a node u_i is to aggregate all attribute vectors of edges connecting u_i. However, it cannot help distinguish multiple distinct edges between two nodes u_i, v_i. In addition, feature information of nodes via multiple hops along edges in G is not captured, which is crucial in constructing effective node features in prior work [7,11,27]. Instead of producing a universal feature vector for node u_i, we construct a feature vector $\mathbf{z}_{u_i}$ for u_i w.r.t. edge e_i. Ideally, $\mathbf{z}_{u_i}$ should contain two parts of feature information, *i.e.*, edge e_i's attributes, as well as features from proximal nodes. First, the attribute vector $\mathbf{x}_i$ of edge e_i is injected into $\mathbf{z}_{u_i}$ so as to not only distinguish it from other edges concerning u_i, but also mitigate the over-smoothing issue, which is inspired by residual connections [11].

Furthermore, we propose to aggregate feature information into $\mathbf{z}_{u_i}$ from proximal nodes pertaining to u_i by leveraging *personalized PageRank* (PPR) [9]. A PPR value $\pi(u, v)$ is the probability that a random walk starting from source node u would end at node v, representing the importance of node v from the perspective of node v. Hence, rather than treating the features of adjacent nodes equally, we can use PPR values as node weights in feature aggregation such that features of important nodes are weighted more heavily. This paradigm has been shown effective in prior GNN models [11]. This is especially useful to capture behaviors such as *layering* in money laundering, which can be described using multi-hop relations in transaction graphs. Distinct from prior models, we only retrieve the nodes with top-k PPR values w.r.t. u_i since we observe that the sum of these k (e.g., 32) nodes' PPR accounts for around 90% of the total PPR w.r.t. the source node in most cases, which implies the remaining nodes are insignificant, and, thus, can be neglected.

Algorithm. Algorithm 2 illustrates the pseudo-code of `PFA`. Initially, `PFA` takes a financial transaction graph G, a decay factor α, an absolute error threshold r_{max} as well as an integer k as inputs, and invokes `ForwardPush` [25] to obtain the approximate top-k PPR values for each node in G, denoted as a set $\mathcal{T}$ (Line 1). $\mathcal{T}(u_i)$ contains k elements, in which each tuple $(w, \hat{\pi}_{u_i}(w))$ represents node w and its approximate PPR w.r.t. u_i. Then, for each node u in G, Algorithm 2 generates an attribute vector $\hat{x}_u$ for u by summing up all attribute vectors of edges connecting node u (Line 2), i.e., $\hat{x}_u = \sum_{e_i \in E(u)} x_i$. After that, `PFA` proceeds to generate feature vectors for the endpoints of each edge. Given edge $e_i \in E$ and its two associated nodes u_i, v_i, we calculate the feature vector of node u_i by the following equation $\hat{z}_{u_i} = \sum_{(w, \hat{\pi}_{u_i}(w)) \in \mathcal{T}(u_i) \setminus (u_i, \hat{\pi}_{u_i}(u_i))} \hat{\pi}_{u_i}(w) \cdot \hat{x}_w$, which means we inject into z_{u_i} the attribute vector $\hat{x}_w$ of node w after weighting it by the approximate PPR value $\hat{\pi}_{u_i}(w)$ of w w.r.t. u_i (Line 4). The feature vector z_{v_i} for another endpoint of edge e_i, i.e., v_i, can be obtained in a similar way (Line 5). Subsequently, we combine the feature vectors of nodes u_i, v_i with edge e_i's attributes as $z_{u_i} = \hat{z}_{u_i} + \hat{\pi}_{u_i}(u_i) \cdot x_i$, and $z_{v_i} = \hat{z}_{v_i} + \hat{\pi}_{v_i}(v_i) \cdot x_i$, which correspond to the feature vectors of nodes u_i, v_i w.r.t. e_i respectively (Lines 6–7). At Line 8, the concatenation of z_{u_i} and z_{v_i}, i.e., $z_{u_i} \parallel z_{v_i}$, yields the feature vector z_i^{α} for edge e_i. Finally, `PFA` returns all feature vectors for edges in E.

A Running Example. We apply a concatenation operator instead of an element-wise addition operator over the feature vectors of endpoints to avoid feature loss. To explain, Fig. 2 illustrates a partial graph containing three nodes $v_1 - v_3$ and two edges e_1, e_2. e_1 is a normal edge, while e_2 is an anomalous edge. x_1 and x_2 correspond to the attribute vectors of edges e_1 and e_2 respectively, which are totally different. Each node is associated with a feature vector, e.g., $\hat{z}_1$ for node v_1. Assume that $\pi(v_1, v_1) = \pi(v_2, v_2) = \pi(v_3, v_3) = 0.5$. According to Algorithm 2, we have $z_1 = \hat{z}_1 + 0.5 \cdot x_1 = [0.5, 0, 0, 0.5]$ and $z_2 = \hat{z}_2 + 0.5 \cdot x_1 = [0.5, 1, 1, 0.5]$ w.r.t. edge e_1, as well as $z_2 = \hat{z}_2 + 0.5 \cdot x_2 = [0, 1.5, 1.5, 1]$ $z_3 = \hat{z}_3 + 0.5 \cdot x_3 = [2, 0.5, 0.5, 1]$ w.r.t. edge e_2. If we apply the element-wise addition to obtain the feature vectors for edges e_1, e_2, we then get $[1, 1, 1, 1]$ for e_1 and $[2, 2, 2, 2]$ for e_2, which are the same after normalization. Recall that e_1 is normal and e_2 is an anomaly, and their attributes are totally different. This implies that using element-wise additions over feature vectors leads to feature loss. In contrast, if we apply the concatenation operator, the issue is resolved.

Analysis. According to [25], `ForwardPush` takes $O\left(n \cdot \frac{1}{r_{max}}\right)$ time to return $\mathcal{T}$ for all nodes in G, where r_{max} is the residue threshold for computing PPR values. At Line 2 of Algorithm 2, we need $O(m \cdot d)$ time to aggregate edge attributes to nodes. Recall that `PFA` iterates over all edges and combine attributes of k nodes for each edge's endpoints with $O(kd)$ time (Lines 3–8). Therefore, the overall time complexity of Algorithm 2 is $O\left(mdk + n \cdot \frac{1}{r_{max}}\right)$. Note that the pre-

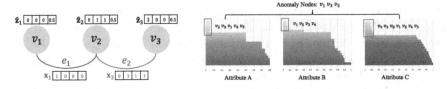

Fig. 2. An example.   **Fig. 3.** Anomaly identification via attributes.

Algorithm 3: AFA

Input: Financial transaction graph G, sampling ratio τ.
Output: Edge feature vectors $\{\mathbf{z}_i^\tau | \forall e_i \in E\}$.

1 **for** $u \in V$ **do** $\hat{\mathbf{x}}_u \leftarrow \sum_{e_i \in E(u)} \mathbf{x}_i$
2 $\mathbf{h} \leftarrow$ HBOS($\{\hat{\mathbf{x}}_u | \forall u \in V\}$);
3 Let V_h be the $\tau \cdot n$ nodes with the largest HBOS values;
4 **for** $u \in V \setminus V_h$ **do** $\mathbf{z}_u \leftarrow \hat{\mathbf{x}}_u$
5 **for** $u \in V_h$, $v \in N(u) \cap V_h$ **do** $\mathbf{z}_u \leftarrow \hat{\mathbf{x}}_u + \hat{\mathbf{x}}_v$
6 **for** $e_i \in E$ **do** $\mathbf{z}_i^\tau \leftarrow \mathbf{z}_{u_i} \parallel \mathbf{z}_{v_i}$

---

computed $\mathcal{T}$ requires $O(nk)$ space. In addition, the total memory consumption of intermediate results yielded in Algorithm 2, *i.e.*, $\hat{\mathbf{x}}_u, \mathbf{z}_u$ for each node $u \in G$ as well as the final output $\mathbf{z}_i^a$ for each edge $e_i \in G$ is bounded by $O(nd)$ and $O(md)$, respectively. Hence, the space complexity of Algorithm 2 is bounded by $O(md + n \cdot (d + k))$.

4.2   AFA: Anomaly Feature Aggregation

Motivation and Basic Idea. Although aggregating features from proximal nodes achieves impressive results in GNN models, it has two major drawbacks, which make its performance limited and suboptimal. The first drawback is that it highly relies on the strong homophily assumption [33], which assumes that nodes connected by edges/paths are more likely to share similar attributes/features as well as labels compared to nodes that are not connected, which is often not true in financial transaction networks. Another severe issue suffered from by this approach is *over-smoothing* [11], *i.e.*, indistinguishable feature representations of nodes with different labels. This is because the feature vectors of nodes in the neighborhood will converge to indistinguishable vectors as the neighborhood-based feature aggregations are constantly applied. This issue is severely aggravated in our case, where anomaly transactions are in the minority, implying that anomaly features in fraud transactions will be overwhelmed by the features of normal transactions. To alleviate the issues mentioned above, we propose another feature aggregation approach, dubbed as AFA. Instead of aggregating features from nodes based on their topological proximities to the given node as in PFA, the basic idea of AFA is to aggregate features from nodes based on their predicted labels. In other words, for each anomaly node, we aggregate features to it from other anomaly nodes so as to amplify these anomaly features in the final feature representations. The key question now is how to identify anomaly nodes in an unsupervised manner since the ground-truth node labels are unavailable in the training phase. Our basic idea is to identify anomaly nodes as per their statistical results on attribute values of nodes, which are fetched from the connected edges. For instance, as illustrated in Fig. 3, for each attribute, we can identify the nodes whose attribute values notably deviate from most of the others, e.g., $\{v_2, v_3, v_1, v_6, v_5\}$, $\{v_1, v_5, v_3, v_4\}$, and $\{v_9, v_2, v_8, v_1, v_5, v_4, v_3\}$ for attributes A, B, and C respectively. Afterwards, we can aggregate the results on all attributes to get the final list of nodes, e.g., $\{v_1, v_3, v_5\}$. Intuitively, such nodes are more likely to contain anomaly information. To implement this idea, we calculate the HBOS (short for Histogram-based Outlier Score [4]) value of each node, i.e., the overall deviation from normal values of all attributes, as its anomaly score.

Algorithm. We display the pseudo-code of AFA in Algorithm 3. AFA starts by taking a financial transaction graph G and a parameter $\tau \in (0, 1)$ as inputs, and then initialize the attribute vector $\hat{x}_u$ for each node $u \in V$ as $\sum_{e_i \in E(u)} x_i$, namely the summation of attribute vectors of edges connecting u (Line 1). After that, Algorithm 3 invokes HBOS with these attribute vectors $\{\hat{x}_u | \forall u \in V\}$ to calculate an anomaly scores for nodes in V (Line 2). Specifically, HBOS constructs static bin-width histograms for each distinct attribute of the input $\{\hat{x}_u | \forall u \in V\}$, i.e., such that numeric values of nodes in terms of this attribute are grouped into b bins, and the heights of bins are normalized to lie in $[0, 1]$. As a result, for each node u and each attribute i, we can obtain the height $hist_i(u)$ of the bin that node u corresponds to in the histograms of attribute i. The returned h of HBOS is a length-n vector, in which each entry signifies the anomaly score for the corresponding node. For each node $u \in V$, its anomaly score h_u is calculated by

$$h_u = \sum_{i=1}^{d} \log \left(\frac{1}{hist_i(u)} \right).$$

Algorithm 3 then proceeds to rank nodes based on the anomaly scores in h and picks the $\tau \cdot n$ nodes with largest anomaly scores as the node anomalies, denoted as V_h (Line 3). Next, for every non-anomaly node u in $V \in V_h$, we use $\hat{x}_u$ as its feature vector (Line 4). Regarding each anomaly node u in V_h, we represent it as $z_u = \hat{x}_u + \sum_{v \in N(u) \cap V_h} \hat{x}_v$, which sums up the attributes of node u as well as its neighbors that are anomalies (Line 5). Similar to PFA, for each edge $e_i \in E$, we generate and return its feature vector $z_i^\top$ as the concatenation of feature vectors of its corresponding node pair u_i, v_i, i.e., $z_{u_i} \| z_{v_i}$ (Line 6).

Analysis. AFA requires aggregating edge attributes to nodes and calculating HBOS scores based on aggregated node attributes, which take $O(md)$ and $O(nd)$ time respectively. Finding $n\tau$-largest entries in n anomaly scores at Line 3 in Algorithm 3 can be implemented by a max-heap, which requires $O(n)$ time for building the heap and $O(n\tau \log (n\tau))$ time for extracting the top-$n\tau$ elements. The time complexity for Lines 4–7 in Algorithm 3 is $O(md)$. In total, AFA's time complexity is $O(md + n\tau \log (n\tau))$, which equals to $O(md)$ when $n\tau$ is considered as a constant. Note that the space consumption in Algorithm 3 is dominated by the feature vectors of edges, and, thus, the overall space time complexity of AFA is $O(md)$.

5 Experiment

5.1 Experimental Setup

Datasets. Following the definition of the financial transaction graph in Sect. 3, we construct three financial networks (i.e., PayPal-X, PayPal-Y, and PayPal-Z) based on three real-world sets of financial transactions generated at PayPal. Table 1 lists the statistics of the three graphs used in our experiments.

Table 1. Data statistics ($K = 10^3, M = 10^6$).

Name	#nodes	#edges	#features	%anomalies
PayPal-X	100K	369K	296	1.4%
PayPal-Y	1M	5M	296	1.2%
PayPal-Z	160M	470M	296	0.8%

We extract a 296-dimensional feature vector for each attributed edge and construct the transaction graph using one month of sampled transaction data. To compare with other baselines that are effective on smaller graphs, we further construct sub-graphs that limit the number of nodes to 100K, 1M: we sort the transactions according to their timestamps, then select the first 100K and 1M users that appear in the transaction list, and filter all transactions related to them. We split all transactions into training, cross-validation, and testing sets using an 8 : 1 : 1 ratio. Note that there may be overlaps for nodes in the training, cross-validation, and testing set. This is reasonable as the transactions from the same users may have different labels. For instance, in account take-over cases, transactions from the same users can be initially benign and later become malicious, which can be distinguished using transaction attribute information. We ensure that the transactions in the cross-validation and testing sets have later timestamps to validate the model's performance on new transactions. Table 1 shows the statistics of the datasets used in our experiments. To construct the financial transaction network, we use the sender and receiver IDs as node IDs in each transaction. Following the definition of the financial transaction graph in Sect. 3, we iterate through all transactions in E and add them as directed attributed edges in G.

Inspired by the sampling method proposed in [8], due to the large volume of transactions and the unbalanced nature of the dataset, we only sample good transactions (i.e., non-anomaly transactions) such that for each pair of nodes, we only retain the latest good transactions. We split the transactions into training and testing sets based on their transaction timestamps.

Baselines and Parameter Settings. We compare our proposed solutions against three categories of methods (10 methods in total) in terms of classification accuracy and empirical efficiency.

Table 2. Anomaly detection performance.

(a) PayPal-X

Algorithm	AUC	PR@1%	F1-Score	AP
Isolation Forest	0.43282	0.00000	0.026707	0.01084
FDGars	0.46034	0.00200	0.02663	0.01192
Player2Vec	0.53873	0.01347	0.03396	0.01560
SemiGNN	0.56313	0.05619	0.07381	0.02915
Logistic Regression	0.86067	0.16232	0.19923	0.14546
XGBoost	0.87784	0.21844	0.25727	0.18102
GCN	0.84134	0.16032	0.19202	0.13477
GraphSAGE	0.86908	0.23246	0.27885	0.21185
Node2Vec	0.87248	0.28257	0.33418	0.25360
GraphConsis	0.88052	0.18249	0.23461	0.16029
PFA	0.88291	0.26453	0.31481	0.24991
AFA	0.90695	0.31864	0.37054	0.32377
DoubleFA	**0.91010**	**0.34469**	**0.41176**	**0.36176**

(b) PayPal-Y

Algorithm	AUC	PR@1%	F1-Score	AP
Isolation Forest	0.45582	0.00086	0.02308	0.00987
FDGars	0.42943	0.00394	0.02310	0.00956
Player2Vec	0.48821	0.00912	0.02310	0.01159
SemiGNN	0.44286	0.00338	0.02310	0.00991
Logistic Regression	0.85057	0.17043	0.18553	0.13516
XGBoost	0.88215	0.21389	0.23135	0.17647
GCN	0.83639	0.16290	0.17742	0.11925
GraphSAGE	0.87952	0.26745	0.28924	0.21649
Node2Vec	0.88268	0.23426	0.25429	0.18842
GraphConsis	0.84883	0.13872	0.19125	0.08857
PFA	0.90531	0.33641	0.36465	0.31356
AFA	0.90918	0.33299	0.35961	0.31569
DoubleFA	**0.90925**	**0.34873**	**0.38143**	**0.33449**

(c) PayPal-Z

Algorithm	AUC	PR@1%	F1-Score	AP
PFA	**0.89749**	0.30510	0.32941	0.27886
AFA	0.89421	0.30407	0.33098	0.27054
DoubleFA	0.89510	**0.31793**	**0.34482**	**0.29061**

- Three non-graph-based methods which input attributes and labels of transactions into Isolation Forest [12], logistic regression and XGBoost [2], respectively.
- One plain graph-based method which learns node embedding vectors via Node2Vec and concatenates node embeddings with the edge attributes as edge feature vectors before inputting them to the classifier.
- Six attributed graph-based methods, including GCN [10], GraphSAGE [7], FDGars [24], Player2Vec [29], SemiGNN [23], and GraphConsis [14], which takes edge attribute vectors as initial edge feature vectors.

For all competitors, we adopt the default parameter settings suggested in their respective papers. In our method `PFA`, we employ a modified parallel version of `ForwardPush` implemented in [25] for the construction of the top-k PPR matrix with parameters $\alpha = 0.2$ and $k = 32$. In `AFA`, to compute `HBOS`, we pick the number of bins to be 10, and label the top 10 percent instances (i.e., $\tau = 0.1$) with the highest `HBOS` as anomaly instances.

5.2 Effectiveness Evaluation

In this set of experiments, we empirically study the effectiveness of our proposed solutions and competing methods in anomaly detection on three datasets, including PayPal-X, PayPal-Y, and PayPal-Z. Table 2 presents the anomaly detection performance on all datasets measured by four commonly used classification metrics, i.e., AUC (area under the curve), PR@1% (precision@1%), F1-score, and AP (average precision). We can first make the following observations from Table 2. As the only unsupervised method, Isolation Forest performs the poorest among all the methods. Supervised methods that only leverage edge attribute information, i.e., Logistic Regression and XGBoost, can outperform some graph-based

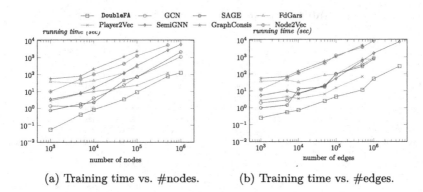

(a) Training time vs. #nodes.   (b) Training time vs. #edges.

Fig. 4. Scalability tests.

methods like GCN. Graph-based baselines GraphSAGE, Node2Vec, and Graph-Consis, perform the best among all the baselines, which suggests the usefulness of combining graph structural information and attribute information. For Graph-SAGE, it can capture neighborhood attribute information through neighborhood aggregation, but it is difficult to capture long-hop relations between nodes. For Node2Vec, it adopts similar ideas from the random walk as well as neighborhood information; thus, it can achieve relatively good performance, but it doesn't explicitly capture anomalies on nodes or edges. For GraphConsis, it considers feature inconsistencies and context inconsistencies but fails to consider long-hop relations. Note that for FDGars, the performance is also very poor due to the graph being obtained by converting transaction edges into nodes wherein their original work the graph is constructed using a line graph. For Player2Vec and SemiGNN, their models focus more on heterogeneous graphs with different relations, which may not be helpful in our settings. Table 2a and 2b report the accuracy results and ablations on the smaller datasets PayPal-X and PayPal-Y. In terms of AUC, compared to the best competitors, DoubleFA achieves an improvement of 0.03, which is around a 3% increase. In terms of F1-Score, DoubleFA achieves an improvement of $0.07 \sim 0.09$, around a $23\% \sim 32\%$ increase. In terms of PR@1%, DoubleFA achieves an improvement of $0.06 \sim 0.08$, around a $22\% \sim 30\%$ increase. In terms of AP, DoubleFA achieves an improvement of $0.11 \sim 0.12$, around a $43\% \sim 55\%$ increase. These metrics indicate that DoubleFA can maintain stable performance for edge anomaly detection. Table 2c reports the results on the dataset that contains 160M users. The purpose of running experiments on this dataset is to validate if our methods can apply to large graphs with massive nodes and edges. In terms of AUC, PR@1%, F1-Score and AP, DoubleFA achieves comparable results as in Table 2a and 2b. This indicates our method's effectiveness and applicability on very large graphs.

5.3 Scalability Evaluation

In terms of scalability, Fig. 4 reports the training time required by our method and other GNN-based methods varied by the number of nodes and the number of edges. The x-axis is the number of nodes/edges in the log scale, and the y-axis is the training time (seconds) in the log scale. All experiments are conducted using a server using Intel Xeon CPU@2.30 GHz with 40 cores and Tesla P100 GPU with 16 GB RAM. For GCN, GraphSAGE, FdGars, Player2Vec, SemiGNN, and GraphConsis, we employ GPUs to conduct experiments since the computational time costs on CPUs are prohibitive. As for our methods and Node2Vec, we utilize CPUs. As shown in Fig. 4, our method consistently outperforms other methods without suffering memory overflow problems. More specifically, compared to the best competitors, our proposed solution `DoubleFA` achieves up to 5.2×, more than 7.7×, and over 27× speed-ups on PayPal-X, PayPal-Y, and PayPal-Z datasets, respectively.

6 Conclusion

This paper proposes a practical framework for anomaly detection in large amounts of financial transactions. In particular, we model a set of financial transactions as an attributed multigraph by regarding each transaction as an edge. Based thereon, we devise two schemes for generating informative feature representations for each edge. Experiments on 3 PayPal transaction networks demonstrate that our proposed framework outperforms baselines in terms of both result accuracy and practical efficiency. As for future work, we will also explore how to extend our method to deal with dynamic networks.

References

1. Brand, M.: Fast low-rank modifications of the thin singular value decomposition. Linear Algebra Appl. **415**(1), 20–30 (2006)
2. Chen, T., Guestrin, C.: XGBoost: a scalable tree boosting system. In: KDD, vol. 13–17-August, pp. 785–794 (2016)
3. Dai, Q., Shen, X., Zhang, L., Li, Q., Wang, D.: Adversarial training methods for network embedding. In: WWW, pp. 329–339 (2019)
4. Goldstein, M., Dengel, A.: Histogram-based outlier score (hbos): a fast unsupervised anomaly detection algorithm. **KI**(1), 59–63 (2012)
5. Goyal, P., Chhetri, S.R., Canedo, A.: dyngraph2vec: Capturing network dynamics using dynamic graph representation learning. Knowl.-Based Syst. **187**, 104816 (2020)
6. Grover, A., Leskovec, J.: Node2vec: scalable feature learning for networks. In: KDD, pp. 855–864 (2016)
7. Hamilton, W.L., Ying, R., Leskovec, J.: Inductive representation learning on large graphs. In: NeurIPS, pp. 1025–1035, June 2017
8. Hu, B., Zhang, Z., Shi, C., Zhou, J., Li, X., Qi, Y.: Cash-out user detection based on attributed heterogeneous information network with a hierarchical attention mechanism. In: AAAI, pp. 946–953 (2019)

9. Jeh, G., Widom, J.: Scaling personalized web search. In: WWW, pp. 271–279 (2003)
10. Kipf, T.N., Welling, M.: Semi-supervised classification with graph convolutional networks. In: ICLR (2017)
11. Klicpera, J., Bojchevski, A., Günnemann, S.: Predict then propagate: graph neural networks meet personalized PageRank. In: ICLR, October 2019
12. Liu, F.T., Ting, K.M., Zhou, Z.H.: Isolation forest. In: ICDM, pp. 413–422 (2008)
13. Liu, Y., et al.: Pick and choose: a GNN-based imbalanced learning approach for fraud detection. In: WWW, pp. 3168–3177. WWW 2021 (2021)
14. Liu, Z., Dou, Y., Yu, P.S., Deng, Y., Peng, H., Peng, H.: Alleviating the inconsistency problem of applying graph neural network to fraud detection. In: SIGIR, pp. 1569–1572 (2020)
15. PayPal Holdings Inc: Annual Report Pursuant to Section 13 or 15(d) of The Securities Exchange Act of 1934. https://www.sec.gov/Archives/edgar/data/1633917/000163339172100018/pypl-20201231.html (2020)
16. Perozzi, B., Al-Rfou, R., Skiena, S.: DeepWalk: Online learning of social representations. In: KDD, pp. 701–710 (2014)
17. Qiu, J., et al.: Netsmf: large-scale network embedding as sparse matrix factorization. In: WWW, pp. 1509–1520 (2019)
18. Ramaswamy, S., Rastogi, R., Shim, K.: Efficient algorithms for mining outliers from large data sets. SIGMOD Record **29**(2), 427–438 (2000)
19. Ravisankar, P., Ravi, V., Raghava Rao, G., Bose, I.: Detection of financial statement fraud and feature selection using data mining techniques. In: DSS, pp. 491–500 (2011)
20. Swersky, L., Marques, H.O., Jörg, S., Campello, R.J., Zimek, A.: On the evaluation of outlier detection and one-class classification methods. In: DSAA, pp. 1–10 (2016)
21. Veličković, P., Casanova, A., Liò, P., Cucurull, G., Romero, A., Bengio, Y.: Graph attention networks. Technical Report (2018)
22. Wang, C., Wang, C., Wang, Z., Ye, X., Yu, P.S.: Edge2vec: edge-based social network embedding. TKDD **14**(4), 1–24 (2020)
23. Wang, D., et al.: A semi-supervised graph attentive network for financial fraud detection. In: ICDM, pp. 598–607 (2019)
24. Wang, J., Wen, R., Wu, C., Huang, Y., Xiong, J.: FDGars: fraudster detection via graph convolutional networks in online app review system. In: WWW, pp. 310–316 (2019)
25. Wang, S., Yang, R., Xiao, X., Wei, Z., Yang, Y.: FORA: simple and effective approximate single-source personalized PageRank. In: KDD, pp. 505–514 (2017)
26. Weber, M., et al.: Anti-money laundering in bitcoin: experimenting with graph convolutional networks for financial forensics. arXiv (2019)
27. Wu, F., Zhang, T., de Souza, A.H., Fifty, C., Yu, T., Weinberger, K.Q.: Simplifying graph convolutional networks. In: arXiv (2019)
28. Yun, S., Jeong, M., Kim, R., Kang, J., Kim, H.J.: Graph transformer networks. In: Wallach, H., Larochelle, H., Beygelzimer, A., d Alché-Buc, F., Fox, E., Garnett, R. (eds.) NeurIPS. vol. 32 (2019)
29. Zhang, Y., Fan, Y., Ye, Y., Zhao, L., Shi, C.: Key player identification in underground forums over attributed heterogeneous information network embedding framework. In: CIKM, pp. 549–558, November 2019
30. Zhang, Z., Cui, P., Li, H., Wang, X., Zhu, W.: Billion-scale network embedding with iterative random projection. In: ICDM, pp. 787–796 (2018)
31. Zhang, Z., Pei, J., Cui, P., Yao, X., Wang, X., Zhu, W.: Arbitrary-order proximity preserved network embedding. In: KDD, pp. 2778–2786 (2018)

32. Zhong, Q., et al.: Financial defaulter detection on online credit payment via multi-view attributed heterogeneous information network. In: WWW, pp. 785–795, April 2020
33. Zhu, J., Yan, Y., Zhao, L., Heimann, M., Akoglu, L., Koutra, D.: Beyond homophily in graph neural networks: Current limitations and effective designs. In: NeurIPS, vol. 2020-December (2020)

The Synergies of Context and Data Aging in Recommendations

Anna Dalla Vecchia[ID], Niccolò Marastoni[ID], Barbara Oliboni[ID], and Elisa Quintarelli[✉][ID]

University of Verona, Strada Le Grazie, 15, Verona, Italy
anna.dallavecchia@studenti.univr.it,
{niccolo.marastoni,barbara.oliboni,elisa.quintarelli}@univr.it

Abstract. In this paper, we investigate the synergies of data aging and contextual information in data mining techniques used to infer frequent, up-to-date, and contextual user behaviours that enable making recommendations on actions to take or avoid in order to fulfill a specific positive goal. We conduct experiments in two different domains: wearable devices and smart TVs.

Keywords: Context-awareness · Data Aging · Data mining · Recommendations

1 Introduction

Mobile technology has become an essential element of modern life. The use of sensors and mobile devices provides a constant and steady stream of new data. Such information can range from wearable devices data, to data coming from IoT home/office devices, and to data related to media consumption habits.

Sensor data is intrinsically temporal in nature, since events happen in succession, and thus temporal data can be analyzed to extract sequences of frequent events (patterns). Looking at a user's historical actions can give insight into their future actions, and events that occurred more recently can be more valuable than older ones for determining new events. In this case, data aging techniques can improve the accuracy of predictions for the future by focusing the attention on recent patterns.

Another aspect that can greatly impact a user's choices is the context in which they find themselves when making those choices. This means that considering additional contextual information could improve the quality of the recommendations. The context, as we refer to it in this paper, encompasses all information that identifies the situation in which a recommendation is made [3]. This contextual data, while providing information that is external to the event, is inherently personalized for each individual user, as users can be influenced by different contextual factors. It therefore cannot be imposed at design time nor will it be equivalent for all users. Contextual information belongs to different categories, among them we can cite the temporal perspective and the one

R. Wrembel et al. (Eds.): DaWaK 2023, LNCS 14148, pp. 80–87, 2023.
https://doi.org/10.1007/978-3-031-39831-5_7

related to the user [9]. One example of a temporal perspective is whether the day is a weekday or on the weekend; one example of a user-related perspective is whether the user is with their family or by themselves.

In this paper we extend our preliminary proposal [8] and we show that aging and contextual data can work together and improve the performance of recommender systems, both in terms of accuracy and computational complexity. In particular, we combine data aging with different contexts, in order to gauge their respective impact on the overall performance of the algorithm.

As general use cases, we consider two different datasets for evaluating our proposal. The first dataset contains logs of data collected by Fitbit devices on physical activity levels and sleeps scores. The second use case is a dataset including logs of TV programs watched by users. In order to enrich the considered data with contextual information, we encode different contextual perspectives according to the considered dataset, including the temporal one which we refine to distinguish between vacation and normal days when possible. As for the TV programs, besides analysing data w.r.t. temporal perspectives, we also use details about the people watching TV, distinguishing between lone child, lone adult, child with family, adult with family.

The paper is organized as follows. Section 2 introduces ALBA, the extension of the LookBack Apriori (LBA) algorithm with the aging mechanism. Section 3 describes how we consider contextual data, and Sect. 4 presents the validation of the proposed approach. Section 5 summarizes the contribution of the paper.

2   ALBA: Adding Aging to LookBack Apriori

In this section, we introduce the Aged LookBack Apriori algorithm (ALBA), an extension of Apriori [4] that increases the importance of recent information through the introduction of an aging mechanism.

ALBA starts from the LookBack Apriori algorithm (LBA) [8] that generates frequent and totally ordered sequence rules from temporal data while considering i) a possibly predefined temporal window τ_w, concatenating each transaction with the previous ones in such a way to be able to look back at $\tau_w - 1$ time units, and ii) the relative time of each item w.r.t. the current time instant. The mined rules have the form $r : I_{-(\tau_w-1)} \wedge \cdots \wedge I_{-2} \wedge I_{-1} \rightarrow I_0 \ [s_r, c_r]$.

It is important to highlight that we have developed our approach without starting from well know algorithms, like GSP [6] because we consider relevant the sequential relationship between each itemset.

In ALBA, differently from LBA, in order to compute the support of the items at each iteration of the Apriori algorithm, the system builds a matrix of dimensions $n \times m$, where n is the number of transactions and m is the number of different items in the considered temporal dataset. The value of the element at row i and column l will be 1 if item l appears in day i, 0 otherwise. With this approach, the support of item l can be easily computed by summing all the elements of its $l - th$ column and dividing the result by n. To penalize the older items, each $i - th$, with $i < \tau_w$, is multiplied by an aging factor $\alpha_i = \frac{i}{n-\tau_w+1}$, where τ_w is the temporal window and n is the number of transactions in the dataset. This guarantees that the items in the temporal window will still be represented with 1, while older items slowly decay but never quite reach 0.

3   Context Modeling

The importance of considering contextual features in recommender systems to improve the relevance of provided suggestions is widely recognized [2,3,9]. Indeed, users' behaviours, preferences, and decisions are often affected by external factors used to characterize the environment or situation they are acting in, e.g. spatio-temporal information, weather, and social conditions, presence of other people. Although there is no standard definition of context, it is commonly accepted that context includes any information that can be used to characterize the situation of an entity, where an entity is anything considered relevant for the interaction between a user and an application [1]. This is mainly composed of highly dynamic features that change their values over time, thus requiring an efficient way to handle them in order to be able to provide fresh and relevant recommendations on the base of the user current context. Since contextual information is not always available, or complete, hierarchical contextual models have been proposed in the literature [5,7], as they allow to represent the context at different levels of granularity.

According to the systematic framework introduced in [9], in this work we consider the contextual modeling paradigm, i.e. the contextual features are integrated into our recommendation model and are considered during the preference computation process. This means that each transaction is composed of itemsets I_{t_i} that contain, besides other items, the item C_{t_i} related to the context at time unit t_i. Context information has to be acquired before being integrated into the preference computation; sometimes it may be gathered from the same smart devices or sensors providing data related to the objects to monitor (e.g. the temporal information can be acquired from a wearable device), but other times it has to be obtained from external sources, as in the case of weather or social information.

4   Evaluation

In this section we present some experiments to validate our approach. We will consider the following aspects: (i) the benefits of data aging to LBA, (ii) the benefits of adding contextual information and (iii) the advantages of modeling contexts with different granularity levels.

The first set of experiments focuses on wearable device data: we analyze logs of physical activity levels and sleep scores extracted from Fitbit devices, enhancing the process with an aging mechanism and contextual information, in order to provide contextual, personalized, and timely recommendations to users. To further evaluate the synergies between the aging mechanism and the relevance of context during the recommendation process we apply our proposal to a Smart TV scenario, where the recommender system suggests the next genre to watch based on historical and contextualized data.

Fitbit. This scenario involves physical activity and sleep score logs obtained from Fitbit devices and it features two datasets: PMdata and Custom. PMdata consists of logs from 16 users and the Custom dataset was collected from 4 users specifically for this study. From both of these datasets we make use of the logs pertaining to "light", "medium" and "heavy" activity, along with the sleep score for each day. These features are recorded by Fitbit as minutes spent in each activity type, thus we discretize them in order to obtain categorical data as described in [8]. During this discretization process the activity levels are further split into 3 sub-levels according to set thresholds, e.g. a heavy activity (HA) can be encoded into three possible labels: $HA : 1, HA : 2$ and $HA : 3$.

Auditel. In this scenario, we study the TV watching habits of 7 users, aged between 5 and 77 years old. The data has been collected by the Auditel company and it consists of logs of TV programs watched, where each data point contains, among other things, the genre of the program itself, a timestamp of the exact start time and end time, and the specific user. An interesting feature of this dataset is the "family" field which indicates the age range of the user ("child" or "adult") and if the user was watching the program alone or with members of the family. The 7 users have been selected from the thousands included in the dataset by virtue of the richness of their logs and familiar context. For these experiments we only consider the data points containing more than 10 min of screen time, filtering out channel surfing and disliked programs. Each data point contains one program viewing and it is encoded by a two-letter label representing one of the five possible genres: documentary, serious, fun, reality, sport.

4.1 Contexts

The first temporal contextual dimension considered in this study is the Part of the week which characterizes each day as either a weekday or part of the weekend. This is used in all the datasets, as the smallest time unit in the Fitbit datasets is the day, so each data point can be easily enriched by the value WD if it represents a day that falls into the weekend. Moreover, Auditel data allows for smaller time granularities, so we add Portion as a second temporal contextual dimension, called TS in this section, which tracks the time slot in which the individual data point (a program viewing) takes place, be it morning, afternoon or night time. As mentioned earlier, the Auditel dataset comes with the "family" feature, so we consider the first individual contextual dimension related to the Type of a group, called here FA, and use it to track four different values: lone child, lone adult, child with family, adult with family. Lastly, we add the Situation dimension, reported as VA, to distinguish the vacation days from normal days. Unfortunately, this is only available on the Custom dataset, as it requires the collection of private data from each user.

4.2   Methodology

We reserve 80% of the logs of each user as a training set to generate the rules and then use the remaining 20% as a test set. We then run the experiments on the training set with multiple configurations of LookBackApriori plus aging and context, and generate a set of rules for each user in the dataset with no specified goal. For the Fitbit dataset we only extract rules that have the sleep score (our target) in the consequent, while on the Auditel dataset the consequent will contain the genre of the last TV program in the frequent itemset.

The test dataset is then divided in multiple queries that are then checked for accuracy against the rules learned in the training phase. For both scenarios we first order the extracted rules by confidence and support, but in the Fitbit experiments we prioritize rules that are "complete", meaning that they contain itemsets for each timestamp. This is done to preserve the feasibility of the original proposal, since we need to be able to suggest what types of activity will lead to better sleep, having a knowledge of the complete history of user's past activities in the Fitbit log.

4.3   Fitbit Validation

To evaluate the impact of aging and context on the PMdata dataset we first ran the experiments with: 1) the original LBA, 2) LBA plus the aging mechanism, 3) LBA with the *WD* context (the only available context for this dataset) and then 4) LBA with both aging and the context. The temporal window has been set to 3, as that is the value that gives the best results while keeping the run time acceptable.

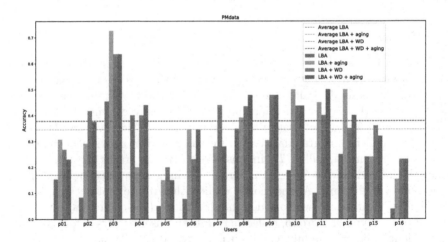

Fig. 1. Accuracies for all PMdata users.

The results can be seen in Fig. 1. From these we can observe that both the aging mechanism and the *WD* context substantially improve the average accuracy of the LBA algorithm when used separately. The combination of *WD* and aging further improves the average accuracy due to a few outliers, but the performance for the individual users is not always better than the previous experiment. The experiments on the Custom dataset were ran with the same configurations as PMdata but with the additional context *VA*, which is only available in this dataset. This adds four configurations: *VA*, the combination of *VA* and *WD*, and both previous configurations with aging.

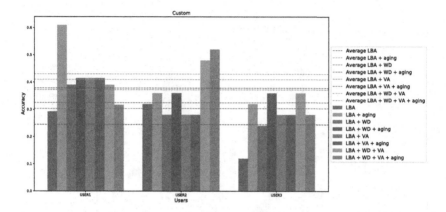

Fig. 2. Accuracies for all Custom users.

The results, shown in Fig. 2, show that the aging mechanism improves LBA average accuracy slightly better than the combination of the *VA* and *WD* contexts. It is also interesting to note that the average accuracy does not tell the whole story, as the experiments on different users in the Custom dataset react differently to data aging and the addition of context. For example, the performance on User 1 is strongly, and positively, affected by the aging mechanism. This is easy to explain, as User 1 started logging data before all the others, resulting in a much bigger dataset that is better suited to aging. We posit that the aging mechanism does not benefit as much those temporal datasets that span shorter periods, and the experiments in the next section further support this proposition. Another example is User 2, which is one of the clearest examples of context synergy in these experiments: the accuracy drops slightly at the addition of the contexts *VA* and *WD* separately, but their combination (with aging) leads to a 63% increase

4.4   Auditel Validation

We run the experiments on the selected users of the Auditel dataset, starting with LBA and then adding the available contexts in all possible configurations. Interestingly, the aging mechanism, while being very effective in the Fitbit experiments, does not improve the results for the Auditel scenario. We suspect that the reasons for this are to be found mainly in the smaller size of the Auditel dataset: since only 3 months of data are available, the aging mechanism is penalizing data that might still be relevant, as it is not particularly old. Another possible reason is the higher granularity of the data: Auditel items always represent events happening in the same day, thus the aging mechanism wrongly penalizes (although not excessively) events that are temporally very close. For this reason the aging mechanism is excluded from the presented experiments, as it would only lead to redundancy.

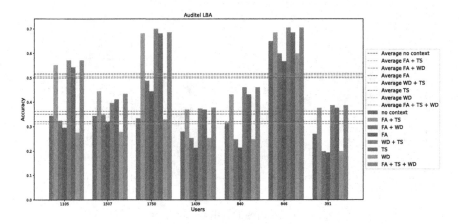

Fig. 3. Accuracies for some Auditel users.

The temporal window for these experiments has been set to 6, as we found that further increasing the value, while leading to better accuracy, incurs in a big running time penalty. The best performing context in this scenario is *TS*, representing the time slot in which the program is televised. As shown in Fig. 3, the addition of *TS* increases the accuracy of the predictions for every user in the dataset. While the boost to the average accuracy is around 38%, the individual rise can be up to 104% (see user 1750 in Fig. 3). This is not surprising, as the nature of TV programs tends to change within the different periods of the day (i.e. serious programs are often at night, while kids program are more common in the afternoon). An interesting finding in this scenario is that the weekday context (*WD*) when used alone, actually decreases the accuracy of the algorithm in every user, while it improves the accuracy if used in conjunction to the *TS* context. The improvement cannot be entirely attributed to the already successful *TS* context, as the results of *TS* + *WD* are superior than those with only *TS*. This is another example of synergy between different contexts. The familiar context (*FA*) is the only context that does not enhance the performance of the recommender system; in fact it often does the opposite, even when paired with more successful contexts.

5   Conclusions and Future Work

In this paper, we combine contextual information and an aging mechanism in data mining techniques to evaluate their impact in the context of recommendation systems. The results of our experiments show that adding context to the recommender system leads to an increase of accuracy in most cases. The aging mechanism proves to be a very positive addition in the Fitbit experiments, improving all results but one, often by quite a large margin. Conversely, aging never improves the results on Auditel data. Overall, it seems like the decision on whether to add contextual data or aging (or neither) to the recommender system needs to be analyzed on a case-to-case basis, as different problems, different scenarios, and different data will require specific solutions.

References

1. Abowd, G.D., Dey, A.K., Brown, P.J., Davies, N., Smith, M., Steggles, P.: Towards a better understanding of context and context-awareness. In: Gellersen, H.-W. (ed.) HUC 1999. LNCS, vol. 1707, pp. 304–307. Springer, Heidelberg (1999). https://doi.org/10.1007/3-540-48157-5_29
2. Adomavicius, G., Mobasher, B., Ricci, F., Tuzhilin, A.: Context-aware recommender systems. AI Mag. **32**(3), 67–80 (2011)
3. Adomavicius, G., Sankaranarayanan, R., Sen, S., Tuzhilin, A.: Incorporating contextual information in recommender systems using a multidimensional approach. ACM Trans. Inf. Syst. **23**(1), 103–145 (2005)
4. Agrawal, R., Srikant, R.: Fast algorithms for mining association rules in large databases. In: Bocca, J.B., Jarke, M., Zaniolo, C. (eds.) Proceedings of VLDB 1994, pp. 487–499 (1994)
5. Bolchini, C., Quintarelli, E., Tanca, L.: CARVE: context-aware automatic view definition over relational databases. Inf. Syst. **38**(1), 45–67 (2013)
6. Harms, S.K., Deogun, J.S.: Sequential association rule mining with time lags. J. Intell. Inf. Syst. **22**(1), 7–22 (2004)
7. Hosseinzadeh Aghdam, M.: Context-aware recommender systems using hierarchical hidden Markov model. Phys. A Stat. Mech. Appl. **518**, 89–98 (2019)
8. Marastoni, N., Oliboni, B., Quintarelli, E.: Explainable recommendations for wearable sensor data. In: International Conference on Big Data Analytics and Knowledge Discovery, pp. 241–246 (2022)
9. Villegas, N.M., Sánchez, C., Díaz-Cely, J., Tamura, G.: Characterizing context-aware recommender systems: a systematic literature review. Knowl. Based Syst. **140**, 173–200 (2018)

Advanced Analytics and Pattern Discovery

Hypergraph Embedding Based on Random Walk with Adjusted Transition Probabilities

Kazuya Nagasato[1(✉)], Satoshi Takabe[1], and Kazuyuki Shudo[2]

[1] Tokyo Institute of Technology, Tokyo, Japan
kazuya.ngst@gmail.com
[2] Kyoto University, Kyoto, Japan

Abstract. In this paper, we consider embedding hypergraphs using random walks. By executing a random walk on a hypergraph and inputting the resulting node sequence into a skip-gram used in natural language processing, a vector representation that captures the graph structure can be obtained. We propose a random walk method with adjustable transition probabilities for hypergraphs. As a result, we argue that it is possible to embed graph features more appropriately. Experimental results show that by tuning the parameters of the proposed method appropriately, highly accurate results can be obtained even for large hypergraphs for machine learning tasks such as node label classification.

Keywords: Hypergraph · Random walk · Embedding

1 Introduction

Graphs are powerful mathematical models for quantitatively dealing with the relationships between people and things. For example, hyperlinks on the World Wide Web, friend networks on social networking services, and the structure of chemical compounds are often represented mathematically by graphs, which have been studied actively in recent years. Among these, recent developments in machine learning methods have attracted attention to embedding methods that represent entire graphs or graph nodes as low-order vectors for graphs, which are unstructured data, and have been applied to graph classification, node classification, and prediction of edges acquired by new nodes.

On the other hand, there are relationships in real networks that interact as a group, and these relationships cannot be represented by ordinary graphs in which an edge connects two nodes. Therefore, a mathematical model called a hypergraph, which is a generalization of graphs, is the target of research. Edges in a hypergraph can contain an arbitrary number of nodes, and higher-order relationships between nodes can be captured. The problem of node embedding for hypergraphs is important issue, and it is applied to recommendation systems, etc.

R. Wrembel et al. (Eds.): DaWaK 2023, LNCS 14148, pp. 91–100, 2023.
https://doi.org/10.1007/978-3-031-39831-5_8

And random walks are useful as a means of capturing node characteristics of graphs and hypergraphs. A random walk on a graph transitions through the nodes that are connected to each other, and thus provides information on the subgraphs of a node's neighbors as a sequence of nodes.

In this paper, we propose a random walk method with adjustable transition probabilities for hypergraphs and apply it to node embedding. Experimental results on node label estimation show that the proposed method can achieve similar estimation results to existing methods and reduce spatial computational complexity.

The remainder of this paper is organized as follows. Section 2 describes related work on node embedding in graphs using random walks and related work on random walks on hypergraphs. Section 3 describes the prior knowledge in this paper, and Sect. 4 describes the proposed method. In Sect. 5, we conduct experiments on a real data set and conclude in Sect. 6.

2   Related Work

In recent years, there has been much research on graph embedding. Spectral embedding [5] is a method based on graph signal processing that produces a vector representation so that the distance of vectors between neighboring nodes is close. There have been many studies on graph neural networks [6], which are derived from this method. However, these methods use $n \times n$ adjacency matrices as input, which are computationally expensive and not suitable for large data sets.

DeepWalk [1] and node2vec [2] are methods for embedding graphs using random walks. These methods obtain a vector representation of each node by inputting a sequence of nodes obtained from a random walk into a skip-gram [4]. These methods are scalable and can perform embedding on large graphs, especially when executed in parallel.

Carletti et al. [3] show that selecting adjacent nodes with equal probability is not a wise decision for random walk transitions in hypergraphs because the connections between nodes belonging to a hyperedge are often stronger than the connections between consecutive individuals. Carletti et al. Therefore, Carletti et al. defined transition probabilities for neighboring nodes based on their weights, which are the number of nodes in the hyper-edge to which they belong minus themselves. Random walks based on this definition can visit many high-order hyperedges [3].

Chitra et al. defined transition probabilities and formulated random walks for hypergraphs with edge-dependent node weights [7]. Chitra et al. show that when node weights are independent of edges in a hypergraph, a random walk on a hypergraph is equivalent to a random walk on a general weighted graph projected onto a clique representation.

Gatta et al. proposed a hypergraph embedding method specifically for music recommendation [19]. They defined a hypergraph data model specific to music recommendation, performs a random walk on the hypergraph, and then inputs the resulting node sequence into a skip-gram to obtain a vector representation of the nodes. The results show that their technique significantly outperforms other state-of-the-art techniques, especially in scenarios where the cold-start problem arise.

3   Preliminaries

3.1   Notation

In this paper, for a undirected hypergraph $\mathcal{H}(V, E)$, let $V = \{1, ..., n\}$ be the set of nodes and $E = \{E_1, ..., E_m\}$ be the set of hyperedges, where n is the number of nodes and m is the number of hyperedges. Each E_α is a subset of V and is equivalent to a general graph, especially when $|E_\alpha| = 2$ holds for any α.

let

$$e_{i\alpha} = \begin{cases} 1 & \text{for } i \in E_\alpha \\ 0 & \text{otherwise} \end{cases} \tag{1}$$

be a hypergraph incidence matrix, the adjacency matrix A of a hypergraph is denoted by $A = ee^T$ and the hyperedge matrix C by $C = e^T e$ where e is a matrix representing the inclusion relationship between hyperedges and nodes, A_{ij} is the number of hyperedges shared by node i and node j, and $C_{\alpha\beta}$ is the number of nodes included in $E_\alpha \cap E_\beta$.

3.2   Hypergraph Projection

In a hypergraph, a hyperedge can contain any number of nodes, and thus has a high degree of freedom, making it difficult to treat the hypergraph as it is. Therefore, there has been a lot of research using a method called Clique expansion, which treats hypergraphs as general unweighted undirected graphs by representing hyperedges as complete graphs (cliques) in which all nodes have edges with each other. However, a clique expansion is an irreversible transformation, which results in a considerable loss of information as a hypergraph. Therefore, in this paper, a random walk on an unweighted undirected graph, which is a clique expansion of a hypergraph, is used as a comparison target.

3.3   Random Walk and Stationary Distribution

In general, if we define a node in a graph as a state and define the transition probability T_{ij} from node i to node j, the transition of a node in the graph can be regarded as a random walk. A stochastic process in which the next state is determined only by the current state is called a Markov chain. An ergodic Markov chain has a stationary distribution p and is obtained by solving

$$p = pT. \tag{2}$$

Equation (2) shows that the stationary distribution of the random walk is a right eigenvector corresponding to eigenvalue 1 of the transition probability matrix T.

3.4   Skip-Gram

We employ skip-gram as a node embedding method [4]. Skip-gram is a neural network that learns a function $f : V \to \mathbb{R}^d$ to obtain a d-dimensional vector representation of nodes, and was designed to maximize word co-occurrence probability in natural language processing. In this study, we perform a random walk on a hypergraph and use the resulting node sequence as input to obtain a vector representation of the nodes that captures the relationships among the nodes.

4   Proposed Method

In this chapter, we propose a random walk whose behavior can be controlled by parameters. By setting parameters that match the dataset and executing a random walk, a set of nodes that capture the graph structure can be obtained. Then, by inputting the obtained nodes into a skip-gram, a vector representation of each node can be obtained.

4.1   Random Walk

If the current node is i, then the adjacent node j is assigned the weight $k_{ij}^H(\beta)$ expressed in Eq. (3).

$$k_{ij}^H(\beta) = \sum_{\alpha}(C_{\alpha\alpha} - 1)^{\beta} e_{i\alpha} e_{j\alpha} \tag{3}$$

This definition is an extension of the previous study [3] by adding the parameter β as a power.

By normalizing so as to impose a uniform choice among the connected hyperedges, we get the expression for the transition probabilities

$$T_{ij} = \frac{k_{ij}^H(\beta)}{\sum_{l \neq i} k_{il}^H(\beta)}. \tag{4}$$

Equation (4) is a function of β. Depending on the value of β, a random walk with various behaviors is realized. $\beta = 1$ is equivalent to the previous study [3], and a behavior of visiting more high-order hyperedges is expected when β is larger than 1. When β is smaller, it is expected to visit more low-order hyperedges. When $\beta = 0$, adjacent nodes are visited with equal probability, which is equivalent to a random walk in a graph with a hyper-edge clique expansion.

From the above, our hypothesis in the proposed random walk can be summarized as follows.

- $\beta > 1$ Extracts many high-order hyperedge structures
- $\beta = 1$ Equivalent to previous studies [3]
- $\beta < 1$ Extracts many low-order hyperedge structures
- $\beta = 0$ Equivalent to a hypergraph with clique expansion

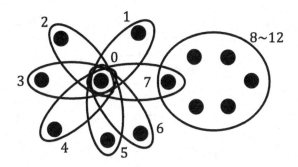

Fig. 1. Artificial hypergraphs. Each number represents a node number.

Also, when the hypergraph is ergodic for a given β, the random walk has a stationary distribution

$$p_j^\infty = \frac{d_j^H(\beta)}{\sum_j d_j^H(\beta)} \tag{5}$$

for all $j = 1, ..., n$, where $d_j^H(\beta)$ is defined as $d_j^H(\beta) = \sum_{l \neq j} k_{jl}^H(\beta)$.

5   Experiment

First, a simple artificial hypergraph is used to verify the behavior of the proposed random walk. Then, we check the accuracy in the label estimation task when the parameters are changed for the three datasets.

5.1   Transition Probabilities in Steady State

For the artificial hypergraph (Fig. 1), the stationary distribution given by Eq. (5) was computed to verify the random walk behavior.

As a result, when $\beta = -1$, the probability of staying at high-order hyperedges was lower and the probability of staying at hubs included in low-order hyperedges was higher than in the random walk in the clique-expanded hypergraph. On the other hand, when $\beta = 2$, the probability of staying at a node included in a high-order hyperedge was higher than $\beta = 1$ in the previous study [3], and the probability of staying at a hub included in a low-order hyperedge was lower (Fig. 2).

Also, examining the probability of staying at each node, higher β induced higher latent probabilities to higher order hyperedges, and vice versa (Fig. 2).

These results indicate that our hypothesis is correct and that we can control the random walk behavior by changing the parameter β.

5.2   Node Label Estimation

In this section, we perform the label estimation task on real data. First, we input the sequence of nodes obtained from the proposed random walk into a

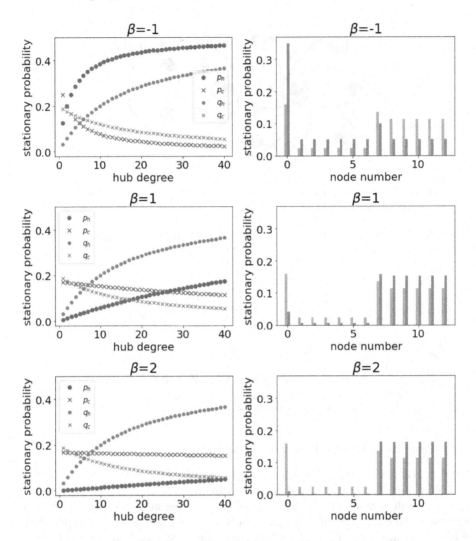

Fig. 2. The left column shows the stationary transition probabilities for node 0 and node 7 when the order of the hub is varied. p_h and q_h represent the stationary transition probabilities for each β at node 0 and node 7, respectively. p_c and q_c represent the stationary transition probabilities for $\beta = 0$ (clique expansion) at node 0 and node 7, respectively. The right column represents the stationary distribution when the degree of node 0 is fixed to 7. The red histograms are for each β, and the gray histograms correspond to $\beta = 0$.(Color figure online)

skip-gram to obtain a vector representation of each node. Then, we perform the task of estimating labels from the vector representation of the nodes using logistic regression. When executing the random walk, the transition probabilities are calculated for each state sequentially, and the walk is performed without explicitly describing the transition probability matrix. This significantly reduces the amount of spatial computation and allows experiments to be performed on large data sets.

The experimental procedure is as follows.

1. A sequence of nodes is obtained from each node in the hypergraph γ times, each with a walk length t, by the proposed random walk.
2. The obtained node sequence is input into a skip-gram to obtain a d-dimensional vector representation for each node.
3. The percentage of supervised data is set to 80% of the number of nodes, and logistic regression is used to estimate the labels of the remaining nodes.
4. (1)–(3) is repeated s times for each β to obtain the mean and standard error of the F1 score.

Table 1. Hypergraphs used in our experiments.

dataset	n	m	avg$|E_i|$	max$|E_i|$	labels
senate-bills	294	29,157	8.0	99	2
contact-primary-school	242	12,704	2.4	5	11
mathoverflow-answers	73,851	5,446	24.2	1,784	1,456

We use the datasets [8–12] shown in Table 1. The senate-bills dataset is a hypergraph, where nodes are US Congresspersons and hyperedges are comprised of the sponsor and co-sponsors of bills put forth in the Senate. The contact-priary-school dataset represents students in proximity. Each hyperedge corresponds to a group of people that were all in proximity of one another at a given time, based on data from sensors worn by students and teachers. Each node is labeled as a teacher or the classroom to which the student belongs. The mathoverflow-answers dataset is a hypergraph where hyperedges are sets of questions answered by users on Math Overflow. Nodes are labeled by the tags used in the questions, and nodes often have multiple labels.

We set $\gamma \in \{8, 16, 32, 64\}$, $t = 20$, $d = 64$, $\beta \in \{-1, 0, 1\}$, $s = 20$ for senate-bills and contact-private-school datasets. For mathoverflow-answers we set $\gamma \in \{8, 16\}$, $t = 20$, $d = 128$, $\beta \in \{-1, 0, 1\}$, $s = 1$. When $\beta = 0$, the embedding technique is equal to the deepwalk in the clique expansion graph, and when $\beta = 1$, the random walk is equal to the [3] in the previous study.

Figure 3 shows the experimental results for the three data sets. Figure 3 shows that the F1 score varies depending on the value of β. In particular, for the senate-bills dataset, setting $\beta = -1$ yields a high F1 score even with a small number of samples, which is significantly higher than that of the existing method $\beta = 0$. The proposed method is also able to obtain results on the mathoverflow-answers dataset, which is a large dataset, due to its ability to reduce the spatial computational complexity, and it correctly answers nearly half of the labels on the $1,456$ label count.

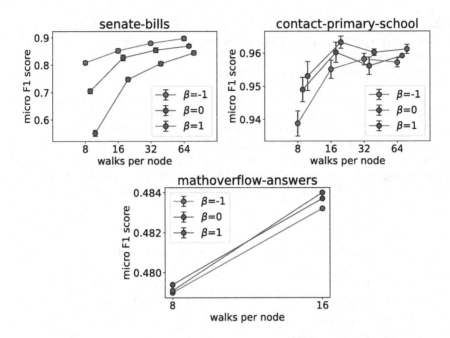

Fig. 3. Experimental results with label estimation for each dataset. Error bars in the figure represent standard errors.

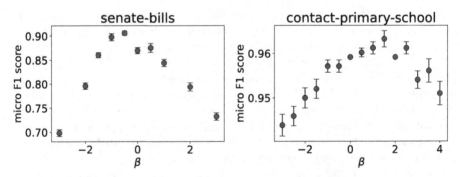

Fig. 4. Dependence of β on F1 score. Error bars in the figure represent standard error.

5.3   Parameter Dependence of F1 Score

In this subsection, we show how the F1 score varies with the parameter β. For two datasets, senate-bills and contact-primary-school, we conducted experiments by moving the value of β from -3 to 3 and from -3 to 4 in 0.5 increments, respectively (Fig. 4). The range of values of β was kept around 0 based on the hypothesis that random walks would not be able to extract relationships between nodes in these cases because random walks would not be able to escape from high-order hyperedges if β is too large or from low-order hyperedges if β is too small. Figure 4 shows that senate-bills and contact-primary-school are convex functions with maximum values around $\beta = -0.5$ and $\beta = 1.5$, respectively. Therefore, it may be possible to optimize the results using the proposed method by appropriately searching for the parameter β according to the data set.

6 Conclusion

In this paper, we propose a random walk with tunable transition probabilities for hypergraphs and apply it to embedding.

By using an artificial hypergraph, we have confirmed that the proposed random walk can freely exhibit behavior by changing the parameters according to our hypothesis. The proposed random walk can embed even large datasets by executing the random walk on the hypergraph, and it shows higher accuracy than existing methods in the label estimation task using real data. The parameters of the proposed method may be optimized for each dataset, and more appropriate embedding can be achieved by optimizing the parameters.

References

1. Perozzi, B., AI-Rfou, R., Skiena, S.: DeepWalk: online learning of social representations. In: KDD, pp. 701–710 (2014)
2. Grover, A., Leskovec, J.: node2vec: scalable feature learning for networks. In KDD, pp. 855–64. ACM, New York (2016)
3. Carletti, T., Battiston, F., Cencetti, G., Fanelli, D.: Random walks on hypergraphs. Phys. Rev. E **101**, 022308 (2020)
4. Mikolov, T., Chen, K., Corrado, G., Dean, J.: Efficient estimation of word representations in vector space. In: ICLR (2013)
5. Tang, L., Liu, H.: Leveraging social media networks for classification. Data Min. Knowl. Disc. **23**(3), 447–478 (2011)
6. Wu, Z., Pan, S., Chen, F., Long, G., Zhang, C., Yu, P.S.: A comprehensive survey on graph neural networks. IEEE Trans. Neural Netw. Learn. Syst. **32** (1), 4–24 (2021)
7. Chitra, U., Raphael, B.J.: Random walks on hypergraphs with edge-dependent vertex weights. In: Proceedings of the 36th International Conference on Machine Learning (ICML) (2019)
8. Chodrow, P.S., Veldt, N., Benson, A.R.: Generative hypergraph clustering: from blockmodels to modularity. Sci. Adv. **7**, eabh1303 (2021)
9. Fowler, J.H.: Connecting the congress: a study of cosponsorship networks. Polit. Anal. **14**, 456–487 (2023)
10. Fowler, J.H.: Legislative cosponsorship networks in the U.S. House and Senate. Soc. Netw. **28**, 454–465 (2006)
11. Stehle, J., et al.: High-resolution measurements of face-to-face contact patterns in a primary school. PLOS ONE **6**, e23176 (2011)
12. Veldt, N., Benson, A.R., Kleinberg, J.: Minimizing localized ratio cut objectives in hypergraphs. In: Proceedings of the ACM SIGKDD International Conference on Knowledge Discovery and Data Mining (KDD) (2020)

13. Gjoka, M., Kurant, M., Butts, C.T., Markopoulou, A.: Walking in Facebook: a case study of unbiased sampling of OSNs. In: Proceedings of the IEEE INFOCOM, pp. 1–9 (2010)
14. Hayashi, K., Aksoy, S.G., Park, C.H., Park, H.: Hypergraph random walks, laplacians, and clustering. In: CIKM 2020, Virtual Event, Ireland, 19–23 October (2020)
15. Zhou, D., Huang, J., Scholkopf, B.: Learning with hypergraphs: clustering, classification, and embedding. In: NIPS 2007, pp. 1601–1608 (2007)
16. Feng, Y., You, H., Zhang, Z., Ji, R., Gao, Y.: Hypergraph neural networks. In: AAAI, pp. 3558–3565 (2019)
17. Belkin, M., Niyogi, P.: Laplacian eigenmaps for dimensionality reduction and data representation. Neural Comput. **15**(6), 1373–1396 (2003)
18. Goyal, P., Ferrara, E.: Graph embedding techniques, applications, and performance: a survey. Knowl. Based Syst. **151**, 78–94 (2018)
19. Gatta, V.L., Moscato, V., Pennone, M., Postiglione, M., Sperli, G.: Music recommendation via hypergraph embedding. IEEE Trans. Neural Netw. Learn. Syst. 1–13 (2022)

Contextual Shift Method (CSM)

Gernot Schmitz[1(✉)], Daniel Wilmes[1(✉)], Alexander Gerharz[1], Daniel Horn[1,2], and Emmanuel Müller[1,2]

[1] TU Dortmund University, 44227 Dortmund, Germany
gernot_schmitz@web.de,
{daniel.wilmes,emmanuel.mueller}@cs.tu-dortmund.de,
{gerharz,dhorn}@statistik.tu-dortmund.de
[2] Research Center Trustworthy Data Science and Security, UA Ruhr,
44227 Dortmund, Germany

Abstract. Explainable AI approaches often create artificial data points to test a given model. Sometimes the created data points are located in areas with low data density, and they are unlikely or even impossible combinations of values. Hence, interpreting the model at those artificial points does not give trustworthy information. This becomes even more relevant the higher the dimensionality of the data. We examine the challenges of creating meaningful, realistic data points, which are essential for many Explainable AI methods. Based on this knowledge, we define a contextual shift as a meaningful artificial data point. The problem of not generating contextual shifts is true for the quantile shift method. We propose the Contextual Shift Method (CSM), which improves the quantile shift method by generating contextual shifts. We show that the CSM reduces the amount of data points created in low data density areas.

Keywords: Explainable AI · Trustworthy AI · Machine Learning · Neural Network

1 Introduction

Explainable artificial intelligence (XAI) is a growing field due to the uprise of increasingly complex artificial intelligence (AI) models taking over increasingly larger parts of our lives. Since the power of AI models increases by their size and their complexity, the crucial aspect of explainability is often lost. Most models themselves are not interpretable and post-hoc methods have to be developed in order to make models understandable to humans to trust and regulate them. [1]

Previous XAI methods include partial dependence plots (PDP) [3], giving a line plot for each dimension of interest, showing the expected prediction for a given value of the dimension of interest when all other dimensions are treated as random variables. This approach highlights the general influence of the dimension of interest on the predictions of the model. Another similar approach is called individual conditional expectation (ICE) [6]. For each data point in a data set, all values except for the one in a given dimension of interest are fixed.

© The Author(s), under exclusive license to Springer Nature Switzerland AG 2023
R. Wrembel et al. (Eds.): DaWaK 2023, LNCS 14148, pp. 101–106, 2023.
https://doi.org/10.1007/978-3-031-39831-5_9

For each existing value in the dimension of interest, a new data point is created and used as input for the model. Therefore, each data point becomes a line showing the influence of the dimension of interest for every single point. Both ICE and PDP test the model by creating new artificial data points that might be unrealistic for the examined data set. This phenomenon is a severe problem for XAI methods, which even latest methods suffer from. Since to the authors' knowledge no other research in this direction exists, in this paper we will investigate which conditions a meaningful artificial data point should fulfill. We use this to improve the quantile shift method (QSM) [5] which is a XAI method that aims to find neighborhoods of classes.

2   Contextual Shifts

The goal of the QSM is to elucidate the relative position of classes in a model in the feature space for a given dimension of interest. This is accomplished by shifting data points in the direction of the dimension of interest and comparing the classification of those points by the model before and after the shift.

We consider a classification model $\hat{f}$ which maps an n-dimensional feature vector $\mathbf{x}$ to its corresponding class $c(\mathbf{x})$: $\mathbf{x} \in \mathbb{R}^n \mapsto c(\mathbf{x}) \in \{1, \ldots, k\}$. Let L denote the set of features and L_l a feature of interest. Then for L_l and a pair $(\mathbf{x}_i, c(\mathbf{x}_i))$ the shifted value $\tilde{x}_{i,l}$ for L_l can be determined via [5]:

$$\tilde{x}_{i,l} = \hat{F}_l^{-1}(\min\{\hat{F}_l(x_{i,l}) + q_l, 1\}), \quad \text{for } q_l \in (0, 1]. \tag{1}$$

Here q_l is the shift in the level of the univariate empirical cumulative distribution function ecdf, $\hat{F}_l$, for feature L_l. q_l gets translated into the level of a concrete feature value by use of the inverse $\hat{F}_l^{-1}$. The usage of the ecdf tries to avoid creating data points that lie in areas where the density of training data is low. The full feature vector $\tilde{\mathbf{x}}_i$ is then changed in that component, that represents a feature of interest, all other components remain unchanged.

In Fig. 1, quality differences of possible shifts are shown. The shift denoted by the red point and the black arrow, shows the problem of extrapolation: the new point was not seen by the model before and is not contained in a dense region of the data. However, the second shift, denoted by the blue arrow is realistic, as it considers the context of the underlying data distribution.

We summarize this as *Challenge 1: Given a data point $\boldsymbol{x}_i$, its shift to $\tilde{\boldsymbol{x}}_i$ might not lie within a dense region of the given data distribution.*

As a second challenge, a meaningful, realistic shift of a data point should not fall prey to a high dimensional data situation. This is a well-known challenge for all methods that operate in high dimensional data: the empty space problem [8], which, put simply, states that with increasing dimensionality, the data space is very sparsely populated. Thus, the distance of the original and the shifted data point become more and more alike and the concept of a high density area becomes more and more meaningless in high dimensions.

We conclude this in *Challenge 2: The higher the dimensionality n of the data, the more likely it is that a shift of a data point does not lie within a high density area.*

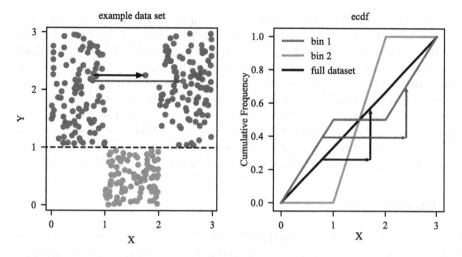

Fig. 1. Example for binning and a contextual shift. The left plot shows an artificial data set split into two bins (orange and blue). Results for a non-contextual (red) and a contextual (green) shift are shown. The right plot shows the belonging ecdf functions, where the black line is the ecdf of all points, while the other two lines denote the ecdfs of the respective bins. (Color figure online)

We define a shift that does not fall prey to both of this challenges as a *contextual shift*, as this type of shift adapts to the context of the underlying data distribution. *Definition 1: A shift $\tilde{\mathbf{x}}_i$ of a data point $\mathbf{x}_i$ is called contextual, if it lies within a high density area.* So in case of a contextual shift there exists a sufficiently large number of data points $\{\mathbf{y}_1, \ldots, \mathbf{y}_m\}$ and a sufficiently small $\epsilon > 0$ such that: $\text{dist}(\tilde{\mathbf{x}}_i, \mathbf{y}_k) \leq \epsilon \quad \forall k \in \{1, \ldots, m\}$. The QSM allows the user to freely choose the parameter q_l according to the data situation at hand. Recall from Equation (1) that q_l determines a shift and has the advantage of circumventing the addition of the same fixed value to all considered data points. However, as stated, a shift in the QSM is likely to be *not* contextual.

3   Contextual Shift Method

As a consequence, our goal is to improve the QSM by generating contextual shifts, while preserving the advantage of finding neighborhoods using the ecdf. We propose to split the data set into multiple bins that do not contain low-density areas that are covered in the ecdf by data lying next to it. Data points are only shifted within their respective bin. To do so, a recursive greedy algorithm, which consists of three steps, is used to split the data set.

First, a subspace of the dimensions of the data set is selected using the *high contrast subspaces* (HiCS) method [7], which is a method to handle data sets with high dimensionality and searches for subspaces in which the data is structured in some way, without assuming any distributions. The subspace containing the dimension of interest with the highest number of dimensions and the best contrast is chosen. From this subspace, the dimension to split the data in will be selected as the dimension that shows the highest contrast in a two-dimensional subspace with the dimension of interest. To find the optimal split value, all possible splits of the data set along the split dimension are created (one for each point), and for each, the difference of the resulting ecdfs along the dimension of interest are compared using the Kolmogorov-Smirnov test. The split with the highest difference in that test will be used to create the final bins. The data set is divided into two subsets and the process is repeated for each. After the binning is complete, the data is shifted analogously to the QSM in all bins separately.

4   Experiments

In order to compare the capabilities of the CSM to the QSM, we visualize the results on three example data sets: the well-known iris data set [2], an artificial data set called *Hole* which was explicitly created to show weaknesses of the QSM and the *soccer* data set collected by Gerharz et al. [4].

For each data set, a simple fully connected feedforward neural network was trained. We will investigate four visualizations for each data set (cp. Fig. 2): The original data set, as well as the class predictions, shifted by QSM and by the CSM. For the iris data set the dimension of interest is petal_length. When comparing the QSM and CSM it can be seen, that the QSM shifts data from the *setosa* class upwards into a previously empty area (non-contextual shift). Using the CSM, the whole *setosa* class is its own bin and no points are shifted out of it (contextual shift). Data points are still shifted from *versicolor* towards *virginica*, which shows, that the CSM also works to find neighborhoods of classes.

The hole data set contains six classes in five dimensions. In dimensions D0 – D3 the first five classes show strong overlap. Only C5 deviates from the pattern. In dimension D4, the classes are ordered according to their number, with C0 having the lowest and C4 having the highest values. The only exception to the ordering is C5, filling a gap between C1 and C2. Therefore a hole exists between C1 and C2, which is covered in the ecdf of D4 by C5.

For the hole data set the shift in D4 leads to some data points ending up in the area where the hole used to be for both methods. The number of points that are shifted into the hole area is strongly reduced using the CSM. New transitions between classes can be observed since some data points skip the hole and are shifted toward the upper classes (contextual shifts).

For the soccer data set, 11 dimensions of performance data on players in the German soccer league (Bundesliga) were collected, and the data were aggregated per player and season. The target variable is the player's playing position, in this paper simplified to the binary classification of "goalkeeper" versus "field player".

Since goalkeepers are separated from the other classes in several dimensions, shifting this data in only one dimension means moving it into areas where there were no data before. Thus, the goalkeeper data points must not be pushed out of their range, which could be prevented by appropriate binning. For both methods

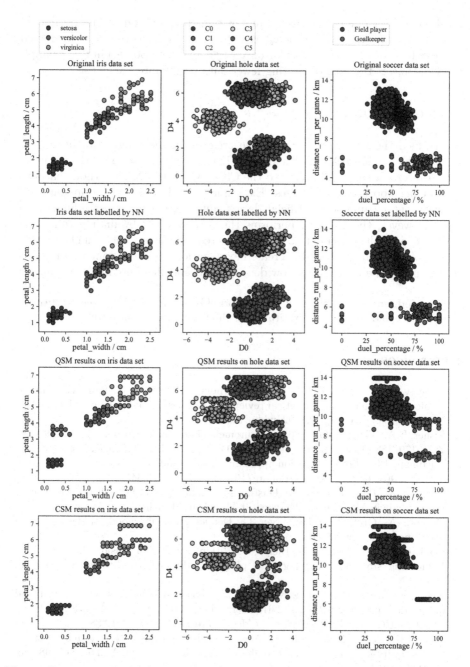

Fig. 2. Comparing results from QSM and CSM. Iris (left), Hole (middle) and soccer (right) data sets are shown. For each data set, four pictures are shown in the following order: The original data set, the class predictions of the training data, the class predictions of the training data shifted by the QSM, and the class prediction of the CSM.

goalkeeper data points are shifted out of their previous area. The results still look vastly different. Overall the QSM shifts 85 goalkeepers out of their previous range, while the CSM does this for only 44 goalkeepers. So also in this example, the CSM meets the goal of shifting fewer data points into low-density areas.

5   Conclusion

We tackled the challenge to avoid the creation of meaningless artificial data points, which modern XAI methods face. For this purpose we gave the definition for a *contextual* shift and proposed the CSM, which is a XAI method that aims to find neighborhoods of classes. Experiments including real-world data sets showed that meaningless artificial data points occur even in low dimensional data. However, the CSM outperforms the related QSM by finding contextual shifts. As a result of contextual shifts, expected transitions between classes were detectable This means that the created explanations are less prone to errors due to unexpected behavior in the low density areas and therefore more meaningful.

Acknowledgements. This work was supported by the Research Center Trustworthy Data Science and Security, an institution of the University Alliance Ruhr.

References

1. Angelov, P.P., Soares, E.A., Jiang, R., Arnold, N.I., Atkinson, P.M.: Explainable artificial intelligence: an analytical review. WIREs Data Mining Knowl. Disc. **11**(5) (2021). https://doi.org/10.1002/widm.1424
2. Fisher, R.A.: The use of multiple measurements in taxonomic problems. Ann. Eugen. **7**(2), 179–188 (1936). https://doi.org/10.1111/j.1469-1809.1936.tb02137.x
3. Friedman, J.H.: Greedy function approximation: a gradient boosting machine. Ann. Statist. **29**(5) (2001). https://doi.org/10.1214/aos/1013203451
4. Gerharz, A.: Positionsbezogene Leistungsdatenanalyse von Fußballspielern. Master thesis, Ludwig-Maximilians-Universität München, München (24052019)
5. Gerharz, A., Groll, A., Schauberger, G.: Deducing neighborhoods of classes from a fitted model. arXiv:2009.05516
6. Goldstein, A., Kapelner, A., Bleich, J., Pitkin, E.: Peeking inside the black box: Visualizing statistical learning with plots of individual conditional expectation. arXiv:1309.6392
7. Keller, F., Muller, E., Bohm, K.: HICS: high contrast subspaces for density-based outlier ranking. In: 2012 IEEE 28th International Conference on Data Engineering, pp. 1037–1048. IEEE (2012). https://doi.org/10.1109/ICDE.2012.88
8. Weber, R., Schek, H.J., Blott, S.: A quantitative analysis and performance study for similarity-search methods in high-dimensional spaces. In: VLDB, vol. 98, pp. 194–205 (1998)

Utility-Oriented Gradual Itemsets Mining Using High Utility Itemsets Mining

Audrey Fongue[1]([✉]), Jerry Lonlac[2], and Norbert Tsopze[1,3]

[1] Department of Computer science, University of Yaoundé I, Yaoundé, Cameroon
{audrey.fongue,norbert.tsopze}@facsciences-uy1.cm
[2] IMT Nord Europe, Institut Mines-Télécom,
University of Lille Centre for Digital Systems, 59000 Lille, France
jerry.lonlac@imt-nord-europe.fr
[3] Sorbonne University, IRD, UMMISCO, 93143 Bondy, France

Abstract. Gradual itemsets capture frequent covariations of attributes of the form "more/less A, more/less B" from a quantitative database. These patterns have gained considerable interest during these years and have been applied in several domains. Various algorithms have been proposed to extract those itemsets efficiently. However, an inherent limitation of the proposed algorithms is that they only evaluate items in terms of increase and decrease. Therefore, all the covariations of items have an equal importance/signification in evaluating the frequency of a gradual itemset. Those algorithms are not appropriate for certain real-world applications where strong covariations which are scarce may be useful. This paper proposes a solution to cope with this limitation with the task of high utility gradual itemsets mining, whose goal is to extract covariations of attributes which generate a high profit for the user. Two algorithms are proposed to mine these patterns efficiently called HUGI (High Utility Gradual Itemsets mining), and HUGI-Merging, which extracts these patterns from both a negative and positive quantitative data separately and merges the obtained results. Experimental results show that the proposed algorithms are efficient and can filter many gradual itemsets to focus only on desired high-utility gradual itemsets.

Keywords: Pattern mining · High utility itemsets · Gradual itemsets

1 Introduction

Gradual itemsets of the form "more/less A, ..., more/less B"' allow expressing covariations between attributes which describe data. They have been the subject of several studies and many algorithms [1,5] have been proposed to extract those patterns. The number of patterns extracted is often high and several measures (support [1], seasonality [4], emergence) have been proposed for the selection of that patterns or some constraints like temporality [5] to reduce the search space. These measures are limited by the fact that they do not allow the user to express his interest on the patterns. For example, in the context of trade, the manager can express his need for patterns that produce more benefits.

© The Author(s), under exclusive license to Springer Nature Switzerland AG 2023
R. Wrembel et al. (Eds.): DaWaK 2023, LNCS 14148, pp. 107–113, 2023.
https://doi.org/10.1007/978-3-031-39831-5_10

Recently, high utility itemsets [2] have been proposed, and many algorithms [7] to extract them also. These patterns allow the expression of other interests of the user on the patterns through the concept of utility. The two forms of patterns (gradual and high utility) allow on one hand to express covariations and on the other hand, to express the interest of the user. To our knowledge, no work in the literature has proposed to extract patterns where covariations of attributes are expressed together with their utility. This type of patterns allows the user to see how a covariation of certain items will impact their interest.

The problem addressed in this paper is defined as follows: "Given a quantitative database and external utilities [8] of the different attributes, find gradual itemsets whose utility is higher than a utility threshold". In contrast with high utility patterns, where utility expresses the interest of the user, gradual patterns express the variation of that interest. Therefore, we propose to extract high utility gradual patterns. To achieve this, the methodological approach consists in transforming the original quantitative database Δ into a new database Δ' that stores the differences between the values of the same attribute in the new transactions. Then, we propose two algorithms called HUGI-Merging and HUGI to extract high utility gradual patterns which operate as follows: 1) the HUGI-Merging algorithm consists in splitting the database Δ' into two parts, one composed of negative values $\Delta'^{\leq}$ and the other part made up of positive values $\Delta'^{\geq}$, then extracts patterns in the different parts separately and merges the results; 2) the HUGI algorithm extracts directly from Δ' high utility gradual patterns. The extraction of itemsets from the different databases is done using a modified version of the EFIM algorithm [8], one of the best algorithms for high utility itemsets mining in term of memory space and time.

The paper is organised as follows: Sect. 2 introduces some notions about high utility itemsets and gradual itemsets. Section 3 focuses on our proposal, and in Sect. 4, the experimental results are presented.

2   Preliminary Definitions

Let $I = \{i_1, \ldots, i_n\}$ a set of items and $T = \{T_1, \ldots, T_m\}$ a quantitative transaction database where $T_z \subseteq I$. Each item i is associated with a positive number $p(i)$, its *external utility* (its relative importance to the user). Let $q(i, T_z)$ be the quantity of item i in the transaction T_z, called the *internal utility* of i in T_z.

Definition 1. *An **itemset** $I_0 = \{i_1, \ldots, i_k\}$ is a set of k distinct items, with $i_j \in I, 1 \leq j \leq k$, and k the size of I_0.*

Definition 2. *An itemset I_0 is said to be a **high-utility itemset** if its utility $u(I_0) = \sum_{I_0 \subseteq T_z (1 \leq z \leq m)} (\sum_{i \in I_0} p(i) \times q(i, T_z))$ is no less than a user-specified minimum utility threshold. Otherwise, I_0 is a low-utility itemset.*

Definition 3. *A **gradual item** is defined in the form of i^*, from an attribute i and a variation $* \in \{\geq, \leq\}$ (increase or decrease) of the values of i. A **gradual itemset** $I_0^g = \{i_1^{*_1}, \ldots, i_k^{*_k}\}$ is a non-empty set of gradual items.*

The support of a gradual itemset amounts to the extent to which the pattern is present in the database. A gradual itemset is frequent in a database if its support is no less than a predefined threshold. Several definitions of the support have been proposed for gradual patterns [1,3]. We consider the definition of the support proposed in [3], which evaluates the support of a gradual itemset as the proportion of pairs of objects that respect the variation of all items in the itemset on the data because it fits more with the idea of searching for gradual itemsets generating a high user profit addressed in this work.

3   High Utility Gradual Itemsets Mining

The goal is to introduce a measure of interest called utility in the extraction of gradual itemsets. The utility measure introduced allows taking into account the user interest during the mining process. In the following, let's denote $u(I_0^g, \Delta)$ as the utility of a gradual itemset I_0^g in the database Δ. When it is clear from context, I_0 is the itemset constituted of the items of gradual itemset I_0^g and $u(I_0, T_z)$ is the utility of I_0 in the transaction T_z. Let $\Delta(I_0^g)$ be the set of transactions couples that verify the constraints expressed by all the gradual items in I_0^g.

Definition 4. (High utility gradual itemset). *A gradual itemset I_0^g is a high utility gradual itemset if its utility $u(I_0^g, \Delta)$ is no less than a user-specified minimum utility threshold minUtil (i.e. $u(I_0^g, \Delta) \geq minUtil$).*
$u(I_0^g, \Delta) = \sum_{(T_x, T_y) \in \Delta(I_0^g) \wedge q(i, T_x) \neq q(i, T_y) \forall i \in I_0} [u(I_0, T_y) - u(I_0, T_x)].$

The utility of a gradual itemset I_0^g is interpreted as the amount of the total interest from the variation of the quantities of the gradual itemset I_0.

   We'll call a positive (resp. negative) gradual itemset, a gradual itemset I_0^g in which the senses of variations of all its items are strictly increasing (resp. decreasing), and mixed itemsets the ones containing both decreasing and increasing items. The extraction of high utility gradual patterns follows two main steps: 1)**Database encoding**. It consists in transforming the original database Δ into a database Δ' such that Δ' has the same attributes as Δ, but the occurrences are the differences between the ones of Δ. 2)**Extraction of high utility gradual itemsets** from the transformed database Δ'. This step is done by applying a modified high utility pattern mining algorithm.

3.1   Database Encoding

To simplify, we suppose that data verify the temporality constraint i.e. the order of appearance of the transactions must be preserved like in [5]. But in contrast with this principle which conserves the senses of variations in the encoded database, we conserve differences between the transactions k and j. For a column l, the values are calculated as follows: $\Delta'^l_{kj} = \Delta^l_j - \Delta^l_k$.

3.2 High Utility Gradual Itemsets Extraction

We propose two approaches which allow expressing the semantics of interest.
First approach: extraction with separation of positives and negatives.
This approach consists in dividing the new database Δ' in two parts: one for
items with positive values $(\Delta^{\geq})$ and the other for items with negative values
$(\Delta^{\leq})$. Extract high utility gradual itemsets from each database using a modified
version of the EFIM algorithm. The obtained results are two sets of patterns: a
set of positive patterns $L^{\geq}$ and a set of negative patterns $L^{\leq}$. The next step is to
combine two patterns (one of each set) $I_0^{g^{\geq}}$ and $I_0^{g^{\leq}}$ such that all items are dis-
tincts to form mixed high utility gradual itemsets. To this end, the modification
in the EFIM algorithm consists in extracting, in addition to high utility patterns, the identifiers of the transactions containing these patterns (extensions).
The extracted identifiers are used to form mixed patterns.

To apply the EFIM algorithm on the database $\Delta^{\leq}$, the values of the latter
are converted to positive values. Algorithm 1 presents the pseudo-code of HUGI-
Merging to do this task. In this algorithm, the function $get\_positive\_items$ (resp.
$get\_negative\_items$) extracts from the encoded database a sub-database $\Delta^{\geq}$
(resp. $\Delta^{\leq}$) containing only positive (resp. negative) items. The function $Merging$
forms mixed itemsets.

*Property 1. (Pruning condition for itemsets from the union of positive and neg-
ative itemsets).* Let I_1^g a gradual itemset such that $u(I_1^g, \Delta^{\geq}) < minUtil$ and
I_2^g a gradual itemset such that $u(I_2^g, \Delta^{\leq}) > -minUtil$. If $I_0^g = I_1^g \cup I_2^g$, then
$-minUtil < u(I_0^g, \Delta) < minUtil$. This allows avoid generating candidates gradual itemsets with low-utility because it leads to low-utility gradual itemsets.

Second approach: extraction from a single database. The itemsets
are extracted from the encoded database. A modified version of EFIM algorithm
(called EFIM_abs) has been proposed for this task.

Algorithm 1: HUGI-Merging

Input: Δ : quantitative database, λ : minimum utility percentage
Output: L : the set of high utility gradual patterns
1 $minUtil = \lambda \times \sum_{T_z \in \Delta} \sum_{i \in T_z} u(i, T_z)$;
2 $\Delta' \leftarrow Encoding(\Delta)$;
3 $\Delta^{\geq} \leftarrow get\_positive\_items(\Delta')$;
4 $\Delta^{\leq} \leftarrow get\_negative\_items(\Delta')$;
5 $L^{\geq} \leftarrow EFIM(\Delta^{\geq}, minUtil)$;
6 $L^{\leq} \leftarrow EFIM(\Delta^{\leq}, minUtil)$;
7 $L^{\geq\leq} \leftarrow Merging(L^{\geq}, L^{\leq})$;
8 $Return\ L^{\geq} \cup L^{\leq} \cup L^{\geq\leq}$;

Algorithm EFIM_abs. In this algorithm, the measures (TWU, sub-tree
utility, remaining utility, Local utility) [8] used are in the absolute values. The
computation of utility is done normally, meaning that we keep the original values
of the internal utilities. The interest in considering rather the absolute values of
measures is to avoid underestimating the utility of itemsets. There can be several

cases: 1) We are only interested in those itemsets I_0^g, such as $u(I_0^g) \geq minUtil$ in order to provide patterns leading to high interest; 2) We are interested in the itemsets I_0^g whose utility respects the relation $u(I_0^g) \leq -minUtil$, so patterns leading to a great loss (decrease) of interest; 3) A union of cases 1 & 2.

4 Experimental Study

We conducted some experiments to evaluate the performances of the proposed algorithms. These experiments were performed on an Intel Core machine i7-7820HQ 2.9GHz CPU with a 8 cores processor and 16GB of RAM. We compared the results obtained to those of T-$Gpattern$ [5] on the number of extracted patterns, execution time and memory space used versus different utility thresholds. The dataset used is Order [6], made for recommendation and contains records of mobile payments in supermarkets and convenience stores. It consists of 418 transactions and 16383 attributes. For our tests, we extracted a dataset from this dataset containing 100 attributes with the same number of transactions.

Figure 1 compares the execution time, the number of extracted patterns and the memory consumed by the algorithms $HUGI$, $HUGI$-$Merging$ and T-$Gpattern$. The utility threshold is used as the support threshold for T-$Gpattern$. The experiments show that: $HUGI$ is significantly faster and outperforms T-$Gpattern$ on the execution times because it filters many non high utility gradual patterns. We notice that $HUGI$-$Merging$ (curve with the red color on the graph) takes more time when the utility threshold dropped below 0.2 because of the merging operation used to combine the results. As mentioned in Sect. 3.2, $HUGI$-$Merging$ first generates several candidates gradual itemsets by joining positive and negative high-utility gradual itemsets and this operation is very computationally expensive regarding memory consumption and execution time. $HUGI$ generates less patterns (see curves AP2: HUGI> and AP2: HUGI< from Fig. 1(b)) than T-Gpattern for the lowest utility values. This is convenient for the user to analyze a small set of patterns. Nevertheless, it should be mentioned that the total number of high utility gradual patterns extracted by the proposed algorithms may be higher than the number of gradual patterns extracted by T-$Gpattern$ (see Fig. 1(b)). In terms of memory usage, $HUGI$-$Merging$ take up more memory space than the other algorithms; however, $HUGI$ uses much less memory than T-$Gpattern$ when the utility threshold was 0.5 or higher.

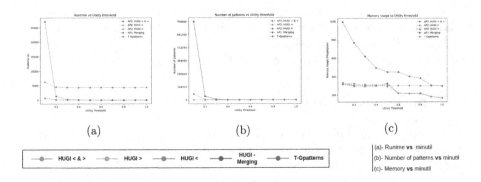

Fig. 1. Comparative evaluation on the dataset *Order*: 100 items, 418 transactions.

5   Conclusion

This paper explores the problem of high utility gradual itemsets mining and proposes two algorithms named HUGI and HUGI-Merging to extract those patterns efficiently. These algorithms exploit the *EFIM* algorithm to extract gradual itemsets whose utility is no less than a user-specified minimum utility threshold. An experimental study with real-life datasets shows that the proposed algorithms are generally faster and extract less gradual patterns than *T-Gpattern* (algorithm for extracting gradual patterns under temporal constraint) by filtering out gradual patterns that are not useful. We compared the proposed algorithms with *T-Gpattern* to see among the gradual patterns extracted, those which are of high utility. For future work, we intend to combine support and utility in order to extract frequent gradual patterns of high utility.

Acknowledgments. The authors would like to thank the French National Centre for Scientific Research (CNRS) for their financial support through the DSCA project FDMI-AMG.

References

1. Di-Jorio, L., Laurent, A., Teisseire, M.: Mining frequent gradual itemsets from large databases. In: IDA, pp. 297–308 (2009)
2. Gan, W., Lin, J.C., Fournier-Viger, P., Chao, H., Tseng, V.S., Yu, P.S.: A survey of utility-oriented pattern mining. IEEE Trans. Knowl. Data Eng. **33**(4), 1306–1327 (2021)
3. Laurent, A., Lesot, M., Rifqi, M.: GRAANK: exploiting rank correlations for extracting gradual itemsets. In: FQAS, pp. 382–393 (2009)
4. Lonlac, J., Doniec, A., Lujak, M., Lecoeuche, S.: Mining frequent seasonal gradual patterns. In: DaWaK, vol. 12393, pp. 197–207 (2020)
5. Lonlac, J., Nguifo, E.M.: A novel algorithm for searching frequent gradual patterns from an ordered data set. Intell. Data Anal. **24**(5), 1029–1042 (2020)

6. Wang, C., Zhang, M., Ma, W., Liu, Y., Ma, S.: Modeling item-specific temporal dynamics of repeat consumption for recommender systems. In: WWW, pp. 1977–1987 (2019)
7. Wu, P., Niu, X., Fournier-Viger, P., Huang, C., Wang, B.: UBP-miner: an efficient bit based high utility itemset mining algorithm. Knowl. Based Syst. **248**, 108865 (2022)
8. Zida, S., Fournier-Viger, P., Lin, J.C.W., Wu, C.W., Tseng, V.S.: EFIM: a fast and memory efficient algorithm for high-utility itemset mining. Knowl. Inf. Syst. **51**(2), 595–625 (2017)

Discovery of Contrast Itemset with Statistical Background Between Two Continuous Variables

Kaoru Shimada$^{(\boxtimes)}$ ⓘ, Shogo Matsuno ⓘ, and Shota Saito ⓘ

Faculty of Informatics, Gunma University, Maebashi 371-8510, Japan
{k.shimada,s.matsuno,shota.s}@gunma-u.ac.jp

Abstract. We previously defined ItemSB as an extension of the concept of frequent itemsets and a new interpretation of the association rule expression, which has statistical properties in the background. We also proposed a method for its discovery by applying an evolutionary computation called GNMiner. ItemSB has the potential to become a new baseline method for data analysis that bridges the gap between conventional data analysis using frequent itemsets and statistical analyses. In this study, we examine the statistical properties of ItemSB, focusing on the setting between two continuous variables, including their correlation coefficients, and how to apply ItemSB to data analysis. As an extension of the discovery method, we define ItemSB that focuses on the existence of differences between two datasets as contrast ItemSB. We further report the results of evaluation experiments conducted on the properties of ItemSB from the perspective of reproducibility and reliability using contrast ItemSB.

Keywords: Knowledge discovery · Big data analytics · Evolutionary computation

1 Introduction

The extraction of frequent itemsets and association rules is widely used as a basic technique in data mining [1]. An association rule expresses the fact that "if an instance (tuple) satisfies P, then it also satisfies Q" in a relational database. It is expressed as "if P then Q ($P \rightarrow Q$)" where P and Q are conditions on database attributes (items). When Q is a class attribute, it can be used for rule-based classification. In the case of class classification using association rules, the expressive form of the rules is readable and considered useful for interpreting the reasons for class classification.

A method has been proposed for directly discovering class association rules that satisfy flexible conditions set by the user without constructing a frequent itemset or going through any manual process, using a network structure and an evolutionary computation method that accumulates results continuously over generations [2]. An extension of the method for determining an attribute combination P such that Q provides a small region in a continuous variable X has also been reported [3].

ⓒ The Author(s), under exclusive license to Springer Nature Switzerland AG 2023
R. Wrembel et al. (Eds.): DaWaK 2023, LNCS 14148, pp. 114–119, 2023.
https://doi.org/10.1007/978-3-031-39831-5_11

In the case of using the expression "if P then Q," if this is an interesting rule, it can indicate that in the set of instances satisfying P (the set of records satisfying P, a small set taken from the whole data), there is a statistical characteristic Q in terms of the whole data. For example, if Q is a class attribute, it can indicate that the ratio of class membership is a characteristic of the population satisfying P. Furthermore, in the case of a small distribution of continuous quantities, the values of the continuous variable X are indicated to show a characteristic distribution in the population satisfying P. Thus, the interpretation of the association rule is to consider a group of cases (a group of records) satisfying P as a single cluster and extract the cluster with the label satisfying P. Furthermore, the basis for extracting the cluster is in Q. In this study, we consider Q in the expression "if P then Q" as an expression using the conventional statistical analysis method and treat a method to discover that "a statistical expression Q is found in a population that satisfies P." Specifically, when Q comprises two continuous variables X and Y in the database, this method reveals that X and Y are correlated in a small population with attribute combination P, even though X and Y are not correlated in the entire database. In this study, we address correlations for two continuous variables, which are expected to realize conditional settings for the ratio and distribution of X and Y to be found and are positioned as a basic method to link large-scale data and conventional statistical analysis methods. We referred to this attribute combination P as ItemSB (Itemsets with statistically distinctive backgrounds) and proposed a method for its discovery [4]. ItemSB has the potential to become a new baseline method for data analysis that bridges the gap between conventional data analysis using frequent itemsets and statistical analysis.

In this paper, to examine the characteristics of ItemSB, we propose a method to discover ItemSB by focusing on the differences between two sets of data and report the results of evaluation experiments on the characteristics of ItemSB. Applying conventional statistical analysis methods to an entire dataset can sometimes be challenging in the analysis of big data. Therefore, to effectively apply statistical analysis methods to big data, statistical analysis should be conducted after establishing a certain type of stratum and dividing the records into subgroups. Typically, this subgrouping is defined a priori by the user; however, it is expected to be determined in an exploratory manner. However, there may be cases where the user wishes to determine a small group in which there is a strong correlation between the two variables of interest. Techniques to understand and apply the characteristics of ItemSB are expected to be the basis of big data mining methods that are characterized by a combination of this subgrouping and statistical analysis.

2   Contrast ItemSB

In this study, we define ItemSB as a set of rules that satisfy the conditions predetermined by the user [4]. Specifically, the set of attribute combinations A_i that appear to have statistical characteristics regarding X and Y in Table 1 is

Table 1. Example of the database.

ID	A_1	A_2	A_3	A_4	A_5	...	X	Y
1	1	0	1	0	0		54.8	2008
2	0	1	1	0	1		40.2	1987
3	1	0	0	1	0		28.4	1964
...	...	...	...	...	...	...	...	...
N	1	0	1	1	1		41.9	2005

determined by expressing the following rule:

$$(A_j = 1) \wedge \cdots \wedge (A_k = 1) \rightarrow (m_X, s_X, m_Y, s_Y, R_{X,Y}), \tag{1}$$

where m_X, s_X, m_Y, and s_Y are the mean values and standard deviations of X and Y, respectively, and $R_{X,Y}$ is the correlation coefficient between X and Y for instances containing the itemset.

We proposed a method for discovering ItemSBs using a genetic network programming (GNP) structure characterized by a directed graph network structure and an evolutionary computation method (GNMiner) [4]. GNMiner is an evolutionary computational method that employs an evolutionary strategy to accumulate the outcomes obtained in each generation. Unlike general evolutionary computation methods, the solution is not the best individual of the last generation; rather, numerous interesting rule candidates are represented and searched for in the individuals in evolutionary computation. Consequently, when an interesting rule is found, it is accumulated in a rule library. Therefore, it has the following characteristics: setting the goodness of fit during the evolutionary operation has little influence on performance; rule discovery can be terminated at any time because it is an accumulative problem-solving method; the flexibility of the individual network structure allows a variety of rules expressions. However, it is limited by its inability to discover all the rules that satisfy the set conditions.

In data mining from dense databases, it is interesting to find differences between two sets of data collected in different settings (A/B) for the same subject, or between two groups (A/B) of data collected for the same subject. When $I = (A_j = 1) \wedge \cdots \wedge (A_k = 1)$ is the same ItemSB obtained from Databases A and B, we define contrast ItemSB as follows: For instance, I^A: $I \rightarrow (m_X^A, s_X^A, m_Y^A, s_Y^A, R_{X,Y}^A)$ in Database A satisfies the condition given by the user. I^B: $I \rightarrow (m_X^B, s_X^B, m_Y^B, s_Y^B, R_{X,Y}^B)$ in database B does not satisfy the same condition. The basic idea of contrast ItemSB discovery was introduced in [4].

In this study, an interesting feature of contrast ItemSB is defined as follows:

$$I^A : I \rightarrow (m_X^A, s_X^A, m_Y^A, s_Y^A, R_{X,Y}^A), \quad I^B : I \rightarrow (m_X^B, s_X^B, m_Y^B, s_Y^B, R_{X,Y}^B),$$

$$R_{X,Y}^A(I) \geq R_{min}, \quad R_{X,Y}^B(I) \geq R_{min}, \tag{2}$$

where R_{min} is a constant provided by the user in advance. In addition, the conditions for *support* (frequency of occurrence) are added to interestingness, such as

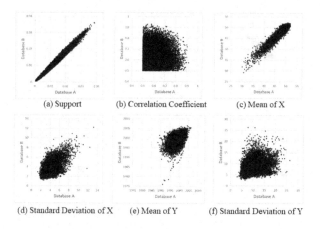

(a) Support   (b) Correlation Coefficient   (c) Mean of X

(d) Standard Deviation of X   (e) Mean of Y   (f) Standard Deviation of Y

Fig. 1. Results of contrast ItemSB discovery.

$$support^A(I) \geq supp_{min}, \quad support^B(I) \geq supp_{min}, \tag{3}$$

where $supp_{min}$ denotes a constant provided in advance by the user. To evaluate the reproducibility and reliability of ItemSB, we used an interest setting different from that in [4].

3   Experimental Results

The YPMSD datasets [5] from the UCI machine learning repository [6] were used to evaluate the proposed method.

The characteristics of the YPMSD dataset are summarized as follows:

- predicting the release year of a song based on audio features;
- 90 attributes without missing values ($T(j)$ ($j = 0, \ldots, 89$). The target attribute value (Y) is the year ranging from 1922 to 2011;
- 463,715 instances for training and 51,630 instances for testing.

We set $X = T(0)$ and discretized continuous attribute values $T(j)$ ($j = 1, \ldots, 89$) to sets of attributes A_{3j-2}, A_{3j-1}, and A_{3j} with values of 1 or 0. Discretization was performed in the same manner as that described in [4]. Two hundred sixty-seven transformed attributes ($A_1, \ldots, A_{267}$) and continuous attributes X and Y were used. The first 100,000 of the 463,715 instances were used as Database A, and 51,630 testing instances were used as Database B. The detailed experimental setup, including the GNP setting, was the same as that described in [4]. The final solution was a set of ItmeSBs with no duplicates obtained after 100 independent rounds of trials.

The following conditions were used for contrast ItemSB discovery:

$$R_{X,Y}^A(I) \geq 0.5, \quad R_{X,Y}^B(I) \geq 0.5, \tag{4}$$

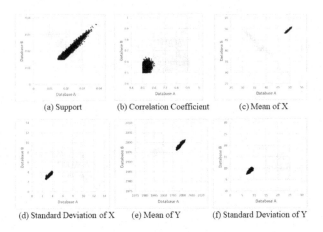

Fig. 2. Results of contrast ItemSB discovery ($supp_{min} = 0.015$).

$$support^A(I) \geq 0.0004, \quad support^B(I) \geq 0.0004. \tag{5}$$

The purpose of this experiment was to gain insight into the reproducibility of ItemSBs found between the two databases such that reliable ItemSB discovery could be performed. The fitness of a GNP individual was defined as

$$F = \sum_{I \in I_c} \{1000 \times (support^A(I) + support^B(I))$$
$$+ 10 \times (R^A_{X,Y}(I) + R^B_{X,Y}(I)) + n_P(I) + c_{new}(I)\},$$

where I_c is the set of suffixes of extracted contrast ItemSBs satisfying (4) and (5) in the GNP individual. $n_P(I)$ is the number of attributes in I, and we set $n_P(I) \leq 6$. $c_{new}(I)$ is a constant for novelty of I.

After 100 rounds of discovery, 133,475 ItemSBs were obtained. The scatter plots of the support value of ItemSB, correlation coefficient, mean values of X and Y, and standard deviations of X and Y for the 133,457 ItemSBs are shown in Fig. 1. Figure 2 shows the scatter plots for the 9,365 ItemSBs satisfying the additional condition $supp_{min} = 0.015$. As shown in Figs. 1 and 2, contrast ItemSB tended to be absent from ItemSB with large support values. Additionally, the values of the correlation coefficients tended to differ between Databases A and B. However, the mean values of X and Y were reproducible. Figure 2 focuses on ItemSB with relatively large support values; however, compared to Fig. 1, the range of the mean values of X and Y is narrower, resulting in a skewed distribution. This indicates that when using ItemSB for classification or stratification by subgroup, it is recommended to use the ItemSB with a relatively high frequency of occurrence. However, ItemSB, which had a comparatively high frequency of occurrence, may not be sufficient to cover all cases. Even if ItemSBs with large correlation coefficients are found, their frequency of occurrence is low, and the correlation coefficient values may not be reproducible when ItemSBs are examined between two datasets collected under the same conditions.

These results suggest that the frequency of occurrence (support) is reproducible and reliable, whereas itemsets are conventionally used based on their frequency of occurrence. The same was considered true for mean values. However, when using complex values such as the correlation coefficient as a statistical feature of ItemSB, evaluating their reliability in the process of discovery is considered an issue.

4 Conclusions

As an extension of the concept of frequent itemsets and a new interpretation of the association rule expression, we defined ItemSB as a set of items with statistical properties as the background and discussed a method of finding ItemSB by applying an evolutionary computation called GNMiner. In this study, we discussed the statistical properties of ItemSB as a basis for big data analysis and its application in forecasting and classification, focusing on setting two continuous variables as the statistical background of ItemSB. Based on experimental results using publicly available data, it is necessary to consider the reproducibility of the characteristics of ItemSB even for small groups with the same combination of items when the correlation coefficient is used as the statistical background of ItemSB. In the future, we aim to establish desirable ItemSB discovery conditions in collaboration with researchers with expertise in the data to which ItemSB is to be applied and to improve the reproducibility and reliability of the ItemSB discovery method in the discovery process.

Acknowledgment. This study was partially supported by JSPS KAKENHI under Grant Number JP20K11964.

References

1. Agrawal, R., Srikant, R.: Fast algorithms for mining association rules. In: Proceedings of 20th International Conference VLDB, pp. 487–499 (1994)
2. Shimada, K., Hirasawa, K., Hu, J.: Class association rule mining with chi-squared test using genetic network programming. In: Proceedings of the IEEE Conference on Systems, Man, and Cybernetics, pp. 5338–5344 (2006)
3. Shimada, K., Hanioka, T.: An evolutionary method for associative local distribution rule mining. In: Perner, P. (ed.) ICDM 2013. LNCS (LNAI), vol. 7987, pp. 239–253. Springer, Heidelberg (2013). https://doi.org/10.1007/978-3-642-39736-3_19
4. Shimada, K., Arahira, T., Matsuno, S.: ItemSB: Itemsets with statistically distinctive backgrounds discovered by evolutionary method. Int. J. Semant. Comput. **16**(3), 357–378 (2022)
5. Bertin-Mahieux, T., Ellis, D.P.W., Whitman, B., Lamere, P.: The million song dataset. In: Proceedings of 12th International Society for Music Information Retrieval Conference, pp. 591–596 (2011)
6. Kelly, M., Longjohn, R., Nottingham, K.: The UCI machine learning repository. https://archive.ics.uci.edu

DBGAN: A Data Balancing Generative Adversarial Network for Mobility Pattern Recognition

Ke Zhang[1]([✉]), Hengchang Liu[2], and Siobhán Clarke[3]

[1] Trinity College Dublin, Dublin D02 R123, Ireland
zhangk2@tcd.ie
[2] The University of Electronic Science and Technology of China,
Chengdu, Sichuan 611731, China
hcliu1984@vip.163.com
[3] Trinity College Dublin, Dublin D02 R123, Ireland
Siobhan.Clarke@tcd.ie

Abstract. Mobility pattern recognition is a central aspect of transportation and data mining research. Despite the development of various machine learning techniques for this problem, most existing methods face challenges such as reliance on handcrafted features (e.g., user has to specify a feature such as "travel time") or issues with data imbalance (e.g., fewer older travelers than commuters). In this paper, we introduce a novel Data Balancing Generative Adversarial Network (DBGAN), which is a specifically designed attention mechanism-based GAN model to address these challenges in mobility pattern recognition. DBGAN captures both static (e.g., travel locations) and dynamic (e.g., travel times) features of different passenger groups, and avoids using handcrafted features that may result in information loss, based on a sequence-to-image embedding method. Our model is then applied to overcome the data imbalance issue and perform mobility pattern recognition. We evaluate the proposed method on real-world public transportation smart card data from Suzhou, China, and focus on recognizing two different passenger groups: older people and students. The results of our experiments demonstrate that DBGAN is able to accurately identify the different passenger groups in the data, with the detected mobility patterns being consistent with the ground truth. These results highlight the effectiveness of DBGAN in overcoming data imbalance in mobility pattern recognition, and demonstrate its potential for wider use in transportation and data mining applications.

Keywords: Mobility pattern recognition · Data imbalance · Kernel embedding · Sequence to image · Generative adversarial networks · Self-attention

---

Supported by CRT-AI.

1   Introduction

Mobility pattern recognition is an important area of research that has gained significant attention in recent years [1,21,33,34], particularly in the field of computer science and transportation. The goal of this research is to identify patterns of human mobility and classify them into different groups, such as older people, students, and regular commuters, based on their travel records. The aim is to understand where different types of passengers are likely to travel and make predictions about their future movements. This information can be used to optimize transportation systems, improve service quality, and make informed decisions about transportation planning.

Current work in this field includes the use of traditional statistical methods, traditional machine learning methods, and deep learning methods. However, there are several limitations that need to be addressed to achieve effective mobility pattern recognition. For example, there is likely to be a data imbalance problem, where the number of samples in different classes is not equal (e.g., fewer older people than commuters). This can lead to biased results and poor classifier performance, as the classifier may be trained on a majority class that does not accurately represent the minority class. To overcome this, oversampling [15] or undersampling[20] techniques can be used, but these techniques can result in loss of information and lead to overfitting. Synthetic data generation techniques, such as SMOTE [5], can also be used to balance the data distribution, but these techniques do not capture the underlying patterns and relationships in the data, which can impact the performance of the classifier.

Another limitation is the difficulty in handcrafting effective features. Mobility pattern recognition requires the extraction of meaningful features from travel records, such as travel locations, travel times, and travel modes. However, handcrafting features can be a challenging task, as the choice of features can greatly impact the performance of the classifier. This challenge can be overcome by using deep learning methods, which can automatically learn features from the data, but these methods can also be computationally expensive and may not be feasible for all applications.

Finally, incorporating both static and dynamic travel information is crucial for achieving accurate mobility pattern recognition, but this can also be a challenge. Both static information, such as locations, and dynamic information, such as travel times, are important for accurately capturing mobility patterns, but combining them can be challenging, as they may have different distributions and may need to be processed differently.

To address these problems, we propose a Data Balancing Generative Adversarial Network called DBGAN for mobility pattern recognition. This framework leverages the power of Reproducing Kernel Hilbert Space (RKHS) embeddings, attention mechanism and Generative Adversarial Networks (GANs). The proposed approach involves transforming the raw travel sequence data into images and using the generator of GANs to generate synthetic images. These images are then classified into different groups by the discriminator, representing different travel patterns. The goal of this approach is to eliminate the need for manual feature engineering and overcome the data imbalance to make it easier to generalize the model to different scenarios.

In this paper, we make the following contributions:

1. Our proposed Generative Adversarial Network (GAN) based framework for mobility pattern recognition leverages the power of Reproducing Kernel Hilbert Space (RKHS) embeddings to convert sequences of travel records into images. This novel approach eliminates the need for handcrafting features, which can be time-consuming and prone to losing some granularity of information. The RKHS embeddings effectively capture rich information about passenger mobility patterns, providing a significant improvement over traditional methods.

2. The proposed GAN-based framework addresses a common challenge in urban mobility pattern recognition: data imbalance. By generating synthetic samples that balance the data distribution, the GAN can enhance the performance of the classifier which is the discriminator, resulting in more robust and accurate classification results, even in scenarios where collecting additional real data is not feasible.

3. Our proposed framework takes into account both static and dynamic features of travel records to improve mobility pattern recognition. We proposed an attention mechanism-based model that effectively leverages the information about the marginal distribution of certain flow attributes and the conditional distribution of the same attribute between two location points.

The rest of this paper is organized as follows: In Sect. 2, we provide a detailed overview of the related work in mobility pattern recognition. In Sect. 3, we give the background about the techniques used in our model. In Sect. 4, we describe the proposed DBGAN framework in detail. In Sect. 5, we present the experimental results and evaluation metrics. Finally, in Sect. 6, we conclude the paper and discuss potential future work.

2   Related Work

Mobility pattern recognition is a crucial area of study in transportation and urban planning. Understanding how people move and behave is essential to making cities better places to live. Researchers have come up with many ways to study mobility patterns and have produced a large amount of information about this subject. Mobility patterns can be useful for understanding people's behavior, seeing behavioral differences between groups, explaining such differences, predicting future behavior, and even detecting anomalous movement trajectories. There are three main ways that researchers study mobility patterns. In this section, we will review some of the most relevant works in this field and discuss their strengths and weaknesses.

The first mainstream approach is clustering, which has recently been widely used for identifying patterns in the movement of individuals or groups. For example, it has been used to identify common routes taken by individuals, or to identify groups of individuals who exhibit similar movement patterns. The process of clustering involves finding the optimal number of clusters and then assigning each data point to the appropriate cluster. This can be done through various techniques such as k-means based clustering [23,36], dense based clustering [33] or hierarchical clustering [31]. The cluster based method can provide valuable insights into mobility patterns and can help inform decision making in areas such as urban planning and transportation management. Clustering is relatively easy to implement, and has a low reliance on labels. However, there are also some disadvantages. For example, it can be sensitive to the initial conditions and the number of clusters chosen. The results can be highly dependent on the similarity metric used. Finally, no prior information is available, making it difficult to provide specific classifications in practice.

Another mainstream approach to mobility pattern recognition is traditional machine learning. This method involves training a model using pre-existing labels, which can then be applied to new data. In the context of mobility pattern recognition, Support Vector Machines (SVM) have been used to classify patterns into various categories such as different modes of transportation or movement patterns [9]. Additionally, popular machine learning algorithms such as Random Forest [24] and Decision Tree [10] have also been employed to perform classification tasks. These traditional machine learning methods provide highly accurate results, but they can be computationally expensive and require significant manual effort to extract meaningful features and rules. Additionally, it can be challenging to ensure that all important temporal (dynamic) and spatial (static) features are extracted. Furthermore, kernel-based methods like SVM can be influenced by the choice of kernel function and regularization parameters, which can impact the model's performance.

The third mainstream approach is deep learning, which uses deep neural networks to analyze human mobility patterns, and has demonstrated promising results in recent years. Kim et al. employed a combination of Deep Neural Network (DNN) and Deep Belief Network (DBN) to investigate the relationship between human mobility and personality [17]. Recurrent Neural Networks (RNN), in particular, are well-suited for analyzing sequential data. Song et al. proposed an RNN-based approach to detect abnormal passengers based on movement trajectories extracted from subway card records [29]. Ke et al. similarly proposed an RNN-based model for recognizing groups with distinct mobility patterns [16]. Kong et al. proposed a hierarchical framework combining with long short-term memory networks (LSTM) and one class support vector machine (OCSVM) to detect the anomalies in subway. Convolutional Neural Networks (CNN) have also been used to understand global human mobility patterns and make predictions about future patterns [25]. These deep learning methods have gained popularity because they can be used to both learn complex patterns and relationships in data, and automatically extract hidden features.

In recent years, there have been numerous studies on mobility pattern recognition with promising results in various scenarios. However, using machine learning methods for this task often comes with the challenge of sensitivity to the input data distribution. While the classification model may perform well with balanced data, its performance can suffer when faced with imbalanced data, which is a common occurrence in real-world scenarios, particularly in tasks such as abnormal passenger detection [9, 29]. To mitigate this issue, Du et al. proposed an undersampling-based ensemble framework [9], while Song et al. employed an oversampling approach by adding additional samples to the minority class [29]. Berke et al. proposed an RNN-based method to generate synthetic mobility data to balance the data distribution [2]. Although these methods have improved the classification performance in their respective models, they also come with limitations such as loss of information granularity and the risk of overfitting. In addition, traditional data-level approaches for generating synthetic instances may not accurately represent the training set. Synthetic data is typically created through duplication or linear interpolation, which fails to generate new, atypical examples that could challenge the classifier's decision boundaries. As a result, the effectiveness of these methods in enhancing performance is limited.

As previously mentioned, existing machine learning methods have limitations, including their dependence on manually engineered features, difficulties in extracting both temporal and spatial information, and challenges in dealing with imbalanced data. These limitations hinder their ability to effectively recognize passenger mobility patterns in different scenarios. In this paper, we propose a Generative Adversarial Network (GAN)-based mobility pattern recognition and classification framework, which leverages the feature extraction power of attention mechanisms to address the weaknesses encountered in related work.

3   Background

Our proposed framework uses Reproducing Kernel Hilbert Space (RKHS) embeddings, attention mechanisms, and Generative Adversarial Networks (GANs), which we first introduce here as background.

3.1   Reproducing Kernel Hilbert Space Embeddings

Determining the probability distribution of a random variable, especially for high-dimensional continuous variables, has proven to be a challenging problem. However, recent advancements in machine learning have resulted in the successful application of RKHS embeddings in various tasks [7, 26]. The key benefit of using RKHS embeddings is that non-linear relationships in the data can be captured by mapping them into a high-dimensional feature space. This allows for the extraction of both static (spatial) and dynamic (temporal) features from raw trajectory data, enabling the training of neural network models.

Our approach uses the concept of Reproducing Kernel Hilbert Space (RKHS) embedding to represent the (conditional) probability distribution of a random variable in a compact and elegant manner [11, 28]. In this section, the (conditional) kernel mean embedding is described. This technique transforms the (conditional) distribution into a high-dimensional feature representation. This is crucial for our proposed GAN-based mobility pattern recognition framework to effectively extract both static (spatial) and dynamic (temporal) features from raw trajectory data.

The kernel mean embedding of a marginal distribution $P(x)$ is defined as the expected value of its feature representation [13]:

$$\mu_X := \mathbb{E}_X\left[\phi(X)\right] = \int_\mathcal{X} \phi(x)dP(x), \tag{1}$$

where $\mathbb{E}_X\left[\phi(X)\right]$ is the expected value of $\phi(X)$ with respect to $P(X)$. The feature mapping, $\phi(X)$, is in a Reproducing Kernel Hilbert Space (RKHS), where the inner product of two feature mappings can be calculated easily using the kernel trick $\langle\phi(x),\phi(x')\rangle\mathcal{H} = k(x,x')$, and $k(,)$ is a positive definite kernel.

We can also use the tensor feature map $\phi(X) \otimes \phi(X)$ to construct a tensor mean embedding as follows:

$$\mu_X^\otimes := \mathbb{E}_X\left[\phi(X) \otimes \phi(X)\right] = \int_\mathcal{X} \phi(X) \otimes \phi(X)dP(x), \tag{2}$$ '

In this paper, we use μ_X as a compact representation of the probability distribution $P(X)$, as the latter can be used as an image and fed into an adversarial neural network for the extraction of features, generation of synthetic data, and classification of mobility patterns.

In addition to the kernel mean embedding, which represents the marginal distribution in a Reproducing Kernel Hilbert Space (RKHS), we are also interested in representing the conditional distribution between two random variables Y and X. This is because the dynamic (temporal) behavior of travelers is a crucial factor in mobility pattern classification. Additionally, RKHS kernel embeddings can be utilized to characterize both the joint and conditional distributions of two random variables.It has been shown in previous research [28] that the conditional kernel embedding $\mathcal{U}_{Y|X}$ of the conditional distribution $P(Y|X)$ can be derived from the following set of equations:

$$\mathcal{U}_{Y|X} = C_{YX}C_{XX}{-1}, \tag{3}$$

where $C_{YX} := \int_{y \times x} \phi(x) \otimes \phi(x)dP(y,x)$, and C_{XX} is defined by replacing Y with X. According to Bochner's theorem [27], the kernel feature mappings $\phi(x)$ and $\phi(y)$ can be derived using the method of Random Fourier Feature Mapping as follows:

$$\phi(x) = \frac{a}{N_\mathcal{H}}\left[cos(\omega_1^T x), sin(\omega_1^T x), ..., cos(\omega_{N_\mathcal{H}} x), sin(\omega_{N_\mathcal{H}} x)\right], \tag{4}$$

where α is the integral of $\mu(\omega)$ and $\omega_1, ..., \omega_{N_\mathcal{H}}$ are sampled randomly from the normalized measure $p = \mu/\alpha$. The Gaussian kernel is used in this paper as follows:

$$k(x,x') = k(x-x') = exp\left\{-(x-x')^T \Sigma^{-1}(x-x')\right\} \tag{5}$$

The multivariate Gaussian distribution $\mathcal{N}(0,\Sigma)$ is denoted by $p(\omega)$. Therefore, the explicit kernel mapping can be constructed by sampling ω_i from $\mathcal{N}(0,\Sigma)$, where $i = 1, 2, ..., N_\mathcal{H}$.

By utilizing the explicit kernel mapping, we estimated the values of $\mu_X^\otimes$ and $\mathcal{U}_{Y|X}$ as $\hat{\mu}_X^\otimes$ and $\hat{\mathcal{U}}_{Y|X}$ respectively through the following steps Equation:

$$\hat{\mu}_X^\otimes = \Phi\Phi^T/n, \tag{6}$$

and

$$\hat{\mathcal{U}}_{Y|X} = \Gamma\Phi^T(\Phi\Phi^T + \lambda nI)^{-1}, \tag{7}$$

where $\Phi = [\phi(x_1), ..., \phi(x_n)]$ and $\Gamma = [\phi(y_1), ..., \phi(y_n)]$ and λ is the regularization parameter and $(x_i, y_i), \forall i$ are the observations.

3.2 Attention Mechanism

Attention mechanisms have gained popularity as they support the deep learning model focusing on the most important parts of the input data, rather than processing the entire input equally, This substantially improves their performance in a wide range of applications including natural language processing [32], computer vision [6], and speech recognition [30].

There are several different types of attention mechanisms, including self-attention, where the attention mechanism is applied to the input features within a single sequence, and multi-head attention, where multiple attention mechanisms are used in parallel to attend to different parts of the input data. Additionally, attention mechanisms can be applied to both sequential and non-sequential data, making them highly flexible and adaptable to a wide range of tasks.

A simple attention mechanism consists of three components: a query, a key, and a value. The query and key are used to calculate an attention score, which determines the relative importance of each element in the sequence. The value is then used to compute the weighted sum of all elements, with the attention score serving as the weight for each element. The output of the attention mechanism is a weighted representation of the input sequence, where the weight of each element reflects its importance in the context of the query.

The attention mechanism is typically implemented as a separate layer in a deep learning model, which is trained in an end-to-end manner along with the rest of the model. The attention layer takes in the intermediate activations from a previous layer and computes a weight or attention map, which represents the importance of each part of the input data. This weight map is then used to weight the activations from the previous layer, allowing the model to focus on the most important parts of the input data.

In this work, we adopt the self-attention mechanism, which has been proven to be effective in various computer vision tasks. The self-attention mechanism works by calculating the attention scores between different parts of the input data, and then using these scores to weigh the contribution of each part to the final output. In our implementation, we apply the self-attention mechanism to the features extracted from the mobility data, and use it to improve the mobility pattern recognition performance.

3.3   Generative Adversarial Network

A GAN is a deep learning architecture that was introduced by Goodfellow et al. in 2014 [12]. It consists of two main components: a generator and a discriminator. The generator is responsible for generating new samples that are similar to the target distribution, while the discriminator is trained to distinguish between real and fake samples. The two components are trained in a game-theoretic framework, where the generator tries to fool the discriminator and the discriminator tries to accurately distinguish between the real and fake samples. Over time, the generator learns to generate more realistic samples that are indistinguishable from the real ones, while the discriminator becomes more and more accurate in its classification. A typical GAN architecture is shown in Fig. 1.

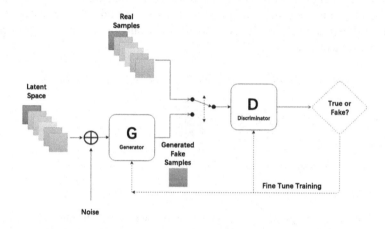

Fig. 1. A Typical Generative Adversarial Network [4].

GANs have been shown to be effective in various applications such as image synthesis [19], data augmentation [3], and domain adaptation [35]. GANs can generate high-quality synthetic data that is similar to the target distribution, which makes them particularly useful for overcoming data imbalance problems in machine learning. By generating additional synthetic samples, GANs can help balance the distribution of the data and improve the performance of machine learning models.

GANs have become an increasingly popular tool in the deep learning community as they can be used to generate high-quality synthetic data and to overcome data imbalance problems in various applications [8,22]. However, training GANs can be challenging, as the loss function is often non-convex and can easily get stuck in suboptimal solutions. Moreover, the minimax game-theoretic framework can be unstable, leading to mode collapse, where the generator produces only a few samples that are indistinguishable from the real ones. In order to overcome these challenges, researchers have proposed various modifications and extensions to the original GAN architecture, such as Wasserstein GANs [14].

4   DBGAN Mobility Pattern Classification Model

The DBGAN mobility pattern recognition method framework is illustrated in Fig. 2. Public transportation systems, such as subways, carry large numbers of passengers on a daily basis, generating vast amounts of travel records. These records contain valuable information about the travel behavior of different types of passengers. By analyzing these travel patterns, it is possible to recognize and categorize different types of passengers based on their unique behaviors. In this study, the relationship between travel state and circumstances state and their effect on passenger behavior is taken into consideration. To concisely capture and represent the probability distributions of these attributes, we utilize RKHS kernel embeddings. In a nutshell, we transform the travel trajectory into an image. By combining two different types of attributes and behaviors (spatial/static and temporal/dynamic), a four-channel image is created and processed by a classifier with self attention based GAN. The classifier outputs the passenger type for each travel trajectory.

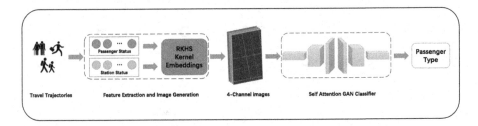

Fig. 2. The proposed sequence-to-image with GAN framework.

4.1   Attributes of Travel Trajectories Utilized for Classification

To ensure the generalizability of our approach, we opt to use raw travel trajectory information, rather than hand-crafted features, as input to our mobility pattern recognition algorithm. In this paper, the raw data utilized is taken from Suzhou, China's subway Automatic Fare Collection (AFC) records. The data consists of various attributes such as the time, location, and transportation mode of each passenger. These attributes provide valuable insights into the travel behavior of different passengers, allowing us to accurately classify their mobility patterns. We created travel trajectories for each passenger using the AFC records. The travel trajectory information encompasses:

1. The travel state sequence built from a passenger's travel trajectory:

$$ts_1, ts_2, ..., ts_n, \tag{8}$$

where n represents the number of time slot during which the passenger's travel state is ts_n.

2. The circumstances state sequence built from the station's crowd density:

$$ss_1, ss_2, ..., ss_n, \tag{9}$$

where n represents the number of time slot during which the crowd density of passenger visited places is ss_n.

The travel state of each passenger is established by determining their location within a functional region and whether they are currently on a subway. This is accomplished by creating a record of their subway travel movements and assigning labels based on the functional areas. The station crowding state is determined based on the crowd density along their travel route.

4.2 Sequences to Images with Kernel Embedding

In this section, we introduce a technique for converting sequential data into an image representation by utilizing RKHS (conditional) mean embedding. The method is applied to early segments of bidirectional flow data, which can be represented as a sequence of information $\{I_1, I_2, ..., I_n\}$. Each element I_j in the sequence represents the travel state or the circumstances state of a passenger in the time slot j.

Our hypothesis posits that each passenger generates a travel sequence with a distinct set of static(spatial) and dynamic(temporal) characteristics, which can be revealed through the analysis of the marginal distribution $p(I_j)$ and the conditional distribution $p(I_{j+1}|I_j)$. While the mean and standard deviation of a distribution provide important information about its location and scale, the shape, particularly local structures, can also be utilized to distinguish between different passengers.

As outlined in Sect. 3, we utilize the RKHS mean embedding method to effectively represent the marginal distribution $p(ts_j)$ and the conditional distribution $p(I_{j+1}|I_j)$.This involves converting the distributions from their original domain to the associated Reproducing Kernel Hilbert Space (RKHS). As a result of this process, we are able to represent a travel sequence as an image with four channels, including $\mu \otimes ts$, $\mathcal{U}_{ts+|ts}$, $\mu \otimes ss$, and $\mathcal{U}_{ss+|ss}$. These channels are derived from the two dimensions of attributes, ts_j and ss_j, present in I_j. For demonstration, we present images depicting the marginal distribution of packet size for a commuter, a student and an older passenger in Fig. 3.

4.3 Classification Using Self Attention-Based Generative Adversarial Network

4-channel images are produced using a kernel embedding approach, and then an attention-based GAN classifier is trained. The generator is trained for synthetic image generation, while the discriminator is used for passenger mobility pattern recognition. Our neural network model was developed using Tensorflow and Keras on a computer with AMD Ryzen 9 5900HX CPU and one Nvidia

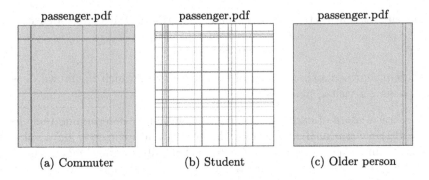

(a) Commuter   (b) Student   (c) Older person

Fig. 3. Images of Commuter, Student and Older passengers.

RTX3070ti GPUs. In this paper, we utilize a self-attention mechanism that includes multiple layers to capture the relationships between different parts of the input feature maps and dynamically adjust their contributions to the output feature maps. Our method employs a GAN architecture, which consists of two main components: a generator and a discriminator. In the first step of our experiment, the generator creates synthetic images from noise inputs while the discriminator assesses the validity of the generated images. The two components are trained in an adversarial manner, with the generator trying to trick the discriminator and the discriminator trying to accurately identify real and generated images. The goal is to train an effective discriminator for later classification using the synthetic data produced by the generator. The generator is comprised of dense layers, convolutional layers, self-attention layers, and batch normalization layers. The output layer uses a tangent activation function. The discriminator is a simple feedforward neural network with dense layers and a sigmoid activation function in the output layer. The network is trained using the Adam optimizer and a cross-entropy loss function. In the second step, the discriminator is further fine-tuned using original images, generated images, and their corresponding labels, processed through a fully connected layer. The loss function used during this step is categorical cross-entropy, and the evaluation metric is accuracy. Our experiments demonstrate that this hybrid approach leads to improved performance and increased robustness of the framework.

5   Evaluation

We conducted our evaluation using a real-world dataset derived from the Automatic Fare Collection (AFC) system of Suzhou, China's subway network. The dataset covers a one-week period, from September 1st to September 7th, 2019 and contains more than 1 million records relating to three types of passenger: commuter, student and older people. The raw attributes include user ID, card type, time stamp, and entry and exit stations for each passenger. Some AFC example records are illustrated in Table 1.

Table 1. AFC example records.

ID	Type	Enter_Time	Enter_Station	Exit_Time	Exit_Station	Update_Time
0	7	9/2/2019 10:20	457	9/2/2019 10:28	456	9/2/2019 11:15
0	7	9/7/2019 13:54	457	9/7/2019 14:15	449	9/7/2019 15:00
0	7	9/7/2019 16:58	449	9/7/2019 17:19	457	9/7/2019 18:15

In this study, our model has two main objectives: mobility pattern recognition and addressing data imbalance. To evaluate the performance of our model, we compare it against two state-of-the-art urban mobility pattern classification methods: Projected K-means(PK-means) [36] and Random Forest (RD) [24]. Additionally, we also compare against three urban anomaly detection methods: Two-Step Support Vector Machine (TS-SVM) [9], Anomalous Trajectory Detection using Recurrent Neural Network (ATD-RNN) [29] and Hierarchical Urban Anomaly Detection (HUAD) [18]. To furthur evaluate the robustness of our model, we divide our dataset into varying ratios to simulate different levels of data imbalance, such as 3:3:4, 1:3:6, and 0.5:1.5:8 for older passengers, students, and commuters respectively. The performance of the proposed framework is evaluated using the overall accuracy metric defined as follows:

$$Overall\_acc = \frac{Number\ of\ correctly\ classified\ passengers}{Number\ of\ all\ passengers}, \qquad (10)$$

We perform 10-fold cross-validation on each of the baselines and our proposed DBGAN model, with average performances illustrated.

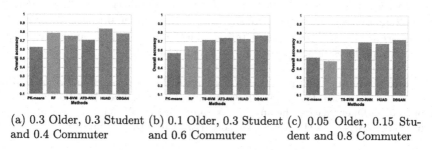

(a) 0.3 Older, 0.3 Student and 0.4 Commuter (b) 0.1 Older, 0.3 Student and 0.6 Commuter (c) 0.05 Older, 0.15 Student and 0.8 Commuter

Fig. 4. Comparing Overall Accuracy of Mobility Pattern Recognition Algorithms with different proportions.

The results in the Fig. 4 show the performance of various algorithms for classifying passengers into three different categories: older travelers, students, and commuters. The figure shows the performance of the algorithms when the data is balanced, as well as when the data becomes increasingly imbalanced.

According to the results illustrated in Fig. 4a, all of the algorithms, except for Projected K-means, perform well on the balanced dataset. This can be attributed to the fact that Projected K-means lacks the information from the passenger labels that is incorporated into the models of the other algorithms. As the data becomes increasingly imbalanced in Fig. 4b and Fig. 4c, as the dataset becomes increasingly imbalanced, the performance of Random Forest drops, while the performance of the anomaly detection methods remains relatively stable. This is likely due to the fact that anomaly detection methods are implemented with techniques, such as undersampling and oversampling, specifically designed to handle imbalanced data.

Our proposed method, DBGAN, shows remarkable stability and robustness across both balanced and imbalanced datasets. Particularly, its performance significantly improves as the data become increasingly imbalanced in comparison to other methods. This is due to the RKHS methods effectively embedding both dynamic and static features and the self-attention mechanism in the GAN extracting and focusing on the most important features, enabling the generator to produce effective synthetic images for training the discriminator. As a result, DBGAN demonstrates significant superiority over the compared methods.

6 Conclusion

We proposed a framework for urban mobility pattern classification and aim to eliminate the need for manual feature engineering. The approach converts the raw travel sequence into images based on RKHS embedding methods and uses self attention based Generative Adversarial Networks (GANs) to classify those images into different passenger mobility patterns.

Experiments have been conducted on real subway Automatic Fare Collection (AFC) data, and the results show significant improvement over existing methods dealing with the imbalanced data. In the future, we plan to improve the accuracy of the model by incorporating more raw travel sequences and to apply the model to other scenarios, such as outdoor public places (e.g., bus and open street) and indoor public places (e.g., airport and train station).

Acknowledgements. This publication has emanated from research supported in part by a grant from Science Foundation Ireland under Grant number 18/CRT/ 6223. For the purpose of Open Access, the author has applied a CC BY public copyright licence to any Author Accepted Manuscript version arising from this submission.

References

1. Ahmed, D.B., Diaz, E.M.: Survey of machine learning methods applied to urban mobility. IEEE Access **10**, 30349–30366 (2022)
2. Berke, A., Doorley, R., Larson, K., Moro, E.: Generating synthetic mobility data for a realistic population with RNNs to improve utility and privacy. In: Proceedings of the 37th ACM/SIGAPP Symposium on Applied Computing, pp. 964–967 (2022)
3. Bird, J.J., Barnes, C.M., Manso, L.J., Ekárt, A., Faria, D.R.: Fruit quality and defect image classification with conditional GAN data augmentation. Sci. Hortic. **293**, 110684 (2022)

4. Chaudhari, P., Agrawal, H., Kotecha, K.: Data augmentation using mg-GAN for improved cancer classification on gene expression data. Soft. Comput. **24**, 11381–11391 (2020)
5. Chawla, N.V., Bowyer, K.W., Hall, L.O., Kegelmeyer, W.P.: Smote: synthetic minority over-sampling technique. J. Artif. Intell. Res. **16**, 321–357 (2002)
6. Chen, C.F.R., Fan, Q., Panda, R.: Crossvit: cross-attention multi-scale vision transformer for image classification. In: Proceedings of the IEEE/CVF International Conference on Computer Vision, pp. 357–366 (2021)
7. Chen, Z., He, K., Li, J., Geng, Y.: Seq2img: A sequence-to-image based approach towards IP traffic classification using convolutional neural networks. In: 2017 IEEE International Conference on Big data (big data), pp. 1271–1276. IEEE (2017)
8. Deng, G., Han, C., Dreossi, T., Lee, C., Matteson, D.S.: IB-GAN: a unified approach for multivariate time series classification under class imbalance. In: Proceedings of the 2022 SIAM International Conference on Data Mining (SDM), pp. 217–225. SIAM (2022)
9. Du, B., Liu, C., Zhou, W., Hou, Z., Xiong, H.: Detecting pickpocket suspects from large-scale public transit records. IEEE Trans. Knowl. Data Eng. **31**(3), 465–478 (2018)
10. Ferreira, P., Zavgorodnii, C., Veiga, L.: edgetrans-edge transport mode detection. Pervasive Mob. Comput. **69**, 101268 (2020)
11. Fukumizu, K., Song, L., Gretton, A.: Kernel Bayes' rule: Bayesian inference with positive definite kernels. J. Mach. Learn. Res. **14**(1), 3753–3783 (2013)
12. Goodfellow, I.J.: On distinguishability criteria for estimating generative models. arXiv preprint arXiv:1412.6515 (2014)
13. Gretton, A., Borgwardt, K.M., Rasch, M.J., Schölkopf, B., Smola, A.: A kernel two-sample test. J. Mach. Learn. Res. **13**(1), 723–773 (2012)
14. Gulrajani, I., Ahmed, F., Arjovsky, M., Dumoulin, V., Courville, A.C.: Improved training of wasserstein gans. In:30th Proceedings of Conference on Advances in Neural Information Processing Systems (2017)
15. Hauser, M.W.: Principles of oversampling a/d conversion. J. Audio Eng. Soc. **39**(1/2), 3–26 (1991)
16. Ke, S., Xie, M., Zhu, H., Cao, Z.: Group-based recurrent neural network for human mobility prediction. Neural Comput. Appl. **34**(12), 9863–9883 (2022)
17. Kim, D.Y., Song, H.Y.: Method of predicting human mobility patterns using deep learning. Neurocomputing **280**, 56–64 (2018)
18. Kong, X., Gao, H., Alfarraj, O., Ni, Q., Zheng, C., Shen, G.: HUAD: hierarchical urban anomaly detection based on spatio-temporal data. IEEE Access **8**, 26573–26582 (2020)
19. Liang, J., et al.: Sketch guided and progressive growing GAN for realistic and editable ultrasound image synthesis. Med. Image Anal. **79**, 102461 (2022)
20. Liu, X.Y., Wu, J., Zhou, Z.H.: Exploratory undersampling for class-imbalance learning. IEEE Trans. Syst. Man Cybernet. Part B (Cybernetics) **39**(2), 539–550 (2008)
21. Loo, B.P., Zhang, F., Hsiao, J.H., Chan, A.B., Lan, H.: Applying the hidden Markov model to analyze urban mobility patterns: an interdisciplinary approach. Chin. Geogra. Sci. **31**, 1–13 (2021)
22. Luo, W., et al.: Fault diagnosis method based on two-stage GAN for data imbalance. IEEE Sens. J. **22**(22), 21961–21973 (2022)
23. Lv, Y., Zhi, D., Sun, H., Qi, G.: Mobility pattern recognition based prediction for the subway station related bike-sharing trips. Transport. Res. Part C: Emer. Technol. **133**, 103404 (2021)

24. Nirmal, P., Disanayaka, I., Haputhanthri, D., Wijayasiri, A.: Transportation mode detection using crowdsourced smartphone data. In: 2021 28th Conference of Open Innovations Association (FRUCT,. pp. 341–349. IEEE (2021)
25. Ouyang, X., Zhang, C., Zhou, P., Jiang, H., Gong, S.: DeepsPace: an online deep learning framework for mobile big data to understand human mobility patterns. arXiv preprint arXiv:1610.07009 (2016)
26. Paruchuri, S.T., Guo, J., Kurdila, A.: Kernel center adaptation in the reproducing kernel Hilbert space embedding method. Int. J. Adapt. Control Signal Process. **36**(7), 1562–1583 (2022)
27. Rudin, W.: Fourier Analysis on Groups. Courier Dover Publications (2017)
28. Song, L., Fukumizu, K., Gretton, A.: Kernel embeddings of conditional distributions: a unified kernel framework for nonparametric inference in graphical models. IEEE Signal Process. Mag. **30**(4), 98–111 (2013)
29. Song, L., Wang, R., Xiao, D., Han, X., Cai, Y., Shi, C.: Anomalous trajectory detection using recurrent neural network. In: Gan, G., Li, B., Li, X., Wang, S. (eds.) ADMA 2018. LNCS (LNAI), vol. 11323, pp. 263–277. Springer, Cham (2018). https://doi.org/10.1007/978-3-030-05090-0_23
30. Song, Q., Sun, B., Li, S.: Multimodal sparse transformer network for audio-visual speech recognition. In: IEEE Transactions on Neural Networks and Learning Systems (2022)
31. Wang, L., Zhang, Y., Zhao, X., Liu, H., Zhang, K.: Irregular travel groups detection based on cascade clustering in urban subway. IEEE Trans. Intell. Transp. Syst. **21**(5), 2216–2225 (2019)
32. Wolf, T., et al.: Huggingface's transformers: state-of-the-art natural language processing. arXiv preprint arXiv:1910.03771 (2019)
33. Yu, C., Li, H., Xu, X., Liu, J., Miao, J., Wang, Y., Sun, Q.: Data-driven approach for passenger mobility pattern recognition using spatiotemporal embedding. J. Adv. Transp. **2021**, 1–21 (2021)
34. Yuan, Y., Raubal, M.: Extracting dynamic urban mobility patterns from mobile phone data. In: Xiao, N., Kwan, M.-P., Goodchild, M.F., Shekhar, S. (eds.) GIScience 2012. LNCS, vol. 7478, pp. 354–367. Springer, Heidelberg (2012). https://doi.org/10.1007/978-3-642-33024-7_26
35. Zhang, M., Wang, H., He, P., Malik, A., Liu, H.: Exposing unseen GAN-generated image using unsupervised domain adaptation. Knowl.-Based Syst. **257**, 109905 (2022)
36. Zhang, S., Yang, Y., Zhen, F., Lobsang, T., Li, Z.: Understanding the travel behaviors and activity patterns of the vulnerable population using smart card data: an activity space-based approach. J. Transp. Geogr. **90**, 102938 (2021)

Bitwise Vertical Mining of Minimal Rare Patterns

Elieser Capillar, Chowdhury Abdul Mumin Ishmam, Carson K. Leung$^{(\boxtimes)}$ [iD],
Adam G. M. Pazdor, Prabhanshu Shrivastava, and Ngoc Bao Chau Truong

University of Manitoba, Winnipeg, MB, Canada
`Carson.Leung@UManitoba.ca`

Abstract. Rare patterns are essential forms of patterns in many real-world applications such as interpretation of biological data, mining of rare association rules between diseases and their causes, detection of anomalies. However, discovering rare patterns can be challenging. In this paper, we present an efficient algorithm for mining minimal rare patterns from sparse and weakly correlated data. The algorithm non-trivially integrates and adapts vertical frequent pattern algorithm VIPER to discover minimal rare patterns in an efficient manner. Evaluation results on our algorithm RP-VIPER show its superiority over existing horizontal rare pattern mining algorithms. Results also highlight the performance improvements brought by our optimized strategies.

Keywords: big data analytics · knowledge discovery · pattern mining · rare patterns · vertical mining · bitwise representation · sparse data

1 Introduction

Nowadays, big data are everywhere. Embedded in these big data are implicit, previously unknown, and potentially useful information (e.g., in the form of patterns), which can be discovered by data science. It makes good use of data mining [1, 2], machine learning, visualization, mathematics, and *statistics*. Among data mining tasks (e.g., clustering [3, 4], classification [5]), association rule mining and frequent pattern mining [6, 7] are popular in various real-life applications (e.g., medical informatics [8–10], transportation analytics [11–13], social analysis [14–16]). In general, frequent pattern mining aims to discover frequently co-occurring items. These frequent patterns can be served as building blocks for antecedent A and consequence C of association rule $A \rightarrow C$, which reveal associative relationships or correlations between items within A and C. It has played an important role in mining other patterns such as emerging [17], sequential [18], periodic [19], and quantitative patterns [20].

Besides frequent pattern mining, *rare pattern mining* [21] can also result in interesting results because they represent infrequent patterns in data. These infrequent patterns are particularly useful in various fields (e.g., biology, medicine, security). For example, in the medical field of pharmacovigilance (which detects, assesses, and studies adverse drug effects), mining for *rare patterns* (*RPs*) can result in the discovery of associating

R. Wrembel et al. (Eds.): DaWaK 2023, LNCS 14148, pp. 135–141, 2023.
https://doi.org/10.1007/978-3-031-39831-5_13

drugs with adverse effects [22]. Provided with a database of drug effects, if "{drug} ∪ {effect A}" is found to be a frequent pattern, and "{drug} ∪ {effect B}" is found to be a RP, this information can be used to determine whether these are desired and expected effects. As another example, in computer network security, given a database of connections to a computer network, mining for RPs can easily isolate uncommon and unusual connections. The resulting list of rare connections can then be further analyzed to determine if any were malicious.

In terms of related works, there were few algorithms for mining rare patterns or minimal rare patterns (mRP). Examples include AprioriRare [22], MRG-Exp [22], and Walky-G [23]. As MRG-Exp and Walky-G were built on the foundation of Apriori-Rare, they performed best on dense and highly correlated data. When it came to sparse and weakly correlated data, their runtime and search space were similar to those of AprioriRare. As a remedy for this, we aim to improve the runtime on finding mRPs in sparse and weakly correlated data. Hence, in this paper, we present an efficient algorithm called RP-VIPER for mining mRPs from sparse and weakly correlated data. The algorithm non-trivially integrates and adapts vertical frequent pattern algorithm VIPER to discover mRPs in an efficient manner. Our *key contributions* of this paper include our design and development of this RP-VIPER algorithm.

The remainder of this paper is organized as follows. The next section provides background and related works. Then, Sect. 3 describes our RP-VIPER algorithm. Section 4 shows evaluation results. Finally, Sect. 5 draws conclusions.

2 Background and Related Works

Definition 1. A pattern X is a *rare pattern* (*RP*) if its support (i.e., frequency) is less than the user-specified minimum support threshold *minsup*—i.e., $\text{sup}(X) < minsup$.

Definition 1. A pattern X is a *minimal rare itemset* (*mRP*) if it is rare but all of its proper subsets are frequent—i.e., $\text{sup}(X) < minsup$, but $\forall W \subset X$ ($\text{sup}(W) \geq minsup$).

A *vertical database* is an item-centric representation of data. It can be considered as a collection of bit vectors (BVs), each of which represent a domain item. The i-th bit of a bit vector representing a domain item x indicates the presence or absence of x in the i-th transaction. Specifically, a "0"-bit reveals that x is absent from the i-th transaction, whereas a "1"-bit reveals that x is present in the i-th transaction. To mine frequent patterns, VIPER [24] represents data vertically by using bit vectors. It mines frequent patterns in a bottom-up, breadth-first fashion. Candidate $(k + 1)$-itemsets (i.e., patterns of cardinality $k + 1$) are generated by applying bitwise AND operations on bit vectors of two frequent k-itemsets. These bitwise operations are cheap.

A *horizontal database* is a transaction-centric representation of data. It can be considered as a collection of transactions, each of which captures co-occurrence of items—represented by their transaction IDs—present in a transaction. AprioriRare [22], MRG-Exp [22], and Walky-G [23] are algorithms for mining mRPs. For example, AprioriRare mines mRPs from a horizontal database in a bottom-up, breadth-first fashion. It performs Apriori [7] as normal, finding all mRPs as a by-product. MRG-Exp also mines mRPs from a horizontal database in a bottom-up, breadth-first fashion. However, it reduces the search space of AprioriRare by looking for generators instead of itemsets. By the nature of generators, if an item is not a frequent generator, then it is an mRP. Similar to MRG-Exp, Walky-G also looks for generators when mining mRPs. However, Walky-G mines mRPs from a vertical database in a depth-first fashion. Among the three, MRG-Exp and Walky-G run faster than AprioriRare partially due to their use of generators. They can efficiently mine mRPs from dense and strongly correlated data. However, their runtimes are similar to that of AprioriRare when mining sparse and weakly correlated data.

3   Our RP-VIPER Algorithm

Due to the nature of sparse and weakly correlated data, frequent patterns usually do not tend to get too large. See Fig. 1, in which Fig. 1(a) captures a dense and highly correlated dataset, whereas Fig. 1(b) captures a sparse and weakly correlated dataset. We observe from their superset lattices that mRPs discovered from these datasets usually lie just past the border between frequent and rare patterns. This observation motivate our current work.

Density=70% (i.e., sparsity=30%)	Density=42% (i.e., sparsity=58%)
BV(a) = [10111 01001]	BV(a) = [11000 11110]
BV(b) = [11111 11100]	BV(b) = [11010 10101]
BV(c) = [01011 11010]	BV(c) = [00011 01111]
BV(d) = [10101 11111]	BV(d) = [00100 00000]
BV(e) = [11111 00011]	BV(e) = [01001 00000]

Fig. 1. (a) A dense dataset with a density of 70% (i.e., sparsity $= 30\%$), and (b) a sparse dataset with a density of 42% (i.e., sparsity $= 58\%$).

Algorithm 1 RP-VIPER

Input: Transactional database (T), max transaction length ($maxXactLen$), $minsup$
Output: List of all minimal rare patterns ($minRare$)
1. $level \leftarrow 1$
2. $minRare \leftarrow$ {infrequent singletons}   // All minimal rare itemsets
3. $Frequents_{level} \leftarrow$ {frequent singletons}   // All frequent 1-itemsets
4. construct vertical bit vectors of each frequent item
5. **while** ($|Frequents_{level}| \geq level+1$ **and** $level+1 \leq maxXactLen$)
6. $Frequents_{level+1} \leftarrow$ generateCandidates($Frequents_{level}$); $level \leftarrow level +1$
7. **return** $minRare$

Our RP-VIPER algorithm is a level-wise, bottom-up, breadth-first vertical algorithm for mining mRPs. Specifically, it first converts a given database to a vertical format if it is not already in a vertical format. It then stores all singleton bit vectors in a hash structure. Afterwards, it performs breadth-first candidate generation and mines mRPs. Pseudo code is shown in Algorithm 1.

In terms of implementation, RP-VIPER takes advantage of a hash structure (hashMap) to capture distinct domain item in a key-value store, where the item is the key and its corresponding bit vector is the value. With this data structure, support of singletons can be easily computed. To generate candidate 2-itemsets, bitwise AND operations are applied to frequent singletons. This step is repeated to generate candidate k-itemsets (for all $k \geq 2$) until (a) no more frequent itemsets can be generated or (b) the level of the largest transaction length is reached. During the candidate generation process, RP-VIPER first calls a subroutine generateCandidate (list) to generate candidates. It, in turn, calls checkSubset (itemset, list) to generate all proper subsets and eliminate one item at a time from the itemset. It then checks its presence in the Frequents list and returns whether or not it is found. True returns are added to the Frequents list if they meet *minsup*. True returns that do not meet *minsup* are saved as an mRP.

4 Evaluation

To evaluate our RP-VIPER, we compared with AprioriRare. Both algorithms were implemented in Java. Experiments were performed on an Intel Core i5-8265U 1.60 GHz CPU on Windows 10 with 8 GB of RAM. We applied both algorithms to four benchmark datasets (which are available at SPMF library[1]):

- T20I6D100K, which is a sparse and weakly correlated dataset with a sparsity of 97.77% with 893 distinct items;
- T25I10D10K, which is a sparse and weakly correlated dataset with a sparsity of 97.33% with 929 distinct items;
- C20D10K, which is a slightly denser and highly correlated dataset with a sparsity of 89.58% with 192 distinct items; and
- C73D10K, which is also a slightly denser and highly correlated dataset with a sparsity of 95.41% with 1592 distinct items.

We measured both runtime and the number of mRPs. The reported runtimes are average of multiple runs. Results show that, when *minsup* increased, the number of mRPs dropped and thus reducing the runtime. The gap between the two algorithms increased when *minsup* decreased. Our RP-VIPER took shorter to mine mRPs than the existing AprioriRare. See Figs. 2 and 3.

---

[1] https://www.philippe-fournier-viger.com/spmf/index.php?link=datasets.php.

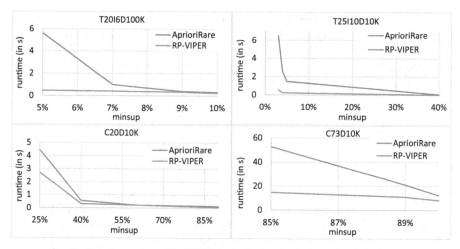

Fig. 2. Runtimes of our RP-VIPER and existing AprioriRare on four datasets.

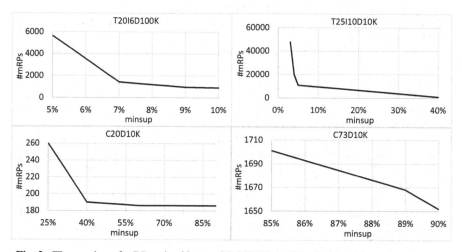

Fig. 3. The number of mRPs mined by our RP-VIPER (and AprioriRare) from four datasets.

5 Conclusions

In this paper, we presented RP-VIPER as a foundation for mining minimal rare patterns in sparse and weakly correlated data. The evaluation results show that, when using sparse datasets regardless of correlation strength, our RP-VIPER outperforms the existing AprioriRare when given low minimum support thresholds. The algorithm is enhanced with an optimization of using a hash structure for all levels of candidate generation (instead of just at the second level). Moreover, the algorithm takes advantage of vertical item-centric representation of data—namely, bitwise representation. By doing so, candidates can be generated by performing bitwise AND operations. As *ongoing and future work*, we explore additional ways to further optimize our RP-VIPER algorithm with an aim to further reduce its runtime in mining minimal rare patterns. We would also like to extend RP-VIPER by incorporate Q-VIPER [25] for mining minimal rare *quantitative* patterns.

Acknowledgement. This work is partially supported by Natural Sciences and Engineering Research Council of Canada (NSERC) and University of Manitoba.

References

1. Aggarwal, C.C.: Data Mining: The Textbook. Springer, Cham (2015). https://doi.org/10.1007/978-3-319-14142-8
2. Han, J., et al.: Data Mining: Concepts and Techniques, 4th edn. MK (2022)
3. Brown, P.O., et al.: Mahalanobis distance based k-means clustering. In: Wrembel, R., Gamper, J., Kotsis, G., Tjoa, A.M., Khalil, I. (eds.) Big Data Analytics and Knowledge Discovery. DaWaK 2022. LNCS, vol. 13428, pp. 256–262. Springer, Cham (2022). https://doi.org/10.1007/978-3-031-12670-3_23
4. Dierckens, K.E., et al.: A data science and engineering solution for fast k-means clustering of big data. In: IEEE TrustCom-BigDataSE-ICESS 2017, pp. 925–932
5. Choudhery, D., Leung, C.K.: Social media mining: prediction of box office revenue. In: IDEAS 2017, pp. 20–29
6. Agrawal, R., et al.: Mining association rules between sets of items in large databases. In: ACM SIGMOD 1993, pp. 207–216
7. Agrawal, R., Srikanth, R.: Fast algorithms for mining association rules. In: VLDB 1994, pp. 487–499
8. de Guia, J., et al.: DeepGx: deep learning using gene expression for cancer classification. In: IEEE/ACM ASONAM 2019, pp. 913–920
9. Fung, D.L.X., Liu, Q., Zammit, J., et al.: Self-supervised deep learning model for COVID-19 lung CT image segmentation highlighting putative causal relationship among age, underlying disease and COVID-19. BMC J. Transl. Med. **19**, 318:1–318:18 (2021). https://doi.org/10.1186/s12967-021-02992-2
10. Leung, C.K., Fung, D.L.X., Hoi, C.S.H.: Health analytics on COVID-19 data with few-shot learning. In: Golfarelli, M., Wrembel, R., Kotsis, G., Tjoa, A.M., Khalil, I. (eds.) DaWaK 2021. LNCS, vol. 12925, pp. 67–80. Springer, Cham (2021). https://doi.org/10.1007/978-3-030-86534-4_6
11. Balbin, P.P.F., et al.: Predictive analytics on open big data for supporting smart transportation services. Procedia Comput. Sci. **176**, 3009–3018 (2020)
12. Leung, C.K., Braun, P., Pazdor, A.G.M.: Effective classification of ground transportation modes for urban data mining in smart cities. In: Ordonez, C., Bellatreche, L. (eds.) DaWaK 2018. LNCS, vol. 11031, pp. 83–97. Springer, Cham (2018). https://doi.org/10.1007/978-3-319-98539-8_7
13. Leung, C.K., Braun, P., Hoi, C.S.H., Souza, J., Cuzzocrea, A.: Urban analytics of big transportation data for supporting smart cities. In: Ordonez, C., Song, I.-Y., Anderst-Kotsis, G., Tjoa, A.M., Khalil, I. (eds.) DaWaK 2019. LNCS, vol. 11708, pp. 24–33. Springer, Cham (2019). https://doi.org/10.1007/978-3-030-27520-4_3
14. Braun, P., Cuzzocrea, A., Jiang, F., Leung, C.-S., Pazdor, A.G.M.: MapReduce-based complex big data analytics over uncertain and imprecise social networks. In: Bellatreche, L., Chakravarthy, S. (eds.) DaWaK 2017. LNCS, vol. 10440, pp. 130–145. Springer, Cham (2017). https://doi.org/10.1007/978-3-319-64283-3_10

15. Leung, C.K., Jiang, F., Poon, T.W., Crevier, P.: Big data analytics of social network data: who cares most about you on Facebook? In: Moshirpour, M., Far, B., Alhajj, R. (eds.) Highlighting the Importance of Big Data Management and Analysis for Various Applications, vol. 27, pp. 1–15. Springer, Cham (2018). https://doi.org/10.1007/978-3-319-60255-4_1

16. Leung, C.K., et al., Personalized DeepInf: enhanced social influence prediction with deep learning and transfer learning. In: IEEE BigData 2019, pp. 2871–2880

17. Dong, G., Bailey, J.: Contrast Data Mining: Concepts, Algorithms, and Applications. Chapman & Hall/CRC, New York (2012)

18. Agrawal, R., Srikant, R.: Mining sequential patterns. In: IEEE ICDE 1995, pp. 3–14

19. Madill, E.W., Leung, C.K., Gouge, J.M.: Enhanced sliding window-based periodic pattern mining from dynamic streams. In: Wrembel, R., Gamper, J., Kotsis, G., Tjoa, A.M., Khalil, I. (eds.) Big Data Analytics and Knowledge Discovery. DaWaK 2022. LNCS, vol. 13428, pp. 234–240. Springer, Cham (2022). https://doi.org/10.1007/978-3-031-12670-3_20

20. Srikant, R., Agrawal, R.: Mining quantitative association rules in large relational tables. In: ACM SIGMOD 1996, pp. 1–12

21. Weiss, G.M.: Mining with rarity: a unifying framework. ACM SIGKDD Explor. **6**(1), 7–19 (2004)

22. Szathmary, L., et al.: Towards rare itemset mining. In: IEEE ICTAI 2007, pp. 305–312

23. Szathmary, L., et al.: Efficient vertical mining of minimal rare itemsets. In: CLA 2012, pp. 269–280

24. Shenoy, P., et al.: Turbo-charging vertical mining of large databases. In: ACM SIGMOD 2000, pp. 22–33

25. Czubryt, T.J., Leung, C.K., Pazdor, A.G.M.: Q-VIPER: quantitative vertical bitwise algorithm to mine frequent patterns. In: Wrembel, R., Gamper, J., Kotsis, G., Tjoa, A.M., Khalil, I. (eds.) DaWaK 2022. LNCS, vol. 13428, pp. 219–233. Springer, Cham (2022). https://doi.org/10.1007/978-3-031-12670-3_19

Inter-item Time Intervals in Sequential Patterns

Thomas Kastner[1,2]([✉]), Hubert Cardot[1], and Dominique H. Li[1]

[1] Université de Tours, LIFAT, EA, 6300 Tours, France
[2] Apivia MACIF Mutuelle, Paris, France
`thomas.kastner@etu.univ-tours.fr`

Abstract. In sequential pattern mining, the time intervals between each pair of items are usually ignored or mined using additional constraints such as gaps. However, a large part of time-related knowledge is lost during the original proposition of sequential pattern mining. In many applications, time-related information is valuable and its preservation is necessary. In this paper, we present two methods for exploiting time intervals between items and evaluate them in a classification task using a real-world customer intent dataset.

Keywords: Pattern mining · Time intervals · Intention prediction

1 Introduction

Creating realistic representations of real-world applications is an essential aspect of data processing. In data mining, sequential pattern is a widely used pattern structure in many domains, such as finance, medicine or e-commerce. Unlike itemset pattern, sequential pattern mining consists in finding patterns from a *sequence database* with temporal order within a list of items. During nearly 30 years of research, two directions of sequential pattern mining have emerged: (1) Mining patterns faster, with models such as PrefixSpan [6] or BIDE [12]; (2) mining patterns with more information. This approach is often based on algorithms like those mentioned above. On the other hand, timestamp is often difficult to process in sequential pattern mining tasks. In literature, we often consider the timestamp of punctual events (items) as the period between two elements, or as the duration of an interval over which the events spreads. The classical sequential pattern mining model does not include timestamps, but several solutions have been developed to solve this problem. In most approaches, constraints are applied during frequent pattern mining, or as a post-processing step. For instance, the *duration constraint* specifies the minimum or maximum time that the patterns must not exceed, the *gap constraint* specifies the number of items between two items in the resulting patterns and the *length constraint* sets the minimum and maximum number of items in the sequence [10].

In this paper, we present a new sequence structure with itemized time intervals instead of timestamps, with which the sequences can be enriched, and for

R. Wrembel et al. (Eds.): DaWaK 2023, LNCS 14148, pp. 142–148, 2023.
https://doi.org/10.1007/978-3-031-39831-5_14

which, the most important, any existing sequential pattern mining algorithms can be applied. Our classification based experimental evaluation shows that the features generated by our proposed sequential pattern structure allow to train more accurate models than original sequences.

The rest of this paper is organized as follows. Section 2 introduces the related work. We describe in Sect. 3 our methods for integrating time intervals within sequences. Section 4 presents an application of our methods on a real customer intent dataset. Finally, we conclude in Sect. 5.

2 Related Work

In sequential pattern mining, the use of timestamps values, when available, has been rather under-exploited in the literature, especially for representing the time intervals between items. Kitakami et al. [8] and Chen et al. [2] extend PrefixSpan pattern mining algorithm by allowing it to use gaps and time intervals. Hirate et al. [7] generalize sequential pattern mining with item interval, gap and time interval handling with constraints. Tran et al. [3] propose to use time interval constraints and utility value to mine high utility sequential patterns. More elaborate methods have been added to the time constraints: MG-PrefixSpan [9] is a derivative from PrefixSpan that enables to deal with the temporal relationships between successive items in the sequential pattern, based on multi-granularities. Guyet et al. [5] adapted the classical Apriori [1] paradigm to propose an efficient algorithm based on a hyper-cube representation of temporal sequences. The algorithm of Gianotti et al. [4] adopts a prefix-projection approach to mine candidate sequences.

Our approach is the first attempt to provide a way to exploit time in patterns without modifying any mining algorithm and without having to apply constraints.

3 Representing Time in Sequences

In this section we present our contributions. We first introduce preliminary notations, then we propose two efficient methods to enrich sequences upstream of classical sequential pattern mining methods.

3.1 Preliminaries

Let $I = i_1, ..., i_n$ be the set of *all items*. A *sequence* is defined as $s = \langle X_1, ..., X_m \rangle$ which corresponds to an ordered list of items from I. A sequence can be associated with a vector of timestamps, denoted as $v_{time} = \langle t_{i_1}, ..., t_{i_m} \rangle$ which corresponds to the time each event of the sequence occurs. A *sequence database* is a large set of sequences, denoted by $\mathcal{S}$.

A sequence s_a of length p is contained in sequence s_b of length $q > p$ if and only if $X_{a_i} \subset X_{b_i} \forall \{i \mid 1 \leq i \leq p\}$: in this case, s_a is a *subsequence* of s_b, denoted as $s_a \sqsubseteq s_b$. Note that a sequence containing p items can have up to $2^q - 1$ distinct subsequences.

For the following methods we consider to compute the interval between each consecutive items of sequences. In this paper, we define the *vector of sequence intervals* as Definition 1.

Definition 1. Let the *interval* between item i_x with timestamp t_{i_x} and item i_{x+1} with timestamp $t_{i_{x+1}}$ be $\delta_{(i_x,i_{x+1})} = t_{i_{x+1}} - t_{i_x}$, the vector of sequence intervals is defined as $v_{inter} = \langle \delta_{(i_1,i_2)}, ..., \delta_{(i_{m-1},i_m)} \rangle$.   □

3.2   Integrating Intervals in Sequences

We consider a sequence $s_a = \langle X_1, ..., X_m \rangle$ composed of m items and the vector

$$v_{inter}^{s_a} = \langle \delta_{i_1,i_2}, ..., \delta_{i_{m-1},i_m} \rangle$$

representing time intervals between items of sequence s_a.

First we map each item of $v_{inter}^{s_a}$ to a discrete partition of $z = \langle I_1, I_2, ..., I_{|z|} \rangle$. Using $v_{inter}^{s_a}$ and an ordered set of thresholds parameters $\alpha = \langle th_1, th_2, ..., th_{|z|-1} \rangle$ with $th_1 < th_2 < ... < th_{|\alpha|}$, each continuous interval $\delta_{(i_{x-1},i_x)} \in v_{inter}^{s_a}$ is mapped to a partition $I_y \in z$ if it validates the following relation:

$$\delta_{(i_{x-1},i_x)} \in I_y \Leftrightarrow th_{y-1} \leqslant \delta_{(i_{x-1},i_x)} < th_y \tag{1}$$

The function $Z : v_{inter}^{s_a} \rightarrow z$, where $z = \{I_1, ..., I_n\}$ discretizes each value of vector $v_{inter}^{s_a}$ into the matching partition of z according to the relation presented above.

Integration as Items (Method 1). We construct a new sequence $s_a{}'$ from the original sequence s_a by inserting intervals partitions from $Z(v_{inter}^{s_a})$ between each observations of s_a as follows:

$$s_a{}' = \langle X_1, Z(i_1), X_2, Z(i_2), ..., X_{m-1}, Z(i_{m-1}), X_m, Z(i_m) \rangle \tag{2}$$

The resulting sequence contains twice as many items as the original sequence and the last interval element $Z(i_m)$ inserted equals zero.

Integration as Items Suffixes (Method 2). This method is derived from the method proposed in the previous section. As for the first method we consider the sequence s_a and the mapping function Z. Instead of adding new items to represent time intervals in the sequence, we add a suffix to each item of s_a that indicates which time interval partition $Z(i)$ has been associated to it. Each item can thus take $|I|^{|z|}$ distinct values.

$$s_a{}'' = \langle X_1 Z(i_1), X_2 Z(i_2), ..., X_{m-1} Z(i_{m-1}), X_m Z(i_m) \rangle \tag{3}$$

4   Experiments

We report in this section that sequences with added time intervals have a better predictive performance than normal sequences. We compare the results of models trained on features created from frequent patterns in the raw and modified sequence databases.

4.1   Datasets and Models

We use a clickstream dataset provided by an e-commerce website, published and described in [11] which has recently been used in the work of [13]. Authors present *seq2pat*, a new research library for sequence-to-pattern generation that allows to turn patterns into features by one hot encoding the presence of each pattern for each sequence. The dataset contains 203,084 click sequences including 8,329 sequences with a purchase (called positive sequences). The sequences are composed of six different types of events (page view, add to cart, purchase,...), and the biggest sequence has 155 items.

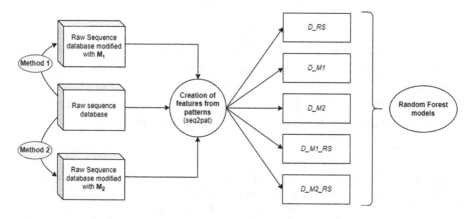

Fig. 1. Overall view of experiments.

We conduct our experiments using the dataset presented previously and *seq2pat* python library. We clean and prepare the dataset using the code provided by [13]. Figure 1 shows the different steps of our analysis. First we apply the two methods presented in the last section to the sequence database using the timestamp vector provided. We choose to model inter-items intervals with three discrete partitions and set the two corresponding thresholds parameters (10 s and 30 s) using the quantiles of the list of all inter-items intervals of the database. It ensures that there is a consistent number of intervals in each partition. This results in three different sequential databases created using *set2pat* to mine frequent patterns. From those three sequential databases we generate features from patterns with *set2pat* (one-hot encoding). Resulting datasets are described as follow: dataset *D_RS* includes features generated from frequent patterns in the raw sequence dataset. *D_M1* includes features generated from frequent patterns in the raw sequence dataset enhanced with method 1. *D_M2* includes features generated from frequent patterns in the raw sequence dataset enhanced with method 2. *D_M1_RS* and *D_M2_RS* include respectively features from *D_RS* + *D_M1* and *D_RS* + *D_M2*. For each dataset we train on AUC a random forest classifier composed of 100 trees, using a ten fold cross validation with *scikit-learn* python library [1].

4.2   Results

Table 1 presents the average results on several metrics for all previously described datasets. When we use only the sequences modified by our methods the performances are not better than with the initial sequences. *D_M1_RS* and *D_M2_RS* have higher scores for all metrics than *D_RS*. This shows the potential of our method to enhance the task of customer intention prediction.

Table 1. Results of the 10-folds cross-validation on a RF using five datasets.

Dataset	AUC	Recall	Precision	F1-score
D_RS	89.32 (±1.91)	75.50 (±4.48)	83.59 (±3.11)	79.26 (±3.11)
D_M1	88.95 (±1.99)	77.80 (±4.28)	83.43 (±3.35)	80.47 (±3.31)
D_M2	85.96 (±2.24)	65.60 (±4.80)	83.68 (±3.60)	73.43 (±3.53)
D_M1_RS	93.06 (±1.87)	**78.70 (±3.77)**	83.78 (±3.20)	**81.12 (±2.99)**
D_M2_RS	**93.68 (±2.24)**	75.90 (±4.25)	**83.84 (±3.07)**	79.61 (±3.05)

When looking at features importance of the model trained with *D_M1*, five of the ten most important features contain interval items that we have constructed. For *D_M2* the notion of time is included in all the features by construction.

Table 2. Advantages and drawbacks of presented methods.

	Advantages	Drawbacks
Method 1	- Independant from pattern mining algorithm	- Double sequence length
	- Sequence enriched by interval data	- Decrease of mean support
	- Fast and simple implementation	- Setting parameters
Method 2	- Independent from pattern mining algorithm	- More distinct items in database
	- Sequence enriched by interval data	- Setting parameters
	- Fast and simple implementation	
	- No increase in sequence length	

5   Conclusion

We presented two methods to include inter-item time intervals into sequences. They have the advantage of being independent from pattern mining algorithms as it is a data preparation step and can be used with almost all existing pattern mining algorithms, including sliding windows based methods. It enhances the sequences using a fast and simple implementation. The resulting sequence database stays interpretable as the mapped time intervals are user defined. Table 2 summarizes the advantages and disadvantages of the presented methods. Experiments show an improvement of performances when using our enhanced sequences on a customer intent classification task. Moreover, our methods do not have a processing time concern as it is a small and simple pre-processing task and it is independent of the pattern mining algorithm used. The integration of temporal knowledge within sequences could improve pattern mining applications, for instance in medicine or in finance. As future work, this method can be adapted to interval items, in other words items that have a duration.

References

1. Machine learning in Python: Scikit-learn. J. Mach. Learn. Res. **12**, 2825–2830 (2011)
2. Chen, Y.L., Chiang, M.C., Ko, M.T.: Discovering time-interval sequential patterns in sequence databases. Expert Syst. Appl. **25**(3), 343–354 (2003)
3. Duong, T.H., Janos, D., Thi, V.D., Thang, N.T., Anh, T.T.: An algorithm for mining high utility sequential patterns with time interval. Cybernet. Inform. Technol. **19**(4), 3–16 (2019)
4. Giannotti, F., Nanni, M., Pedreschi, D.: Efficient mining of temporally annotated sequences. In: Proceedings of the 2006 SIAM International Conference on Data Mining, pp. 348–359. SIAM (2006)
5. Guyet, T., Quiniou, R.: Mining temporal patterns with quantitative intervals. In: 2008 IEEE International Conference on Data Mining Workshops, pp. 218–227. IEEE, Pisa, Italy (December 2008)
6. Han, J., et al.: Prefixspan: mining sequential patterns efficiently by prefix-projected pattern growth. In: Proceedings of the 17th International Conference on Data Engineering (2001)
7. Hirate, Y., Yamana, H.: Generalized sequential pattern mining with item intervals. J. Comput. **1**(3), 51–60 (2006)
8. Kitakami, H., Kanbara, T., Mori, Y., Kuroki, S., Yamazaki, Y.: Modified prefixspan method for motif discovery in sequence databases. In: PRICAI 2002 (2002)
9. Li, N., Yao, X., Tian, D.: Mining temporal sequential patterns based on multi-granularities. In: IJCCC (2014)
10. Pei, J., Han, J., Wang, W.: Mining sequential patterns with constraints in large databases. In: Proceedings of the Eleventh International Conference on Information and Knowledge Management, pp. 18–25 (2002)
11. Requena, B., Cassani, G., Tagliabue, J., Greco, C., Lacasa, L.: Shopper intent prediction from clickstream e-commerce data with minimal browsing information. Sci. Rep. **10**(1), 1–23 (2020)

12. Wang, J., Han, J.: Bide: Efficient mining of frequent closed sequences. In: Proceedings. 20th International Conference on Data Engineering, pp. 79–90. IEEE (2004)
13. Wang, X., Hosseininasab, A., Colunga, P., Kadıoğlu, S., van Hoeve, W.J.: Seq2pat: sequence-to-pattern generation for constraint-based sequential pattern mining. In: Proceedings of the AAAI Conference on Artificial Intelligence, vol. 36 (2022)

Fair-DSP: Fair Dynamic Survival Prediction on Longitudinal Electronic Health Record

Xin Huang[1(✉)], Xiangyang Meng[1], Ni Zhao[3], Wenbin Zhang[2], and Jianwu Wang[1]

[1] University of Maryland, Baltimore County, MD, USA
xinh1@umbc.edu
[2] Michigan Tech University, Houghton, MI, USA
[3] Johns Hopkins University, Baltimore, MD, USA

Abstract. Scarce medical resources and highly transmissible diseases may overwhelm healthcare infrastructure. Fair allocation based on disease progression and fair distribution among all demographic groups is demanded by society. Surprisingly, there is little work quantifying and ensuring fairness in the context of dynamic survival prediction to equally allocate medical resources. In this study, we formulate individual and group fairness metrics in the context of dynamic survival analysis with time-dependent covariates, in order to provide the necessary foundations to quantitatively analyze the fairness in dynamic survival analysis. We further develop a fairness-aware learner (Fair-DSP) that is generic and can be applied to a dynamic survival prediction model. The proposed learner specifically accounts for time-dependent covariates to ensure accurate predictions while maintaining fairness on the individual or group level. We conduct quantitative experiments and sensitivity studies on the real-world clinical PBC dataset. The results demonstrate that the proposed fairness notations and debiasing algorithm are capable of guaranteeing fairness in the presence of accurate prediction.

Keywords: Dynamic survival analysis · Fairness · Deep learning · EHR

1 Introduction

Fairness in AI-based decision-making systems has drawn a lot of attention and becomes a very important topic in AI community [12]. A recent study [9] found the current clinical prediction models are systematically under-performing and biased against black patients, leading to unfair treatment of a particular type of cancer that disproportionately impacts the patients. In the early days of COVID-19 which spawned world-widely in a short time, many hospitals and medical institutes suffer from a shortage of medical resources (E.g., ventilators, Anti-SARS-CoV-2 Monoclonal Antibodies). Making fair medical resource allocations

© The Author(s), under exclusive license to Springer Nature Switzerland AG 2023
R. Wrembel et al. (Eds.): DaWaK 2023, LNCS 14148, pp. 149–157, 2023.
https://doi.org/10.1007/978-3-031-39831-5_15

and treating patients fairly in prioritization of ventilator or ICU admission across demographic groups or patients render an urgent need to promote fairness in medical care and AI-based clinical prediction models.

The longitudinal record in an Electronic Health Records (EHR) System enables doctors to trend labs over multiple encounters for a more holistic and longitudinal overview of their patient's health [3, 16]. Survival analysis enables us to understand the relationship between events of interest and the covariates so that we can assess the competing risk of events of interest. Cox proportional hazards model, a regression statistical model, investigates the association between the survival time of patients and one or more predictor variables and prioritizes the process of providing care to the ones in need of treatments [4]. However, the Cox model does not consider the time-varying covariates when fitting for individual and practical treatment questions. It only utilizes partial of repeated measurements, the last available measurement.

This has led to an active area of research in applying deep neural networks in developing prognostic models in survival analysis for time-varying risks or covariates. DeepHazard [8], a flexible neural network, was proposed to make survival predictions for time-varying risks. Dynamic-DeepHit learns the estimated joint distributions of survival times and competing events by incorporating the available longitudinal data comprising various repeated measurements [2], but fair treatment on a different group of patients (gender, race...) or individuals remains unexplored. Studies [5, 16, 17] show the AI-based survival predictions can generate biases for or against certain individuals or demographic groups, so it is important to understand the quantification of fairness in the dynamic survival prediction (e.g. prioritization of ventilator or ICU admission) with time-varying predictor variables and develop fairness-aware dynamic survival model.

In this paper, we propose a generic fairness-aware dynamic survival prediction (Fair-DSP) framework for longitudinal EHR to reduce bias and provide a fair prioritization process, in order to promote fair treatment of certain individuals or demographic groups. More specially, the novelty of this research comes from three aspects: (1) We formulate individual and group fairness metrics in the context of dynamic survival analysis with time-dependent covariates, thus providing the necessary foundations to quantitatively analyze the fairness in dynamic survival analysis. (2) We further develop a fairness-aware learner that is generic and can be applied to a dynamic survival prediction model. (3) Quantitative experiments and sensitivity studies on real-world clinical datasets demonstrate that the proposed fairness notations and debiasing algorithm are capable of guaranteeing fairness in the presence of accurate prediction.

2   Related Work

Understanding the relationship between covariates and the distribution of survival time is fundamental in survival analysis. Traditionally, several semiparametric survival models such as Cox Model and Additive Hazard Model have been extensively used in many fields including healthcare, bio-medicine, and economics [4]. By modeling the hazards as a linear combination of covariates, they suffer very limited applicability in many read-word applications. In recent years, survival analysis models have been augmented with deep neural networks to overcome that limitations [1]. DeepSurv [6] trains the neural network by making use of the classical Cox partial likelihood and bases the analysis on the proportional hazard assumptions. DeepHit [7] treats the survival time as discrete and uses a deep neural network to learn the distribution of survival time directly and smoothly handles competing risks with cause-specific neural layers. However, those traditional survival models don't incorporate the temporal changes of the covariates, where the patients could have repeated measurements during the treatment. Dynamic-DeepHit [2] proposes to measure the influence of each covariate on risk predictions and the temporal importance of longitudinal measurements, with the aim of identifying covariates that are influential for different competing risks. It employs an RNN subnetwork with temporal attention to capture the representation of covariates and uses cause-specific subnetwork to predict survival outcomes for competitive risks.

Amid the popularity of survival models, it becomes intrinsic to ensure the fairness of the survival prediction for demographic groups or individuals. Our work situates in this under-explored research area to tackle the fair dynamic survival predictions. Fairness in machine learning[5] has gained a lot of interest in recent years, been applied in streaming learning, graph neural network, and mobility learning. Keya et al develop a pioneer work in fair survival analysis by applying fairness regularizer in the popular proportional Cox Model [1]. Zhang et al. develop survival universal random forests (SURF) to jointly consider fair data encoding and discrimination reduction by introducing a new fairness-oriented splitting criterion to select the potential fair splitting candidates in the presence of censorship [5,16]. To the best of our knowledge, our work is the first in developing a fair dynamic survival model in the risk predictions on longitudinal survival data with repeated measurements and dynamic risks.

3   Methodology

Figure 1 depicts the framework of our generic fairness-aware dynamic survival system. Firstly the fairness-aware dynamic survival model makes predictions of risks that will be used to compute the expected probability of having the event k at the time point t. We then train the neural network by leveraging a combination of loss functions that incorporates model loss and individual/group fairness constraints, to optimize the parameters of a fairness-aware dynamic survival network. Finally, the outputs from the trained model will be evaluated by individual and group fairness measurements, respectively.

3.1   Fair Dynamic Survival Model

Inspired by Dynamic-DeepHit [2], the proposed fair dynamic survival model (Fair-DSP) jointly learns the dynamic risk prediction and fairness constraints to achieve fair dynamic risk prediction for demographic groups or individuals.

The model employs an RNN subnetwork with temporal attention to capture the representation of covariates and uses a cause-specific subnetwork to predict the survival outcomes for competitive risks. The loss function of dynamic survival model L consists of the log-likelihood of joint TTE (time to event) distribution (L_{tte}), concordance-based ranking loss (L_{rank}), and step ahead prediction loss of the RNN modules ($L_{prediction}$), as shown:

$$L = L_{tte} + L_{rank} + L_{prediction} \tag{1}$$

The loss function L_{total} of fairness aware dynamic survival model is composed of the survival loss L and fairness regularizer loss $L_{fairness}$, as formulated in 2, in which λ is a trade-off parameter which adjusts a balance between prediction accuracy and fairness.

$$L_{total} = L + \lambda * L_{fairness} \tag{2}$$

We choose individual fairness (F_I) and group fairness (F_G) respectively as the fairness constraints in regularizing our dynamic survival prediction model. We introduce individual and group fairness quantification that fits the longitudinal survival data in the following sections.

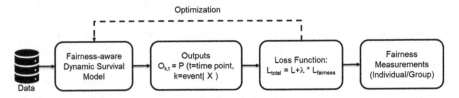

Fig. 1. Framework of the proposed fair dynamic survival prediction (Fair-DSP).

3.2   Individual Fairness

Existing statistical formulations of fairness can be categorized into two main families, individual fairness and group fairness, that evaluate fairness at the individual level or at the group level, respectively [10,11]. Individual fairness [10] aims to ensure that similarly individuals are treated similarly, rendering more fine-grained fairness constraints than any group-notion fairness.

To fit individual fairness in dynamic survival prediction, we use $L1$ distance to measure the distance between two prediction output risks of any pair of patients, and euclidean distance to measure the distance between any two individuals at the first measurement time point. The model's output risk is the expected probability of having a particular event k at a specific time t. We define individual fairness in dynamic survival prediction as follows:

$$F_I = \sum_{i=1}^{N} \sum_{j=i+1}^{N} Max(0, |O_{k,t}(X_i) - O_{k,t}(X_j)| \tag{3}$$

$$-\alpha * D(X_{i0}, X_{j0}))$$

where $O_{k,t}(X)$ is the prediction output risks of having the event k at the time point t, given the features X of a patient, and $D(X_{i0}, X_{j0})$ is a distance metric (e.g., euclidean distance) to measure the distance between any two individuals at the first time point, α is a scale value that adjusts with the impacts of the distance on the first measurement. We choose to use the first-time point because all patients have at least one initial measurement when admitted to the cohort.

3.3   Group Fairness

Group fairness also called demographic parity or statistical parity is one of the most well-known criteria for fairness [11]. Intuitively, it means the predicted risks of patients from two demographic groups (e.g., genders or races) must be equal. Formally, we define group fairness, mathematically represented as F_G, as the maximum deviation of discriminative abilities across different demographic groups of the model:

$$F_G = max_{a \in A} |E[O_{k,t}(a)] - E[O_{k,t}(x)]| \tag{4}$$

$E[O_{k,t}(a)]$ is the expected output probabilities of having the event k at the time point t, A is the set of sensitive values in the protected attributes (e.g., genders or races).

4   Experiments

We use C-index (concordance index) and Brier Score to measure the performance of our Fair-DSP model.

– C-index [15]: It is a rank order statistic for predictions against true outcomes. It is assumed that patients who lived longer should have higher risks than patients who lived less long, $R_i > R_j, t_i < t_j$, R represents the assigned risk while and t represents time to event. The higher the C-index is, the better the result is.
– The Brier Score [14]: It measures the weighted mean squared difference between the predicted probability of possible outcomes for sample i and the actual outcome. The lower the Brier Score is, the better the result is.

We evaluate the performance of our system on the dataset *pbcseq: Mayo Clinic Primary Biliary Cirrhosis, sequential data* [13]. This dataset contains the follow-up laboratory data for each study patient as well as the time to events information. The dataset contains 312 randomized patients' measurements, with 15 covariates (features). This dataset also contains time-to-event(survival) data, which is time-to-events, and labels indicating the type of event.

Table 1. Performance comparison on real-world dataset.

Dynamic-Deephit [2]			
T=156 (weeks)	$\Delta T=12$	$\Delta T=36$	$\Delta T=60$
C-Index	**0.948882**	**0.929326**	0.91657
BRIER-SCORE	0.082197	0.104873	0.117934
Group Fairness	0.064016	0.064859	0.06711
Individual Fairness	0.018866	0.0173635	0.048403
Deep-Hazard [8]			
C-Index	0.923410	0.913249	0.907654
BRIER-SCORE	0.082567	0.114674	0.129876
Group Fairness	0.065012	0.066543	0.068822
Individual Fairness	0.019987	0.019443	0.050422
Fair-DSP with F_G regularizer			
C-Index	0.939297	0.928629	**0.921565**
BRIER-SCORE	**0.080842**	**0.102076**	0.114453
Group Fairness	**0.061551**	**0.058288**	**0.066522**
Fair-DSP with F_I regularizer			
C-Index	0.942492	0.923575	0.920046
BRIER-SCORE	0.081845	0.104302	**0.110512**
Individual Fairness	**0.002008**	**0.0025195**	**0.019375**

4.1 Quantitative Analysis

Table 1 summarizes the experiment results when comparing the proposed Fair-DSP to the state of art dynamic survival models Dynamic-DeepHit [2] and Deep-Hazard [8]. Recall that the lower the fairness value is, the better fairness it achieves. By introducing individual fairness or group fairness as a regularizer in our fair dynamic survival model, our approach achieves more fair prediction with lower fairness distances in all cases, with better C-Index/Brier Score in some cases or only slight degradation of the C-Index or Brier Score in certain cases, thus demonstrates the effectiveness of our approach in enforcing fairness in dynamic survival predictions. Compared with group fairness regularizer, individual fairness regularizer performs better as individual fairness is generally a stronger fairness notion on the individual level.

In addition, we evaluated our system on different λ values for both individual fairness and group fairness regularizers. Figure 2 and Fig. 3 show how C-index and individual fairness changes as we change λ, respectively. From leftmost to rightmost, the value of λ decreases so that it gives less weight to the fairness contribution to the total loss, which leads to a higher C-index and higher fairness distance (implies less fair prediction). They both show that fine-tuning the λ can achieve a good trade-off between predictive accuracy and fairness.

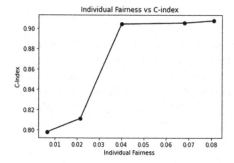

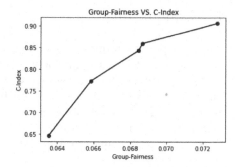

Fig. 2. Individual fairness vs C-index with the change of λ.

Fig. 3. Group fairness vs C-index with the change of λ.

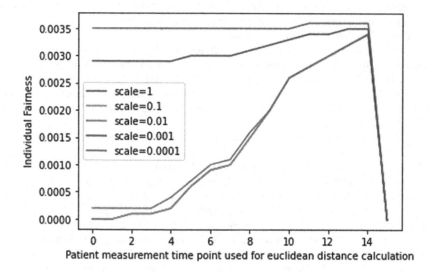

Fig. 4. Sensitivity test with different scale α values.

4.2 Sensitivity Study

We performed a sensitivity study on the model in terms of individual fairness constraints vs. patient measurement time points for calculating euclidean distance, as shown in Fig. 4. It shows individual fairness for different scale factors, with $\alpha=1$, $\alpha=0.1$, $\alpha=0.01$, $\alpha=0.001$, and $\alpha=0.0001$. The result demonstrates the increase of the fairness distance in all five scale values by taking the later measurement time points as the euclidean distance penalty for individual fairness, thus implying the necessity of using the first measurement time point. The similar trend across the five scale values demonstrates the robustness of our fair dynamic survival prediction. Note the sudden drop in the last measurement is caused by the very limited data that only 3 patients have 16 measurements.

5 Conclusions

In this paper, we propose a generic fair dynamic survival prediction system for longitudinal EHR with time-dependent measurements. Extensive experiments and sensitivity studies on the real-world clinical PBC dataset demonstrate that the proposed fairness notations and debiasing model are capable of guaranteeing fairness in the presence of accurate prediction. Our future work is to evaluate our method on more longitudinal clinical datasets and gain more insights into fairness-aware dynamic survival analysis.

References

1. Keya, K.N., Islam, R., Pan, S., Stockwell, I., Foulds, I.: Equitable allocation of healthcare resources with fair survival models. In: Proceedings of the 2021 SIAM International Conference on Data Mining (SDM), pp. 190–198. Society for Industrial and Applied Mathematics (2021)
2. Lee, C., Yoon, J., Van Der Schaar, M.: Dynamic-DEEPHIT: a deep learning approach for dynamic survival analysis with competing risks based on longitudinal data. IEEE Trans. Biomed. Eng. **67**(1), 122–133 (2020)
3. Huang, X., et al.: A Generic knowledge based medical diagnosis expert system. In: The 23rd International Conference on Information Integration and Web Intelligence (2021)
4. Cox, D.R.: Regression models and life tables (with discussion). J R Statist Soc B **34**, 187–220 (1972)
5. Zhang, W., Weiss, J.: Fair Decision-making Under Uncertainty. In: ICDM (2021)
6. Katzman, J., Shaham, U., Cloninger, A., Bates, J., Jiang, T., Kluger, Y.: DeepSurv: personalized treatment recommender system using a Cox proportional hazards deep neural network. BMC Med. Res. Methodol. **18**(1), 24 (2018)
7. Lee, C., Zame, W., Yoon, I., van der Schaar, M.: DeepHit: a deep learning approach to survival analysis with competing risks. In: Thirty-Second AAAI Conference on Artificial Intelligence (2018)
8. Rava, D., Bradic, J.: DeepHazard: neural network for time-varying risks. ArXiv, abs/2007.13218 (2020)
9. Chen, C., Wong, R.: Black patients miss out on promising cancer drugs-propublica. 2018 (2019)
10. Dwork, C., Hardt, M., Pitassi, T., Reingold, Q., Zemel, R.: Fairness through awareness. In: Proceedings of the 3rd Innovations in Theoretical Computer Science Conference, pp. 214–226 (2012)
11. Kamiran, F., Calders, T.: Data preprocessing techniques for classification without discrimination. Knowl. Inf. Syst. **33**, 1–33 (2012)
12. Barocas, S., Hardt, M., Narayanan, A.: Fairness in machine learning. Knowledge and Information Systems, Nips tutorial **1**, 2 (2017)
13. Esquivel, C.Q., et al.: Transplantation for primary biliary cirrhosis. Gastroenterology 94, 1207–1216 (1988)
14. Graf, E., Schmoor, C., Schumacher, M.: Assessment and comparison of prognostic classification schemes for survival data. Stat. Med. **18**(17–18) (1999)

15. V. Raykar, H. Steck, B. Krishnapuram, C. Dehing-Oberije, and P. Lambin. On ranking in survival analysis: Bounds on the concordance index. Proceedings of the Twenty-First Annual Conference on Neural Information Processing Systems, Vancouver, British Columbia, Canada, December 3–6, 2007, 1209–1216
16. Zhang, W., Weiss, J.: Longitudinal Fairness with Censorship. In: Proceedings of the 36th AAAI Conference on Artificial Intelligence (AAAI), online (2022)
17. Rahman, M., Purushotham, S.: Fair and Interpretable Models for Survival Analysis of the 28th ACM SIGKDD Conference on Knowledge Discovery and Data Mining (KDD 2022), pp. 1452–1462 (2022)

Machine Learning

DAT@Z21: A Comprehensive Multimodal Dataset for Rumor Classification in Microblogs

Abderrazek Azri[1]([✉])[ID], Cécile Favre[1], Nouria Harbi[1], Jérôme Darmont[1][ID], and Camille Noûs[2]

[1] Université de Lyon, Lyon 2, UR ERIC, 5 avenue Pierre Mendès France, F69676 Bron Cedex, France
{a.azri,cecile.favre,nouria.harbi,jerome.darmont}@univ-lyon2.fr
[2] Université de Lyon, Lyon 2, Laboratoire Cogitamus, Lyon, France
camille.nous@cogitamus.fr

Abstract. Microblogs have become popular media platforms for reporting and propagating news. However, they also enable the proliferation of misleading information that can cause serious damage. Thus, many efforts have been taken to defeat rumors automatically. While several innovative solutions for rumor detection and classification have been developed, the lack of comprehensive and labeled datasets remains a major limitation. Existing datasets are scarce and none of them provide all of the features that have proven to be effective for rumor analysis. To mitigate this problem, we propose a big data-sized dataset called DAT@Z21, which provides news contents with rich features including textual contents, social context, social engagement of users and spatiotemporal information. Furthermore, DAT@Z21 also provides visual contents, i.e., images, which play a crucial role in the news diffusion process. We conduct exploratory analyses to understand our dataset's characteristics and analyze useful patterns. We also experiment various state-of-the-art rumor classification methods to illustrate DAT@Z21's usefulness, especially its visual components. Eventually, DAT@Z21 is available online at https://git.msh-lse.fr/eric/dataz21.

Keywords: Social networks · Rumors · Datasets · Multimodal learning

1 Introduction

The explosive growth of social media platforms has led to the generation and proliferation of a large amount of data that reaches a wide audience. The openness and unrestricted method of sharing information on microblogging platforms fosters information spread regardless of its credibility. Misinformation, which is also known as rumors, may cause severe damages in the real-world. For example, during the US presidential elections, it was estimated that over 1 million tweets were related to "Pizzagate," a piece of fake news from Reddit, which led to a real shooting[1].

---

[1] https://en.wikipedia.org/wiki/Pizzagate_conspiracy_theory.

R. Wrembel et al. (Eds.): DaWaK 2023, LNCS 14148, pp. 161–175, 2023.
https://doi.org/10.1007/978-3-031-39831-5_16

To mitigate this problem, automatic rumor detection on social media has recently started to attract considerable attention. Many innovative and significant solutions have been proposed in the literature to defeat rumors (see a comprehensive survey in [2,13]). However, these techniques remain dependent on the availability of comprehensive and community-driven rumor datasets, which have become a major road block.

Most of the publicly available datasets contain only news contents or linguistic features [12,14,20]. Apart from their individual limitations, the common drawback of those datasets is the lack of social context and information other than the text of news articles. In addition to news contents, a few datasets also contain social context, such as user comments and reposts on social media platforms [11,16,19]. Although these engagement-driven datasets are valuable for fake news detection, they have mostly not covered any user profiles, except FakeNewsNet [16].

Detecting rumors on social media is a very challenging task because fake news pieces are intentionally written to mislead consumers, which makes it difficult to spot fake news from news content only. Thus, we need to explore information in addition to news content, such as the social engagements and social behavior of users on social media [16].

The existing datasets cannot address this challenges. Thus, in this study we collect a comprehensive and large-scale repository with multidimensional information, such as textual and visual content, spatiotemporal information, social engagements, and the behaviors of users. The main contributions of this work are summarized as follows:

- We aim to construct and release a comprehensive rumor dataset. DAT@Z21 provides multidimensional information on news articles and social media posts, while our dataset includes rich features such as textual, visual, spatiotemporal, and network information. We provide the methodological details on how it is built.
- We perform various exploratory analyses (data statistics and distributions) on news articles and on the social diffusion (tweets) of the dataset to understand their key properties and characteristics.
- We conduct extensive experiments on the rumor classification task using the DAT@Z21 data and several baselines to validate its usefulness, especially its visual components. Baselines are obtained using either single-modal or multimodal information of tweets, and we utilize either traditional machine learning algorithms or deep learning models.

The rest of this paper is organized as follows. We first survey related works in Sect. 2. We detail the data collection process, and we present exploratory statistics, analysis and data distribution in Sect. 3. Experiments using DAT@Z21 dataset to classify rumor credibility are designed and conducted in Sect. 4. We conclude this paper and shed some light on future work in Sect. 5.

2   Related Works

In this section, we briefly review the related datasets. These datasets can be grouped into those that were recently designed for fake health news related to the pandemic COVID-19, and datasets for fake news and rumors in general.

2.1   Fake Health News Datasets

With the emergence of the COVID-19 pandemic, political and medical rumors have increased to a level where they have created what is commonly referred to as the global infodemic. In a short amount of time, many COVID-19 datasets have been released. Most of these datasets either have no annotations at all, they employ automated annotations using transfer learning or semi-supervised methods, or are not specifically designed for COVID-19 misinformation.

The earliest and most noteworthy dataset depicting the COVID-19 pandemic at a global scale was contributed by the John Hopkins University [5]. The authors developed an online real-time interactive dashboard[2], which was first made public in January 2020. This dashboard lists the daily number of positive cases, the number of cured patients, and the mortality rates at a country and state/province level.

In terms of datasets collected for COVID-19 infodemic analysis and detection, examples include CoAID [3], which contains automatic annotations for tweets, replies, and claims for fake news; ReCOVery [21], which is a multimodal dataset annotated for tweets sharing reliable versus unreliable news, annotated via distant supervision; FakeCovid [15], which is a multilingual cross-domain fake news detection dataset with manual annotations; and [4], which is a large-scale Twitter dataset that is also focused on fake news. A survey of the different COVID-19 datasets can be found in [9,18].

2.2   Fake News Datasets

A high-quality dataset plays an extremely important role in the task of rumor classification. However, the lack of labeled fake news datasets is a major bottleneck when building an effective detection system for online misleading information. Generally, news data with ground truth are gathered by either expert journalists, fact-checking websites or crowdsourcing. Although, there is no consensual benchmark for fake news detection [17], some publicly available resources are worth mentioning. Most of them only contain news contents, and more particularly textual content information.

The BuzzFeedNews[3] dataset contains URLs from posts produced by nine verified Facebook publishers over a week close to the 2016 U.S. election from September 19 to 23 and September 26 and 27. Each post and linked article is manually

---

[2] https://coronavirus.jhu.edu/map.html.
[3] https://www.buzzfeed.com/.

fact-checked and annotated for veracity by BuzzFeed journalists. It contains 1,627 articles-826 mainstream, 356 left-wing, and 545 right-wing articles.

LIAR[4] [20] dataset is collected from fact-checking website PolitiFact[5]. It has 12.8K human labeled short statements collected from PolitiFact and the statements are labeled into six categories ranging from completely false to completely true as pants on fire, false, barely true, half-true, mostly true, and true. It is useful to find misinformation in short statements, but cannot be applied for complete news articles.

The BS Detector[6] is a web crawler with knowledge about fake news websites, which was used to build a dataset by monitoring these websites. Labels are output by the BS detector not by human annotators.

NELA-GT-2018 [12] provides 714k news articles in general topics from 194 news producers. The labels are obtained from eight assessment sites. Among the 194 news sources, 40 of them are not found any labels from the assessments sites.

FA-KES [14] is a fake news dataset that focuses on the Syrian war. It contains 804 articles that labeled as real or fake. However, the labels were annotated based on a database from the Syrian Violations Documentation Center with a cluster algorithm, so the reliability of these labels may be of concerns.

Apart from their own limitations, the common drawback of the data listed here is the lack of social context and information other than the text of news articles. Besides the news content, other researchers have also collected user engagement features of the news, such as user engagement on online social media.

FakeNewsNet[7] [16] is collected from fact-checking websites PolitiFact and GossipCop. News contents and the related ground truth labels are crawled from these two websites. The authors collected social engagement through Twitter's Advanced Search API. This dataset contains 1,056 articles from PolitiFact and 22,864 articles from GossipCop. Each article is labeled as fake or true. Overall, the dataset nearly contains two million tweets .

CREDBANK[8] [11] contains about 1000 news events whose credibility are labeled by 30 annotators who are sourced from Amazon mechanical Turk. This dataset contains 60 million tweets that were posted between 2015 and 2016.

Although these datasets are valuable for fake news detection, they mostly have not cover any user profiles except FakeNewsNet. Moreover, only one dataset includes visual data.

The existing datasets often only focus on the textual content of news articles, and little attention has been paid to visual content of both news articles and user engagement of news on social media. However, fake images on media posts can easily go viral on social platforms, and cause serious social disruptions.

Unlike this category of datasets, we propose in this paper a new dataset: DAT@Z21. It not only includes both true and fake news articles but also their

---

[4] https://www.cs.ucsb.edu/william/data/liardataset.zip.
[5] https://www.politifact.com/factchecks/.
[6] https://www.kaggle.com/mrisdal/fake-news.
[7] https://github.com/KaiDMML/FakeNewsNet.
[8] http://compsocial.github.io/CREDBANK-data/.

diffusion on social media. It contains all possible features of interests, including multi-modal information (textual, visual and spatiotemporal information), and user engagement data (user profile, tweets, retweets, replies, likes) as shown in Table 1 that compares our new dataset to existing ones. In addition, DAT@Z21 can be updated automatically to fetch the latest news article and social media content.

Table 1. Comparison with Existing Fake News Datasets

Features Datasets	News Content		User Engagement					Spatiotemporal data	
	textual	Visual	User profile	Tweet	Retweet	Reply	Attached image	Spatial	Temporal
BuzzFeedNews	✓	—	—	—	—	—	—	—	—
LIAR	✓	—	—	—	—	—	—	—	—
BS Detector	✓	—	—	—	—	—	—	—	—
NELA-GT-2018	✓	—	—	—	—	—	—	—	—
FA-KES	✓	—	—	—	—	—	—	—	—
FakeNewsNet	✓	✓	✓	✓	✓	—	—	✓	✓
CREDBANK	✓	—	✓	✓	—	—	—	✓	✓
DAT@Z21	✓	✓	✓	✓	✓	✓	✓	✓	✓

3   Data Collection

In this section, we present the process that we followed to build DAT@Z21. We show how we collected data with a valuable ground truth. We will also show how we collected news articles and messages that were posted on social media. The overall data collection process can mainly be divided into three stages, as shown in Fig. 1: (1) collecting news articles with ground truth labels; (2) preparing tweets collection; and (3) collecting content generated on social media. The following subsections will describe these stages in more detail.

3.1   News Articles and Ground Truth Collection

We start by collecting as of August 30, 2021 all verified rumor news articles, with reliable ground truth labels from the fact-checking website *PolitiFact*[9]. Politi-Fact is a well-known fact-checking and rumor debunking website that contains many types of reports. For each checked rumor, professional analysts provide a conclusion of the rumor followed by a full description, source, origin, supporting/opposing arguments of the rumor story, as well as the truth label. Each statement is divided into six categories of veracity: true, mostly true, half-true, mostly false and "pants on fire" for outrageously false statements.

We use the source URLs of web pages that publish the news articles, provided by PolitiFact professionals, to crawl the related news contents. After identifying the web pages, we access and retrieve them using Politifact API. We then pull

---
[9] www.PolitiFact.com/factchecks/.

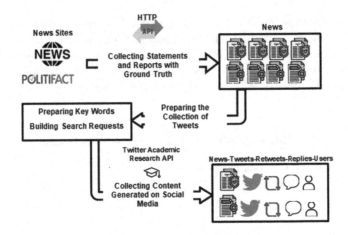

Fig. 1. Data Collection pipeline for DAT@Z21

data out from the content of these pages using *Beautiful Soup Python library*[10]. To guarantee a high quality ground truth, and reduce the number of false positive and false negative in the news labels of our dataset, we select only news articles that are explicitly tagged as true, false, or grossly false. We are particularly interested in the information related to each news article, which are:

1. Article ID: We assign a unique identifier to each news article;
2. Article Title: The title of the news article, which might well summarize or give clues about the content of the article;
3. Article Author: The author(s) of the news article;
4. Publication Date: The date of publication of the article;
5. Article URL: The URL to the news article;
6. Article Body Text: The textual content of the news article;
7. Article Image: The URL to the image(s) attached to the article;
8. Article Label: The original ground truth of the article credibility, either as True or Fake.

Figures 2 and 3 show some of the textual features of news content generated from both the title and the body text. From Fig. 2, we can note that the number of words in articles follows a scaling long-tail distribution, with a mean value of $\simeq 2,513$ and an average of $\simeq 2,418$. Furthermore, Fig. 3 shows the word cloud of all articles. Because the news articles are statements, reports and claims, which could be potentially a fake or real rumors, they naturally frequently share some vocabularies, such as "said" ($\#= 11,131$), "state" ($\#= 10,112$), "stated" ($\#= 9,373$), and "according" ($\#= 9,208$).

We can also capture the source of the rumor from the titles of the news articles. In PolitiFact, this source is usually a statement from a politician but it could also be a message posted on a social media or a blog. Figure 4 illustrates

---

[10] https://pypi.org/project/beautifulsoup4/.

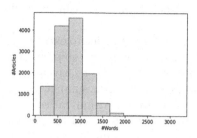

Fig. 2. Word Count Distribution

Fig. 3. Word Cloud

the top 10 sources for both real and fake news articles. With respect to real news, we can notice that "Facebook posts", "Viral images" and "Blog posting" are the three main sources of fake news articles.

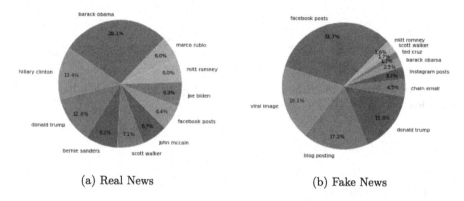

(a) Real News

(b) Fake News

Fig. 4. Top 10 News Sources

3.2 Preparing the Tweets Collection

In this stage, we track the diffusion of the news articles on the Twitter microblog, by collecting tweets of users that discuss or interact with these events. Because we aim to build a multimedia dataset with images, we only keep news articles with images attached.

We start by constructing two sets of the most representative keywords from (1) the title and (2) the body text of each article. This way of proceeding is motivated, on the one hand, by the fact that the titles used by the journalists are not always significant and, on the other hand, by the intention to capture all the keywords of the news articles appearing either in the title or in the body of the text.

For the title, we develop a natural language processing pipeline to extract key words by removing special tokens and common stop words. In this step, we kept at most six key words to avoid over general queries. Given the relatively large size of the body text, we first use the TextRank extractive text summarization algorithm [10]. The advantage of using an extractive algorithm rather than an abstractive one is to keep the original keywords of the text and the general meaning. We then apply our natural language processing pipeline to clean up the text. Note that we keep the named entities in both sets of keywords, which plays an important role for the semantics of the search queries. For example, Fig. 6 shows the keywords extracted from the title and the body text of a news article.

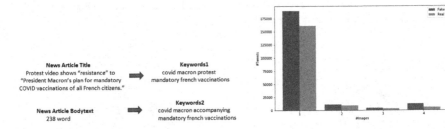

Fig. 5. An Example of Keyword Extraction

Fig. 6. Image Distribution

3.3   Tweets Collection

We have prepared two keyword sets that serve as queries to search for matching tweets. We next use the *Twitter Academic research API*[11] with the Full-archive search endpoint to collect the complete history of public tweets from March 2006 (Twitter creation) until September 07, 2021 (for the current version of the dataset). With a 1024 character limit on the query rules, this endpoint also supports a more advanced query language (Boolean) to help yield more precise, complete, and unbiased filtered results. Thanks to these features, we were able to create a single search query that combines the two keyword lists that were prepared earlier for each news article.

The Full-archive search endpoint returns matching tweets with complete information such as their IDs, text, date and times of being created. It also enables us to pull valuable details about users (e.g., their IDs) and their social activity (e.g.,

---

[11] https://developer.twitter.com/en/products/twitter-api/academic-research/product-details.

the number of retweets/followers/friends) and media content (e.g., images, places, and poll metadata) using the new expansion parameters. We also consider that all the tweets that discuss a particular news article will bear the truth value (i.e., the label of the article) because it contributes to the diffusion of a rumor (true or fake), even if the tweet denies or remains skeptical regarding the veracity of the rumor. The statistical details of DAT@Z21 are shown in Table 2.

Table 2. Dataset Statistics

Characteristics	Real	Fake	Overall
All News articles	5,671	7,213	12,884
#With images	1,765	2,391	4,156
All Tweets	1,209,144	1,655,386	2,864,530
#With images	179,153	216,472	395,625
#Images	211,447	271,419	482,866
#Tweets as retweets	4,734	5,502	10,236
#Tweets as replies	37,593	55,937	93,530
#Tweets as likes	2,803	4,453	7,256
#Tweets with location	3,823	4,497	8,320
#Users	98,967	115,669	194,692
Average #Followers	107,786	286,101	205,353
Average #Followees	2,929	3,153	3,051

Unlike most existing datasets that are focused on only text content, we aim to build a multimedia dataset that includes images. We collect the original tweet texts and attached images. Thus, from the 2,864,530 collected tweets, we remove text-only tweets, duplicated tweets, and images to obtain 395,625 tweets with attached images. This represents a significant part ($\simeq 14\%$). In addition, among these tweets, almost 10% share more than one image, which shows that visual information is widely used on social media. Figure 6 shows the distribution of true and fake tweets with the number of included images.

For the textual content of our dataset, we analyze the topic distribution of fake and real tweets. From Fig. 7, we can observe that the topic distribution of fake tweets is slightly different from that of the real tweets. Indeed, both distributions share most of the frequent words related to political events, such as "Donald", "trump", "Hillary", "clinton", and "election" or to the Covid-19 health crisis, such as "health", "care", "Coronavirus", and "death". The accordance of real and fake news topic distributions ensures that the fake news detection models trained on our dataset are not topic classifiers.

For users involved in spreading tweets, we find that the number of users spreading fake and true news is greater than the total number of distinct users in the dataset. This indicates that a user can engage in spreading both fake and real news. As for the users social interaction, the distribution of their followers

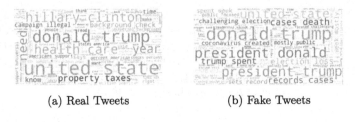

(a) Real Tweets   (b) Fake Tweets

Fig. 7. Topic Distribution of Fake and Real Tweets

and followees count are presented in Fig. 8. We can observe that the followees and followers count of users spreading fake and real tweets follows a distribution that resembles a power law, which is commonly observed is social network structures.

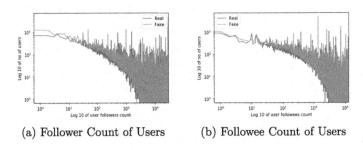

(a) Follower Count of Users   (b) Followee Count of Users

Fig. 8. Followers and Followees Count Distribution of Fake and Real Tweets

To understand how users interact with each other, we have created an interaction graph using the Python package Networkx [6] that represent three types of interactions between two users: retweets, replies, and quoted tweets. Figure 9 shows the largest connected component for our original graph for users spreading real and fake tweets. We note that both users display echo-chamberness. In addition, users posting fake tweets tend to create tightly knit groups characterized by relatively highly-dense ties.

Our dataset includes spatiotemporal information that depicts the temporal user engagement for news articles, which provides the necessary information to further investigate the utility of using spatiotemporal information to detect fake news. Figure 10 depicts the geolocation distribution of users engaging in posting tweets. Given that it is often rare for the location to be explicitly provided by the users in their profiles, we consider the locations information attached with tweets.

Note that to comply with Twitter's Terms of Service[12], the full contents of user social engagements and network are not able to be directly published. Instead, we only publicly release the collected tweet IDs, so that the tweets can be

---

[12] https://developer.twitter.com/en/developer-terms/agreement-and-policy.

(a) Real Tweets   (b) Fake Tweets

Fig. 9. Social Connections Graph for Real and Fake Tweets

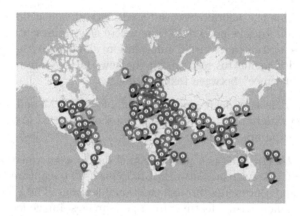

Fig. 10. Spatial Distribution of Users posting Fake and Real Tweets

rehydrated for non-commercial research use, but we also provide the instructions for obtaining the tweets using the released IDs for user convenience. More details can be seen in https://git.msh-lse.fr/eric/dataz21. We also maintain and update annually the repository to ensure its usability.

4   Rumor Classification Using DAT@Z21

In this section, we conduct comparative experiments on the rumor classification task using the DAT@Z21 dataset. These experiments aim to show the usefulness of our dataset with different kinds of methods. We include both monomodal methods and multimodal state-of-the-art methods as baselines to predict the credibility of tweets. We specify these methods in Sect. 4.1. We then provide the experiments settings in Sect. 4.2. Finally, we present the performance analysis for these methods in Sect. 4.3.

4.1   Baselines

For monomodal methods, we consider the following algorithms:

NB (Naive Bayes). We use the doc2vec embedding model to represent texts and feed the representations to a Gaussian Naive Bayes algorithm.

LR (Logistic Regression). We use the word2vec embedding model to represent texts and feed the representations to the model.

RF (Random Forest). We use the Term Frequency-Inverse Document Frequency (TF-IDF) to represent texts and feed them to the model.

Text-CNN [8]. This is a convolutional neural network for text classification, we feed the word embedding into a Convolution1D which will learn filters. We then add a max pooling layer.

For multimodal methods, we consider the following state of the art baselines:

SAFE [22][13]. This is a neural-network-based method that utilizes news multimodal information for fake news detection, where news representation is learned jointly by news textual and visual information along with their relationship (similarity). SAFE facilitates recognizing the news falseness in its text, images, and/or the "irrelevance" between text and images.

att-RNN [7]. This is a deep model that employs LSTM and VGG-19 with attention mechanism to fuse textual, visual and social-context features of tweets. We set the hyper-parameters as in [7].

deepMONITOR [1]. This is a multi-channel deep model where first a Long-term Recurrent Convolutional Network (LRCN) is used to capture and represent text semantics and sentiments through emotional lexicons. The VGG19 model is used to extract salient visual features from post images. Image features are then fused with the joint representations of text and sentiment to classify messages.

4.2   Experiment Settings

We randomly divided the dataset into training and testing subsets, with a ratio of 0.7:0.3. We use the scikit-learn library from Python to implement the traditional algorithms with default settings and we do not tune parameters. We use the Keras library from Python to implement multimodal methods. We use TF-IDF with 1000 features, word2vec and doc2vec embeddings with 200 vector size to represent texts. Because the dataset has a slightly unbalanced distribution between false and true classes ($\simeq 1.17 : 1$), we use the following metrics to evaluate the performance of rumor classification algorithms: PR AUC (Area Under Precision-Recall Curve), ROC AUC(Area Under Receiver Operating Characteristic Curve), precision, recall, and F_1-score. Finally, we run each method three times and report the average score in Table 3.

---

[13] https://github.com/Jindi0/SAFE.

Table 3. Classification Performance Using the DAT@Z21 Dataset

Type	Methods	PR AUC	ROC AUC	Precision	Recall	F_1
Monomodal methods	LR	0.778	0.799	0.729	0.777	0.752
	NB	0.496	0.561	0.593	0.480	0.531
	RF	0.843	0.846	0.754	0.783	0.764
	Text-CNN	0.861	0.866	0.759	0.774	0.771
Multimodal methods	att-RNN	0.880	0.871	0.805	0.753	0.777
	SAFE	0.937	0.935	0.851	0.857	0.854
	deepMONITOR	0.984	0.982	0.901	0.948	0.924

4.3 Experimental Results

From Table 3, we can see that among the four simple models, the Text-CNN is the best performing model and it achieved the best scores for four out of five comparison metrics, including PR AUC, ROC AUC, precision, and F_1-score. For multimodal techniques, deepMONITOR outperforms other models, with the best scores in all comparison metrics. Multimodal methods generally perform better than monomodal methods because they incorporate signals from other message modalities then the text content like visual, sentiment or social context information, which enables them to better capture contextual information.

Because the dataset is slightly unbalanced in favor of fake labels, the models tend to generate slightly more false positive cases. However, the PR AUC (which is an appropriate metric for unbalanced datasets), recall and F_1-score values are satisfactory. Figure 11 illustrates the PR curve for the monomodal and multimodal methods. Given that the proportion of true and fake information is likely to be even more unbalanced in real world datasets, practical detection solutions need to handle this type of imbalance more effectively.

With these experiments, we show that the DAT@Z21 dataset allows to explore various methods dealing with different types of features, by comparing their efficiency. According to the obtained results, we see that the visual content is very important since these methods obtain better results. Thus, a dataset containing all these features was required.

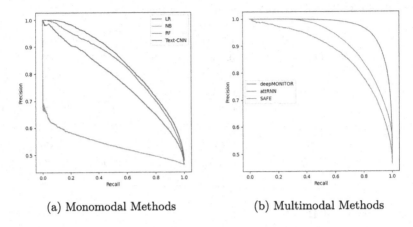

(a) Monomodal Methods   (b) Multimodal Methods

Fig. 11. PR curve for Monomodal and Multimodal Methods

5 Conclusion and Perspectives

In this work, we have constructed a comprehensive multimodal dataset DAT@Z21 for rumor classification, which contains textual, visual, spatiotemporal, user engagement, and social platform posts. We have described how we collected the dataset. To show the distinctive features between rumors and facts, we have provided a brief exploratory data analysis of news articles and the corresponding messages from the Twitter social platform. In addition, we have demonstrated the usefulness and the relevance of our dataset through a rumor classification task over several monomodal and multimodal state-of-the-art methods using the data of DAT@Z21.

The DAT@Z21 dataset can be extended by including (1) news articles and tweets from other languages than English, and (2) other reliable news sources(e.g., other fact-checking or organizations and platforms that assess the reliability and bias of news sources) to create a large, centralized set of ground truth labels. We hope that researchers will find DAT@Z21 useful for their own research and that together we contribute to combat rumors.

References

1. Azri, A., Favre, C., Harbi, N., Darmont, J., Noûs, C.: Calling to CNN-LSTM for rumor detection: a deep multi-channel model for message veracity classification in microblogs. In: Dong, Y., Kourtellis, N., Hammer, B., Lozano, J.A. (eds.) ECML PKDD 2021. LNCS (LNAI), vol. 12979, pp. 497–513. Springer, Cham (2021). https://doi.org/10.1007/978-3-030-86517-7_31
2. Cao, J., Guo, J., Li, X., Jin, Z., Guo, H., Li, J.: Automatic rumor detection on microblogs: a survey. arXiv preprint arXiv:1807.03505 (2018)
3. Cui, L., Lee, D.: CoAID: COVID-19 healthcare misinformation dataset. arXiv preprint arXiv:2006.00885 (2020)

4. Dai, E., Sun, Y., Wang, S.: Ginger cannot cure cancer: battling fake health news with a comprehensive data repository. In: Proceedings of the AAAI ICWSM, vol. 14, pp. 853–862 (2020)
5. Dong, E., Du, H., Gardner, L.: An interactive web-based dashboard to track COVID-19 in real time. Lancet. Infect. Dis **20**(5), 533–534 (2020)
6. Hagberg, A., Swart, P., S Chult, D.: Exploring network structure, dynamics, and function using networkX. Tech. rep., Los Alamos National Lab. (LANL), Los Alamos, NM (United States) (2008)
7. Jin, Z., Cao, J., Guo, H., Zhang, Y., Luo, J.: Multimodal fusion with recurrent neural networks for rumor detection on microblogs. In: ICM 2017, pp. 795–816. ACM (2017)
8. Kim, Y.: Convolutional neural networks for sentence classification. In: Proceedings of EMNLP, pp. 1746–1751. ACL, Doha, Qatar (2014)
9. Latif, S., et al.: Leveraging data science to combat COVID-19: a comprehensive review. IEEE Trans. AI **1**(1), 85–103 (2020)
10. Mihalcea, R., Tarau, P.: TextRank: bringing order into text. In: Proceedings of the 2004 Conference on EMNLP, pp. 404–411 (2004)
11. Mitra, T., Gilbert, E.: CREDBANK: a large-scale social media corpus with associated credibility annotations. In: Ninth AAAI ICWSM (2015)
12. Nørregaard, J., Horne, B.D., Adalı, S.: NELA-GT-2018: a large multi-labelled news dataset for the study of misinformation in news articles. In: Proceedings of the AAAI ICWSM, vol. 13, pp. 630–638 (2019)
13. Pathak, A.R., Mahajan, A., Singh, K., Patil, A., Nair, A.: Analysis of techniques for rumor detection in social media. Procedia Comput. Sci. **167**, 2286–2296 (2020)
14. Salem, F.K.A., Al Feel, R., Elbassuoni, S., Jaber, M., Farah, M.: Fa-kes: A fake news dataset around the Syrian war. In: Proceedings of the AAAI ICWSM, vol. 13, pp. 573–582 (2019)
15. Shahi, G.K., Nandini, D.: FakeCovid-a multilingual cross-domain fact check news dataset for COVID-19. arXiv preprint arXiv:2006.11343 (2020)
16. Shu, K., Mahudeswaran, D., Wang, S., Lee, D., Liu, H.: FakeNewsNet: a data repository with news content, social context and spatialtemporal information for studying fake news on social media. arXiv preprint arXiv:1809.01286 (2018)
17. Shu, K., Sliva, A., Wang, S., Tang, J., Liu, H.: Fake news detection on social media: a data mining perspective. ACM SIGKDD Explor. Newsl. **19**(1), 22–36 (2017)
18. Shuja, J., Alanazi, E., Alasmary, W., Alashaikh, A.: COVID-19 open source data sets: a comprehensive survey. Appl. Intell. **51**, 1–30 (2020)
19. Tacchini, E., Ballarin, G., Della Vedova, M.L., Moret, S., de Alfaro, L.: Some like it hoax: automated fake news detection in social networks. arXiv preprint arXiv:1704.07506 (2017)
20. Wang, W.Y.: "Liar, Liar Pants on Fire": A new benchmark dataset for fake news detection. arXiv preprint arXiv:1705.00648 (2017)
21. Zhou, X., Mulay, A., Ferrara, E., Zafarani, R.: Recovery: A multimodal repository for COVID-19 news credibility research. In: Proceedings of the 29th ACM CIKM, pp. 3205–3212 (2020)
22. Zhou, X., Wu, J., Zafarani, R.: SAFE: similarity-aware multi-modal fake news detection. In: Lauw, H.W., Wong, R.C.-W., Ntoulas, A., Lim, E.-P., Ng, S.-K., Pan, S.J. (eds.) PAKDD 2020. LNCS (LNAI), vol. 12085, pp. 354–367. Springer, Cham (2020). https://doi.org/10.1007/978-3-030-47436-2_27

Dealing with Data Bias in Classification: Can Generated Data Ensure Representation and Fairness?

Manh Khoi Duong[✉][iD] and Stefan Conrad[iD]

Heinrich Heine University, Universitätsstraße 1, 40225 Düsseldorf, Germany
{manh.khoi.duong,stefan.conrad}@hhu.de

Abstract. Fairness is a critical consideration in data analytics and knowledge discovery because biased data can perpetuate inequalities through further pipelines. In this paper, we propose a novel pre-processing method to address fairness issues in classification tasks by adding synthetic data points for more representativeness. Our approach utilizes a statistical model to generate new data points, which are evaluated for fairness using discrimination measures. These measures aim to quantify the disparities between demographic groups that may be induced by the bias in data. Our experimental results demonstrate that the proposed method effectively reduces bias for several machine learning classifiers without compromising prediction performance. Moreover, our method outperforms existing pre-processing methods on multiple datasets by Pareto-dominating them in terms of performance and fairness. Our findings suggest that our method can be a valuable tool for data analysts and knowledge discovery practitioners who seek to yield for fair, diverse, and representative data.

Keywords: fairness · bias · synthetic data · fairness-agnostic · machine learning · optimization

1 Introduction

Data analytics has grown in popularity due to its ability to automate decision-making through machine learning. However, real-world data can contain biases that produce unfair outcomes, making fairness in data pipelines involving machine learning a pressing concern. Fairness in machine learning typically deals with intervening algorithms providing equitable outcomes regardless of protected characteristics such as gender, race, or age group.

The existing related works can be divided into three categories [5,8,20]. The first category of methods are pre-processing methods, which aim to reduce bias in the data. Examples of such methods include data augmentation and data

This work was supported by the Federal Ministry of Education and Research (BMBF) under Grand No. 16DHB4020.

balancing [2]. The second category of methods are in-processing methods, which aim to enforce fairness constraints during the training procedure [15]. Examples of in-processing methods include regularization techniques and constrained optimization [31]. The last category are post-processing methods that allow the improvement of fairness after training by correcting the outputs of the trained model [14].

The goal of this paper is to introduce a pre-processing method that achieves fairness by including generated data points. This is done by utilizing a statistical model that learns the distribution of the dataset, enabling the generation of synthetic samples. Additionally, a discrimination measure is employed to evaluate the fairness when incorporating the generated data points. Our method treats the discrimination measure as a black-box, making it able to optimize any discrimination measure defined by the user. We refer to this property of our algorithm as *fairness-agnostic*. This makes it suitable for cases where a specific fairness notion is required.

For the experimentation, multiple datasets known to be discriminatory were used. The experiments were performed by firstly loading the datasets and then pre-processing them using different pre-processing techniques. The pre-processed datasets were then fed into several classifiers. The performance of each classifier was then evaluated in terms of performance and fairness to assess the effectiveness of the pre-processing methods. Our experiments have empirically shown that our technique effectively lessens discrimination without sacrificing the classifiers' prediction qualities. Moreover, it is compatible with any machine learning model. Of the pre-processors tested, none were able to meet all of these conditions. The scope and application of our method is not necessarily limited to tabular data and classification tasks, even though experiments were conducted on them. The method is more broadly suitable for supervised learning tasks where the data, label, and protected attribute are available. Only the appropriate discrimination measures have to be derived for the right task. Generally, our primary contributions are:

- The introduction of a novel pre-processing technique that can optimize any given fairness metric by pre-selecting generated data points to include into the new fair dataset.
- We carry out a comprehensive empirical study, comparing our method against three widely recognized pre-processors [9,13,31], using multiple datasets commonly found in fairness literature.
- We present interesting and valuable properties, such as the empirical evidence that our method consistently improved fairness in comparison to the unprocessed data.

2   Related Work

Many pre-processing algorithms in literature alter the dataset to achieve fairness [4,9,31]. Because the methods simply return a fair dataset, they can be used with any estimator. However, such approaches cannot be used with ease: They often require a parameter setting that sets how aggressive the change should be. As the approaches differ in their methodology, it is hard to interpret the parameter's setting and their unexpected effects on the data. Data alteration methods also have a higher risk of producing data that do not resemble the original data distribution in any ways.

Other approaches return a weight for each sample in the dataset that the estimator should account for when fitting the data [1,13]. While the approaches seem promising [1,13], they require estimators to be able to handle sample weights. A way to account for this is to replicate samples based on their sample weights. However, this is not computationally scalable for larger datasets or for larger differences between the sample weights.

Another related approach is removing data samples that influence estimators in a discriminatory way [28]. Nevertheless, this approach does not seem feasible for smaller datasets.

Differently from related works, we present an algorithm that does not come with the above mentioned drawbacks. Further, our approach is able to satisfy any fairness notion that is defined for measuring discrimination or bias in the dataset. While the work of Agarwal et al. [1] also features this property, the fairness definitions must be formalizable by linear inequalities on conditional moments. In contrast, our work requires the fairness definitions to quantify discrimination in a numeric scale where lower values indicate less discrimination. This can be as simple as calculating the differences of probabilistic outcomes between groups.

While there exist works that train fair generative models to produce data that is fair towards the protected attribute on images [7,24,27] or tabular data [12,23], our approach can be seen as a framework that employs generative models and can therefore be used for any data where the protected attribute is accessible. Specifically, our research question is not *"How can fair generative models be constructed?"*, we instead deal with the question *"Using any statistical or generative model that learns the distribution of the dataset, how can the samples drawn from the distribution be selected and then included in the dataset such that fairness can be guaranteed?"*. Other works that generate data for fairness include generating counterfactuals [26] and generating pseudo-labels for unlabeled data [6].

3   Measuring Discrimination

In this section, we briefly present *discrimination measures* that assess the fairness of data. For that, we make use of following notation [5,8,20]: A *data point* or *sample* is represented as a triple (x, y, z), where $x \in X$ is the *feature*, $y \in Y$ is the ground truth *label* indicating favorable or unfavorable outcomes, and $z \in Z$ is the *protected attribute*, which is used to differentiate between groups. The sets X, Y, Z typically hold numeric values and are defined as $X = \mathbb{R}^d$, $Y = \{0, 1\}$, and $Z = \{1, 2, \ldots, k\}$ with $k \geq 2$. For simplicity, we consider the case where protected attributes are binary, i.e., $k = 2$. Following the preceding notation, a *dataset* is defined as the set of data points, i.e., $\mathcal{D} = \{(x_i, y_i, z_i)\}_{i=1}^n$. Machine learning models $\phi : X \times Z \to Y$ are trained using these datasets to predict the target variable $y \in Y$ based on the input variables $x \in X$ and $z \in Z$. We call the output $\hat{y} := \phi(x, z)$ *prediction*.

Based on the work of [32], we derive *discrimination measures* to the needs of the pre-processing method in this paper. To make our algorithm work, a *discrimination measure* must satisfy certain properties which we introduce in the following.

Definition 1. *A discrimination measure is a function $\psi : \mathbb{D} \to \mathbb{R}^+$, where $\mathbb{D}$ is the set of all datasets, satisfying the following axioms:*

1. *The discrimination measure $\psi(\cdot)$ is bounded by [0, 1]. (Normalization)*
2. *Minimal and maximal discrimination are captured with 0, 1 by $\psi(\cdot)$, respectively.*

The first and second axiom together assure that the minimal or maximal discrimination can be assessed by this measure. Furthermore, through normalization it is possible to evaluate the amount of bias present and its proximity to the optimal solution. As achieving no discrimination is not always possible, i.e., $\psi(\mathcal{D}) = 0$, we consider lower discrimination as better and define a fairer dataset as the one with the lower discrimination measure among two datasets.

Literature [2, 5, 8, 19, 20, 32] on fairness-aware machine learning have classified fairness notions to either representing group or individual fairness. We subdivide the most relevant fairness notions into two categories which are *dataset* and *prediction notions* and derive discrimination measures from it as suggested by [32]. From now on, we denote x, y, z as random variables describing the events of observing an individual from a dataset $\mathcal{D}$ taking specific values.

Dataset notions typically demand the independency between two variables. When the protected attribute and the label of a dataset are independent, it is considered fair because it implies that the protected attribute does not influence or determine the label. An example to measure such dependency would be the *normalized mutual information* (NMI) [29] where independency can be concluded if and only if the score is zero. Because it is normalized as suggested by the name, it is a discrimination measure.

Definition 2 (Normalized mutual information). *Let $H(\cdot)$ be the entropy and $I(y; z)$ be the mutual information [25]. The normalized mutual information score is defined in the following [30]:*

$$\psi_{NMI}(\mathcal{D}) = 2 \frac{I(y; z)}{H(y) + H(z)}.$$

Statistical parity [15,31] and *disparate impact* [9] are similar notions that also demand independency, except they are specifically designed for binary variables. Kang et al. [16] proved that zero mutual information is equivalent to statistical parity. To translate statistical parity to a discrimination measurement, we make use of differences similarly to Žliobaitė [32].

Definition 3 (Statistical parity). *Demanding that each group has the same probability of receiving the favorable outcome is statistical parity, i.e.,*

$$p(y = 1 \mid z = 1) = p(y = 1 \mid z = 0).$$

Because we want to minimize discrimination towards any group, we measure the absolute difference between the two groups to assess the extent to which the dataset fulfills statistical parity. This is also known as (absolute) statistical disparity (SDP) [8]. A value of 0 indicates minimal discrimination:

$$\psi_{SDP}(\mathcal{D}) = |p(y = 1 \mid z = 1) - p(y = 1 \mid z = 0)|. \tag{1}$$

Because disparate impact [9] essentially demands the same as statistical parity but contains a fraction, dividing by zero is a potential issue that may arise. Therefore, its use should be disregarded [32]. Note that dataset notions can also be applied to measure the fairness on predictions by exchanging the data label with the prediction label.

Parity-based notions, fulfilling the *separation* or *sufficiency* criterion [2], require both prediction and truth labels to evaluate the fairness. Contrary to the category before, measuring solely on datasets is not possible here. Despite this, it is still essential to evaluate on such measures to account for algorithmic bias. Here, the discrimination measure takes an additional argument, which is the prediction label $\hat{y}$ as a random variable. According fairness notions are, for example, *equality of opportunity* [10], *predictive parity* [2], and *equalized odds* [2].

Definition 4 (Equalized odds). *Equalized odds is defined over the satisfaction of both* equality of opportunity *and* predictive parity *[10],*

$$p(\hat{y} = 1 \mid y = i, z = 1) = p(\hat{y} = 1 \mid y = i, z = 0) \; \forall i \in \{0,1\},$$

where equality of opportunity is the case of $i = 1$ and predictive parity is the case of $i = 0$, correspondingly. Making use of the absolute difference, likewise to SDP (1), we denote the measure of equality of opportunity as $\psi_{EO}(\mathcal{D}, \hat{y})$ and predictive parity as $\psi_{PP}(\mathcal{D}, \hat{y})$.

To turn equalized odds into a discrimination measure, we can calculate the average of the absolute differences for both equality of opportunity and predictive parity. This is referred to as *average odds error* [3]:

$$\psi_{ODDS}(\mathcal{D}, \hat{y}) = \frac{\psi_{EO}(\mathcal{D}, \hat{y}) + \psi_{PP}(\mathcal{D}, \hat{y})}{2}. \tag{2}$$

4   Problem Formulation

Intuitively, the goal is to add an amount of synthetic datapoints to the original data to yield for minimal discrimination. With the right discrimination measure chosen, it can be ensured that the unprivileged group gets more exposure and representation in receiving the favorable outcome. Still, the synthetic data should resemble the distribution of the original data. The problem can be stated formally in the following: Let $\mathcal{D}$ be a dataset with cardinality n, let $\tilde{n}$ be the number of samples to be added to $\mathcal{D}$. The goal is to find a set of data points $S = \{d_1, d_2, \ldots, d_{\tilde{n}}\}$ that can be added to the dataset, i.e., $\mathcal{D} \cup S$ with $S \sim P(\mathcal{D})$, that minimizes the discrimination function $\psi(\mathcal{D} \cup S)$. Hence, we consider the following constrained problem:

$$\begin{aligned} \min \quad & \psi(\mathcal{D} \cup \mathcal{S}) \\ \text{subject to} \quad & \mathcal{S} \sim P(\mathcal{D}) \\ & |\mathcal{S}| = \tilde{n}. \end{aligned} \tag{3}$$

The objective (3) suggests that the samples d_i that are added to the dataset $\mathcal{D}$ are drawn from $P(\mathcal{D})$. To draw from $P(\mathcal{D})$, a statistical or generative model P_G that learns the data distribution can be used. Therefore generating data samples and bias mitigation are treated as sequential tasks where the former can be solved by methods from literature [22]. Because the discrimination measure ψ can be of any form, the optimization objective is treated as a black-box and is solved heuristically.

5   Methodology

Our algorithm relies on a statistical model, specifically the Gaussian copula [22], to learn the distribution of the given dataset $P(\mathcal{D})$. Gaussian copula captures the relationship between variables using Gaussian distributions. While assuming a Gaussian relationship, the individual distributions of the variables can be any continuous distribution, providing flexibility in modeling the data.

Still, the type of model for this task can be set by the user as long as it can sample from $P(\mathcal{D})$. Because discrimination functions are treated as black-boxes, the algorithm does not require the derivatives of ψ and optimizing for it leads to our desired *fairness-agnostic* property: It is suitable for any fairness notion that can be expressed as a discrimination function. Our method handles the size constraint in Eq. (3) as an upper bound constraint, where a maximum of $\tilde{n}$ samples are added to $\mathcal{D}$.

Our method, outlined in Algorithm 1, begins by initializing $\hat{\mathcal{D}}$ with the biased dataset $\mathcal{D}$. Then $\hat{n}$ is set as a multiplicative $r > 1$ of the original dataset's size. Lastly in the initialization, the distribution of $P(\mathcal{D})$ is learned by a generative model P_G. The algorithm then draws m samples from the generative model P_G which are referred to as the set of candidates C. The next step is decisive for the optimization (Line 9): The candidate which minimizes the discrimination most when included in the dataset $\hat{\mathcal{D}}$ is added to $\hat{\mathcal{D}}$. The steps of drawing samples and adding the best candidate to the dataset is repeated till $\hat{\mathcal{D}}$ has a cardinality of $\hat{n}$ or the discrimination is less than the fairness threshold ϵ. Because ϵ is set to 0 by default, the algorithm can stop earlier before the dataset reaches its requested size if the discrimination cannot be further reduced, i.e., $\psi(\hat{\mathcal{D}}) = 0$. Because calculating $\psi(\hat{\mathcal{D}} \cup \{c\})$ (Line 9) does not involve retraining any classifier and solely

---

Algorithm 1. Pseudocode of MetricOptGenerator

Input: $\mathcal{D}, r = 1.25, m = 5, \epsilon = 0$
Output: $\hat{\mathcal{D}}$

   Initialization:
1: $\hat{\mathcal{D}} \leftarrow \mathcal{D}$
2: $\hat{n} \leftarrow \lfloor r \cdot |\mathcal{D}| \rfloor$
3: $P_G \leftarrow$ learn distribution of $P(\mathcal{D})$
   Generating fair samples:
4: **for** $i = 1$ to $\hat{n} - |D|$ **do**
5:    **if** $(\psi(\hat{\mathcal{D}}) \leq \epsilon)$ **then**
6:       **return** $\hat{\mathcal{D}}$
7:    **end if**
8:    $C \leftarrow$ sample m candidates from P_G
9:    $\hat{\mathcal{D}} \leftarrow \hat{\mathcal{D}} \cup \{\text{argmin}_{c \in C}\ \psi(\hat{\mathcal{D}} \cup \{c\})\}$
10: **end for**
11: **return** $\hat{\mathcal{D}}$

---

evaluates the dataset, this step is practically very fast. In our implementation, we generate a set of synthetic data points prior to the for-loop, eliminating the sampling cost during the optimization step. We refer to Appendix A for the proof outlining the polynomial time complexity of the presented method.

6 Evaluation

To evaluate the effectiveness of the presented method against other pre-processors in ensuring fairness in the data used to train machine learning models, we aim to answer following research questions:

- **RQ1** What pre-processing approach can effectively improve fairness while maintaining classification accuracy, and how does it perform across different datasets?
- **RQ2** How stable are the performance and fairness results of classifiers trained on pre-processed datasets?
- **RQ3** How does pursuing for statistical parity, a data-based notion, affect a prediction-based notion such as average odds error?
- **RQ4** Is the presented method fairness-agnostic as stated?

To especially address the first three research questions, which deal with effectiveness and stability, we adopted the following experimental methodology: We examined our approach against three pre-processors on four real-world datasets (see Table 1). The pre-processors we compare against are *Reweighing* [13], *Learning Fair Representation* [31] (LFR), and *Disparate Impact Remover* [9] (DIR). The data were prepared such that categorical features are one-hot encoded and rows containing empty values are removed from the data. We selected sex, age, race, and foreign worker as protected attributes for the respective datasets. Generally, the data preparation was adopted from AIF360 [3].

Table 1. Overview of datasets.

Dataset	Protected Attribute	Label	Size	Description
Adult [17]	Sex	Income	45 222	Indicates individuals earning over $50 000 annually
Bank [21]	Age	Term Deposit	30 488	Subscription to a term deposit
COMPAS [18]	Race	Recidivism	6 167	Arrested again for a new offense within a period of 2 years after initial arrest
German [11]	Foreign Worker	Credit Risk	1 000	Creditworthiness of loan applicants

All hyperparameter settings of the pre-processsors were kept as they are, given the implementation provided by AIF360 [3]. For the case of LFR, we empirically had to lower the hyperparameter of optimizing for fairness. It was initially set too high which led to identical predictions for all data points. For our approach, we set $r = 1.25$ which returns a dataset consisting of additional 25% samples of the dataset's initial size. The discrimination measure chosen was the absolute difference of statistical parity (1), which all other methods also optimize for. Further, we set $m = 5$ and $\epsilon = 0$ as shown in Algorithm 1.

The experimental methodology for a single dataset is visualized in Fig. 1 as a pipeline. The given dataset is firstly split into a training (80%) and test set (20%). Afterwards, the training set is then passed into the available pre-processors. Then, all debiased data are used to train several classifiers. We employed three different machine learning algorithms—*k-nearest neighbors* (KNN), *logistic regression* (LR), and *decision tree classifier* (DT)—to analyze the pre-processed datasets and the original, unprocessed dataset for comparison. The unprocessed dataset is referred to as the baseline. Finally, the performance and fairness is evaluated on the prediction of the test set. It is noteworthy to mention that the test sets were left untouched to demonstrate that by pre-processing the training data, unbiased results can be achieved in the prediction space even without performing bias mitigation in the test data. Due to stability reasons (and to handle **RQ2**), we used Monte Carlo cross-validation to shuffle and split the dataset. This was done 10 times for all datasets. The results from it set the performance-fairness baseline. While our optimization focuses on SDP, we address **RQ3** by assessing the error of average odds. To answer **RQ4**, we refer to Sect. 6.2 for the experimentation and discussion.

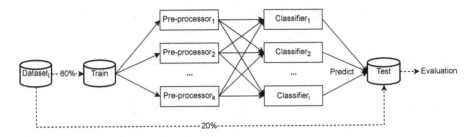

Fig. 1. Experimental methodology visualized as a pipeline

6.1 Comparing Pre-processors

Table 2 presents the performance-fairness test results of pre-processors on different datasets (**RQ1**). For the discrimination, the table displays SDP and average odds error of the predictions on the test sets. To assess the classifier's performances, we used *area under the receiver operating characteristic curve* (AUC). An estimator that guesses classes randomly would produce an AUC score of 0.5. Here, higher scores imply better prediction performances. Means and standard deviations of the Monte Carlo cross-validation results are also displayed to evaluate the robustness (**RQ2**). We note that all classifiers except of KNN were able to handle sample weights in training, which are required for Reweighing. Therefore Reweighing was not able to mitigate bias in KNN and performed as well as the baseline in contrast to other approaches including ours.

Because all pre-processors aim to reduce statistical disparity (or the equivalent formulation), we compare the SDP scores between the pre-processors: In most cases, our approach produced Pareto optimal solutions with respect to both SDP and AUC. Generally, only Reweighing and our approach appear to consistently improve fairness without sacrificing notable prediction power. In direct comparison, LFR improved the fairness at most across all experiments but at the same time sacrifices prediction quality of all classifiers to such a great extent that the predictions become essentially useless. In experiments where LFR attained standard deviations of 0 across all scores (Table 2b, 2d), we investigated the pre-processed data and found that LFR had modified almost all labels to a single value. As a result, the estimators were unable to classify the data effectively, as they predicted only one outcome. The results of DIR are very inconsistent. DIR sometimes even worsens the fairness, as seen in the COMPAS and German datasets, where SDP and average odds error are increased in most settings. This situation arises when there is an excessive correction of the available discrimination for the unprivileged group, leading to discrimination against the privileged group. If the discrimination measures are defined such that the privileged or unprivileged groups do not matter (similarly to this paper), reverse discrimination would not mistakenly occur by our approach. This extra property renders our method more suitable for responsible use cases.

When comparing the average odds error rates (**RQ3**), our approach has successfully reduced algorithmic bias without aiming for it under nearly all experiments. The increase in the average odds error rate (mean), albeit negligible, was observed only when training DT on the Banking data and LR in the German dataset. In all other ten model and dataset configurations, our approach did reduce the error rate without particularly optimizing for it. This can be expected in practice as the independency of the label with the protected attribute (SDP) is a sufficient condition for average odds.

Table 2. The tables displays each classifier's mean test performance and discrimination when trained on different pre-processed training sets. The best performing statistic for each classifier is marked in bold. Minimal standard deviations are marked bold, too. All values displayed are percentages.

(a) Adult

Model	Preprocessor	AUC mean	std	SDP mean	std	AVG Odds Error mean	std
	DIR	81.28	0.53	19.65	1.04	24.50	1.27
	LFR	50.18	1.20	**0.18**	**0.46**	**0.16**	**0.37**
DT	Our	78.91	0.52	9.68	1.19	10.55	1.26
	Original	**81.35**	0.52	19.77	0.96	24.66	1.21
	Reweighing	78.95	**0.48**	4.96	1.22	1.37	0.76
	DIR	75.35	0.86	20.94	2.35	22.31	3.32
	LFR	51.80	4.96	**1.14**	3.26	**0.70**	2.02
KNN	Our	75.26	**0.60**	18.84	2.91	19.80	3.54
	Original	**75.53**	0.85	21.09	**2.16**	22.33	3.05
	Reweighing	**75.53**	0.85	21.09	**2.16**	22.33	3.05
	DIR	80.12	0.59	17.84	0.46	22.80	0.49
	LFR	55.35	8.64	**1.33**	3.05	**0.80**	2.02
LR	Our	76.96	0.52	3.60	0.82	1.30	0.63
	Original	**80.13**	0.59	17.75	**0.45**	22.71	**0.48**
	Reweighing	77.29	**0.51**	4.63	0.57	1.90	0.71

(b) Bank

Model	Preprocessor	AUC mean	std	SDP mean	std	AVG Odds Error mean	std
	DIR	67.66	1.24	3.30	2.03	7.81	3.66
	LFR	50.00	**0.00**	**0.00**	**0.00**	**0.00**	**0.00**
DT	Our	72.68	0.96	10.36	3.75	6.93	3.42
	Original	**72.94**	1.15	10.69	1.70	6.49	2.97
	Reweighing	72.81	1.16	9.58	1.99	5.93	3.07
	DIR	81.42	0.82	8.43	3.69	6.56	3.16
	LFR	50.00	**0.00**	**0.00**	**0.00**	**0.00**	**0.00**
KNN	Our	**86.98**	0.66	9.05	3.55	5.00	3.74
	Original	**86.98**	0.65	9.05	3.55	5.00	3.75
	Reweighing	**86.98**	0.65	9.05	3.55	5.00	3.75
	DIR	91.48	0.42	3.90	1.08	3.60	2.59
	LFR	50.00	**0.00**	**0.00**	**0.00**	**0.00**	**0.00**
LR	Our	91.25	0.51	6.72	2.24	2.87	2.00
	Original	**92.14**	0.34	7.00	2.85	4.37	2.57
	Reweighing	92.11	0.36	5.82	2.47	3.37	1.81

(c) COMPAS

Model	Preprocessor	AUC mean	std	SDP mean	std	AVG Odds Error mean	std
	DIR	70.75	**0.53**	23.58	4.70	22.06	4.71
	LFR	50.26	3.97	**8.32**	21.12	**8.05**	21.17
DT	Our	70.91	0.85	10.55	4.31	8.63	4.06
	Original	70.76	0.82	21.16	**3.64**	19.67	3.77
	Reweighing	70.35	0.97	10.22	4.02	8.98	**2.80**
	DIR	**65.78**	2.88	21.74	7.64	20.58	7.03
	LFR	53.58	6.15	**2.29**	**3.67**	**3.00**	**4.20**
KNN	Our	65.13	**1.49**	12.56	7.98	11.94	7.48
	Original	64.84	2.52	15.62	7.88	14.55	8.16
	Reweighing	64.84	2.52	15.62	7.88	14.55	8.16
	DIR	**72.28**	0.54	23.20	3.41	21.31	3.77
	LFR	56.94	9.07	**1.95**	3.44	**2.55**	3.57
LR	Our	71.78	0.68	2.31	**1.51**	5.38	**1.79**
	Original	72.08	**0.48**	21.74	3.76	20.01	4.06
	Reweighing	71.52	0.76	3.89	2.46	5.61	2.17

(d) German

Model	Preprocessor	AUC mean	std	SDP mean	std	AVG Odds Error mean	std
	DIR	61.15	4.02	23.22	13.70	28.23	14.30
	LFR	50.00	**0.00**	**0.00**	**0.00**	**0.00**	**0.00**
DT	Our	**63.08**	3.47	16.61	10.90	26.78	11.09
	Original	62.76	3.95	15.07	11.29	27.12	10.83
	Reweighing	62.71	4.65	17.53	15.03	33.15	6.73
	DIR	**55.42**	4.44	18.70	9.28	23.91	7.47
	LFR	50.00	**0.00**	**0.00**	**0.00**	**0.00**	**0.00**
KNN	Our	54.21	4.34	12.95	2.95	13.81	3.26
	Original	54.08	3.86	16.99	4.63	18.27	4.95
	Reweighing	54.08	3.86	16.99	4.63	18.27	4.95
	DIR	78.05	2.15	20.05	11.84	29.79	14.16
	LFR	50.00	**0.00**	**0.00**	**0.00**	**0.00**	**0.00**
LR	Our	77.60	1.73	15.49	10.46	31.00	9.50
	Original	**78.10**	1.67	16.79	11.10	29.83	10.54
	Reweighing	78.05	1.88	16.42	11.10	30.32	10.56

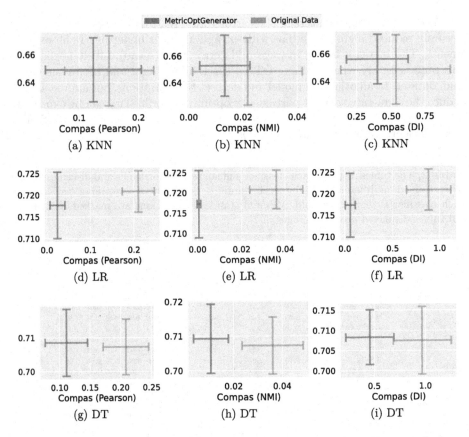

Fig. 2. Results of optimizing different discrimination objectives with our method on the COMPAS dataset. Objectives are ordered by columns, classifiers by rows. The y-axis displays AUC.

6.2   Investigating the Fairness-Agnostic Property

To demonstrate the fairness-agnostic property of our algorithm (**RQ4**), we evaluated our method against the baseline dataset on multiple measures and examine whether the objective was improved (see Fig. 2). The COMPAS dataset was used for this experiment. The chosen objectives are: the absolute value of Pearson's ρ, NMI (2), and the objective of disparate impact (DI) as given by [9]. All other experimental settings remained the same as described prior, except that other pre-processing methods were not used.

It can be observed that all discrimination measures were lowered significantly. Generally, our method was able to optimize on any fairness notion, as evidenced here and Sect. 6.1. It was even able to outperform algorithms that were specifically designed for a single metric, demonstrating its adaptability.

7   Conclusion

Machine learning can be utilized for malicious purposes if estimators are trained on data that is biased against certain demographic groups. This can have an incredibly negative impact on the decisions made and the groups that are being discriminated against.

The presented pre-processing method in this work is a sampling-based optimization algorithm that firstly uses a statistical model to learn the distribution of the given dataset, then samples points from this distribution, and determines which one to add to the data to minimize the discrimination. This process continues until the predefined criteria set by the user are satisfied. The method can optimize any discrimination measure as it is treated as a black-box, making it more accessible for wider use cases.

The results of our experiments demonstrate that our technique is reliable and significantly reduces discrimination while not compromising accuracy. Although a few other methods performed similarly in a few experiments, they were not compatible with certain estimators or even added bias to the original data. Because fairness was improved among the experiments and our method adds samples, it indicates that representativeness can be achieved with our method. Our research underscores the importance of addressing bias in data and we hope to contribute such concerns in data analytics and knowledge discovery applications.

8   Discussion and Future Work

The results of our approach demonstrate that it is possible to achieve fairness in machine learning models using generated data points. Despite our approach showing promise, it is important to acknowledge that our results rely heavily on the quality of the statistical model used to generate synthetic data. For tabular data, Gaussian copula [22] seems to be a good choice.

In future work, we aim to explore the potential of our method in making pre-trained models fairer with our method. While retraining large models using debiased datasets may not always be feasible from a cost-effective perspective, our approach allows using generated data to fine-tune the model for fairness, which provides a more efficient alternative.

Additionally, our evaluation deals with datasets where the protected attribute is a binary variable, which leaves some use cases untreated. Neglecting to recognize non-binary groups can lead to overlooking those who are most in need of attention. Similarly, research on dealing with multiple protected attributes at the same time could be done. This is to make sure that no protected group is being disadvantaged. Previous studies have touched on this subject [1, 4, 32], but we hope to reformulate these issues as objectives that work with our approach.

A   Proof of Time Complexity

Theorem 1 (Time complexity). *If the number of candidates m and fraction r are fixed and calculating the discrimination $\psi(\mathcal{D})$ of any dataset $\mathcal{D}$ takes a linear amount of time, i.e., $\mathcal{O}(n)$, Algorithm 1 has a worst-case time complexity of $\mathcal{O}(n^2)$ where n is the dataset's size when neglecting learning the data distribution.*

Proof. In this proof, we will focus on analyzing the runtime complexity of the for-loop within our algorithm as the steps before such as learning the data distribution depends heavily on the used method. The final runtime of the complete algorithm is simply the sum of the runtime complexities of the for-loop that is focus of this analysis and the step of learning the data distribution.

Our algorithm firstly checks whether the discrimination of the dataset $\hat{\mathcal{D}}$ is already fair. The dataset grows at each iteration and runs for $\lfloor rn \rfloor - n = \lfloor n(r-1) \rfloor$ times. For simplicity, we use $n(r-1)$ and yield,

$$\sum_{i=0}^{n(r-1)-1} n+i = \sum_{i=1}^{n(r-1)} n+i+1$$

$$= \sum_{i=1}^{n(r-1)} n + \sum_{i=1}^{n(r-1)} i + \sum_{i=1}^{n(r-1)} 1$$

$$= n^2(r-1) + \frac{(n(r-1))^2 + (n(r-1)+1)}{2} + n(r-1) \in \mathcal{O}(n^2),$$

making the first decisive step for the runtime quadratic.

The second step that affects the runtime is returning the dataset that minimizes the discrimination where each of the m candidates $c \in C$ is merged with the dataset, i.e., $\psi(\hat{\mathcal{D}} \cup \{c\})$. The worst-case time complexity of it can be expressed by

$$\sum_{i=1}^{n(r-1)} m(n+i) = m \cdot \sum_{i=1}^{n(r-1)} n+i = m \cdot \left(\sum_{i=1}^{n(r-1)} n + \sum_{i=1}^{n(r-1)} i \right)$$

$$= m \cdot \left(n^2(r-1) + \frac{(n(r-1))^2 + (n(r-1))}{2} \right) \in \mathcal{O}(n^2),$$

which is also quadratic. Summing both time complexities makes the overall complexity quadratic. □

Although the theoretical time complexity of our algorithm is quadratic, measuring the discrimination, which is a crucial part of the algorithm, is very fast and can be assumed to be constant for smaller datasets. Conclusively, the complexity behaves nearly linearly in practice.

In our experimentation, measuring the discrimination of the Adult dataset [17], which consists of 45 222 samples, did not pose a bottleneck for our algorithm.

References

1. Agarwal, A., Beygelzimer, A., Dudík, M., Langford, J., Wallach, H.: A reductions approach to fair classification. In: International Conference on Machine Learning, pp. 60–69. PMLR (2018)
2. Barocas, S., Hardt, M., Narayanan, A.: Fairness and Machine Learning. fairmlbook.org (2019). http://www.fairmlbook.org
3. Bellamy, R.K.E., et al.: AI fairness 360: an extensible toolkit for detecting, understanding, and mitigating unwanted algorithmic bias. CoRR arxiv:1810.01943 (2018)
4. Calmon, F., Wei, D., Vinzamuri, B., Natesan Ramamurthy, K., Varshney, K.R.: Optimized pre-processing for discrimination prevention. In: Guyon, I., et al. (eds.) Advances in Neural Information Processing Systems, vol. 30. Curran Associates, Inc. (2017). https://proceedings.neurips.cc/paper/2017/file/9a49a25d845a483fae4be7e341368e36-Paper.pdf
5. Caton, S., Haas, C.: Fairness in machine learning: a survey. arXiv preprint arXiv:2010.04053 (2020)
6. Chakraborty, J., Majumder, S., Tu, H.: Fair-SSL: building fair ML software with less data. arXiv preprint arXiv:2111.02038 (2022)
7. Choi, K., Grover, A., Singh, T., Shu, R., Ermon, S.: Fair generative modeling via weak supervision. In: III, H.D., Singh, A. (eds.) Proceedings of the 37th International Conference on Machine Learning. Proceedings of Machine Learning Research, vol. 119, pp. 1887–1898. PMLR, 13–18 July 2020
8. Dunkelau, J., Leuschel, M.: Fairness-aware machine learning (2019)
9. Feldman, M., Friedler, S.A., Moeller, J., Scheidegger, C., Venkatasubramanian, S.: Certifying and removing disparate impact. In: proceedings of the 21th ACM SIGKDD International Conference on Knowledge Discovery and Data Mining, pp. 259–268 (2015)
10. Hardt, M., Price, E., Srebro, N.: Equality of opportunity in supervised learning. In: Advances in Neural Information Processing Systems 29 (2016)
11. Hofmann, H.: German credit data (1994). https://archive.ics.uci.edu/ml/datasets/Statlog+%28German+Credit+Data%29
12. Jang, T., Zheng, F., Wang, X.: Constructing a fair classifier with generated fair data. In: Proceedings of the AAAI Conference on Artificial Intelligence, vol. 35, pp. 7908–7916 (2021)
13. Kamiran, F., Calders, T.: Data preprocessing techniques for classification without discrimination. Knowl. Inf. Syst. **33**(1), 1–33 (2012)
14. Kamiran, F., Karim, A., Zhang, X.: Decision theory for discrimination-aware classification. In: 2012 IEEE 12th International Conference on Data Mining, pp. 924–929. IEEE (2012)
15. Kamishima, T., Akaho, S., Asoh, H., Sakuma, J.: Fairness-aware classifier with prejudice remover regularizer. In: Flach, P.A., De Bie, T., Cristianini, N. (eds.) ECML PKDD 2012. LNCS (LNAI), vol. 7524, pp. 35–50. Springer, Heidelberg (2012). https://doi.org/10.1007/978-3-642-33486-3_3
16. Kang, J., Xie, T., Wu, X., Maciejewski, R., Tong, H.: InfoFair: information-theoretic intersectional fairness. In: 2022 IEEE International Conference on Big Data (Big Data), pp. 1455–1464. IEEE (2022)
17. Kohavi, R.: Scaling up the accuracy of Naive-Bayes classifiers: a decision-tree hybrid. In: KDD 1996, pp. 202–207. AAAI Press (1996)

18. Larson, J., Angwin, J., Mattu, S., Kirchner, L.: Machine bias, May 2016. https://www.propublica.org/article/machine-bias-risk-assessments-in-criminal-sentencing

19. Makhlouf, K., Zhioua, S., Palamidessi, C.: Machine learning fairness notions: bridging the gap with real-world applications. Inf. Process. Manage. **58**(5), 102642 (2021). https://doi.org/10.1016/j.ipm.2021.102642. https://www.sciencedirect.com/science/article/pii/S0306457321001321

20. Mehrabi, N., Morstatter, F., Saxena, N., Lerman, K., Galstyan, A.: A survey on bias and fairness in machine learning. ACM Comput. Surv. (CSUR) **54**(6), 1–35 (2021)

21. Moro, S., Cortez, P., Rita, P.: A data-driven approach to predict the success of bank telemarketing. Decis. Support Syst. **62**, 22–31 (2014)

22. Patki, N., Wedge, R., Veeramachaneni, K.: The synthetic data vault. In: 2016 IEEE International Conference on Data Science and Advanced Analytics (DSAA), pp. 399–410, October 2016. https://doi.org/10.1109/DSAA.2016.49

23. Rajabi, A., Garibay, O.O.: TabfairGAN: fair tabular data generation with generative adversarial networks. Mach. Learn. Knowl. Extr. **4**(2), 488–501 (2022)

24. Sattigeri, P., Hoffman, S.C., Chenthamarakshan, V., Varshney, K.R.: Fairness GAN: generating datasets with fairness properties using a generative adversarial network. IBM J. Res. Dev. **63**(4/5), 1–3 (2019)

25. Shannon, C.E.: A mathematical theory of communication. Bell Syst. Tech. J. **27**(3), 379–423 (1948)

26. Sharma, S., Henderson, J., Ghosh, J.: CERTIFAI: a common framework to provide explanations and analyse the fairness and robustness of black-box models. In: Proceedings of the AAAI/ACM Conference on AI, Ethics, and Society, pp. 166–172 (2020)

27. Tan, S., Shen, Y., Zhou, B.: Improving the fairness of deep generative models without retraining. arXiv preprint arXiv:2012.04842 (2020)

28. Verma, S., Ernst, M.D., Just, R.: Removing biased data to improve fairness and accuracy. CoRR arXiv:2102.03054 (2021)

29. Vinh, N.X., Epps, J., Bailey, J.: Information theoretic measures for clusterings comparison: is a correction for chance necessary? In: Proceedings of the 26th Annual International Conference on Machine Learning, pp. 1073–1080 (2009)

30. Witten, I.H., Frank, E., Hall, M.A., Pal, C.J., DATA, M.: Practical machine learning tools and techniques. In: Data Mining, vol. 2 (2005)

31. Zemel, R., Wu, Y., Swersky, K., Pitassi, T., Dwork, C.: Learning fair representations. In: International Conference on Machine Learning, pp. 325–333. PMLR (2013)

32. Žliobaitė, I.: Measuring discrimination in algorithmic decision making. Data Min. Knowl. Disc. **31**, 1060–1089 (2017)

Random Hypergraph Model Preserving Two-Mode Clustering Coefficient

Rikuya Miyashita$^{(\boxtimes)}$, Kazuki Nakajima, Mei Fukuda, and Kazuyuki Shudo

Tokyo Institute of Technology, Tokyo, Japan
`miyashita.r.ac@m.titech.ac.jp`

Abstract. Real-world complex systems often involve interactions among more than two nodes, and such complex systems can be represented by hypergraphs. Comparison between a given hypergraph and randomized hypergraphs that preserve specific properties reveal effects or dependencies of the properties on the structure and dynamics. In this study, we extend an existing family of reference models for hypergraphs to generate randomized hypergraphs that preserve the pairwise joint degree distribution and the degree-dependent two-mode clustering coefficient of the original hypergraph. Using empirical hypergraph data sets, we numerically show that the extended model preserves the properties of the node and hyperedge as designed.

Keywords: Hypergraph · Two-mode clustering coefficient · Configuration model

1 Introduction

Networks are often used to represent complex systems that consist of nodes and pairwise interactions (i.e., edges) among the nodes [4]. On the other hand, real-world complex systems often involve interactions among more than two nodes [2]. For example, in email networks, there are multiple senders and receivers of a single e-mail [7]; in co-authorship networks, there are more than two coauthors for a single paper [11,15]. Such complex systems can be represented as hypergraphs consisting of a set of nodes and hyperedges, where each hyperedge contains an arbitrary number of nodes.

Randomized networks that preserve specified properties are often used for analyzing the structure and dynamics of empirical networks [8,12,14]. In general, by comparing the structure or dynamics between a given network and randomized networks that preserve specified properties of the original network, we investigate how the preserved properties affect the structure or dynamics of interest of the original network [5]. The dK-series [6,8,14] is a family of models to generate such randomized networks. Given a network, the dK-series generates randomized networks that preserve up to the degree of each node, the pairwise joint degree distribution, and the degree-dependent clustering coefficient. A recent study proposed the hyper dK-series, which is an extension of

© The Author(s), under exclusive license to Springer Nature Switzerland AG 2023
R. Wrembel et al. (Eds.): DaWaK 2023, LNCS 14148, pp. 191–196, 2023.
https://doi.org/10.1007/978-3-031-39831-5_18

the dK-series to the case of hypergraphs [10]. The hyper dK-series preserves up to the degree distribution of the node, the pairwise joint degree distribution of the node, the degree-dependent redundancy coefficient of the node, and the size distribution of the hyperedge in the original hypergraph.

In this study, we extend the hyper dK-series to preserve the pairwise joint degree distribution of the node and the degree-dependent two-mode clustering coefficient of the node, proposed in Ref. [13]. Using empirical hypergraph data sets, we numerically show that the extended hyper dK-series preserves the properties of a given hypergraph as designed. This paper is an extended version of our previous work published as an extended abstract [9]. This paper presents definitions and notations in Sect. 2 and further analysis and consideration of the experimental results in Sect. 4 and Fig. 1.

2   Preliminaries

We represent an unweighted hypergraph that consists of a set of nodes $V = \{v_1, \ldots, v_N\}$ and a set of hyperedges $E = \{e_1, \ldots, e_M\}$, where N is the number of nodes and M is the number of hyperedges. Then, we denote by $G = (V, E, \mathcal{E})$ the bipartite graph that corresponds to the given hypergraph, where $\mathcal{E}$ is a set of edges in the bipartite graph. An edge (v_i, e_j) is connected to each node v_i and each hyperedge e_j if and only if v_i belongs to the hyperedge e_j in the hypergraph. We assume that G does not contain multiple edges.

We denote by k_i and s_j the degree of node v_i (i.e., the number of hyperedges to which v_i belongs) and the size of hyperedge e_j (the number of nodes that belong to the hyperedge e_j), respectively. We also denote by $\bar{k}$ and $\bar{s}$ the average degree of the node and the average size of the hyperedge, respectively.

We use the joint degree distribution and the average degree of the nearest neighbors of nodes with degree k, which were defined in Ref. [10]. We denote by $P(k, k')$ the joint degree distribution of the node [10]. We denote by $k_{\mathrm{nn}}(k)$ by the average degree of the nearest neighbors of nodes with degree k [10].

We define the two-mode clustering coefficient of each node v_i [13]:

$$c_i = \frac{(\text{number of closed 4-paths centered on node } v_i)}{(\text{number 4-paths centered on node } v_i)}.$$

We denote by $\bar{c}$ the average two-mode clustering coefficient of the node. We define $c(k)$ the degree-dependent two-mode clustering coefficient of the node as

$$c(k) = \frac{1}{N(k)} \sum_{i=1, k_i=k}^{N} c_i,$$

where $N(k)$ is the number of nodes with degree k.

3   Extending the Hyper dK-Series to the Case of $d_v = 2.5+$

The hyper dK-series generates a bipartite graph that preserves the joint degree distributions of the node in the subgraphs of size $d_v \in \{0, 1, 2, 2.5\}$ or less and

Table 1. Data sets. N: number of nodes, M: number of hyperedges, $\mathcal{M}$: number of edges in the corresponding bipartite graph, $\bar{k}$: average degree of the node, $\bar{s}$: average size of the hyperedge, $\bar{c}$: average two-mode clustering coefficient of the node, $\bar{l}$: average shortest path length between nodes.

Data	N	M	$\mathcal{M}$	$\bar{k}$	$\bar{s}$	$\bar{c}$	$\bar{l}$	Refs.
email-Enron	143	1,512	4,550	31.82	3.01	0.68	2.08	[3,7]
NDC-classes	628	816	5,688	9.06	6.97	0.31	3.53	[3]
primary-school	242	12,704	30,729	126.98	2.42	0.70	1.73	[3,16]

the size distributions of the hyperedge in the subgraphs of size $d_e \in \{0,1\}$ or less in the given bipartite graph [10]. We extend the model to the case of $d_v = 2.5+$ and $d_e \in \{0,1\}$ to generate a randomized bipartite graph that preserves the joint degree distribution and the degree-dependent two-mode clustering coefficient of the node in addition to the average or distribution of the hyperedge's size.

In the model with $d_v = 2.5+$ and $d_e \in \{0,1\}$, we first generate a randomized bipartite graph with $d_e = 2$ and given d_e using the original hyper dK-series. Then, we repeat the rewiring process for the generated bipartite graph [10]. We select a pair of edges, (v_i, e_j) and $(v_{i'}, e_{j'})$, in the bipartite graph such that $i \neq i'$, $j \neq j'$, and $k_i = k'_i$ uniformly at random. We replace (v_i, e_j) and $(v_{i'}, e_{j'})$ by $(v_i, e_{j'})$ and $(v_{i'}, e_j)$ if and only if the normalized L_1 distance defined as

$$D_{2.5+} = \frac{\sum_{k=1}^{M} |c'(k) - c(k)|}{\sum_{k=1}^{M} c(k)}$$

decreases, where $c'(k)$ represents the degree-dependent two-mode clustering coefficient of the node for the hypergraph after rewiring the edge-pair. The rewiring procedure preserves the pairwise joint degree distribution of the node and the size of each hyperedge. We repeat the rewiring attempts $R = 500\mathcal{M}$ times.

4   Experiments

We apply the extended hyper dK-series to three empirical hypergraphs. The email-Enron hypergraph is an email network [3,7], where nodes are email addresses and hyperedges are sets of all addressees of senders and receivers of each email. The NDC-classes hypergraph is a drug network [3], where nodes are class labels and hyperedges are sets of class labels applied to each drug. The primary-school hypergraph is a contact network [3,16], where nodes are people and hyperedges are sets of people who contact each other face-to-face. Table 1 shows the properties of the largest connected component for the data sets.

Table 2 shows the distance in five properties between the original hypergraph and the hypergraphs generated by the hyper dK-series with $d_v \in \{0,1,2,2.5,2.5+\}$ and $d_e \in \{0,1\}$. We calculated the Kolmogorov-Smirnov distance between the cumulative distributions of the degree distribution of the node

Table 2. Distance between the empirical hypergraphs and those generated by the reference models.

Data	(d_v, d_e)	$P(k)$	$k_{nn}(k)$	$c(k)$	$P(s)$	$P(l)$
email-Enron	$(0, 0)$	0.434	0.420	0.781	0.158	0.595
	$(1, 0)$	0.000	0.202	0.206	0.160	0.491
	$(2, 0)$	0.000	0.012	0.207	0.160	0.429
	$(2.5, 0)$	0.000	0.013	0.222	0.160	0.416
	$(2.5+, 0)$	0.000	0.013	0.023	0.160	0.368
	$(0, 1)$	0.406	0.412	0.772	0.000	0.647
	$(1, 1)$	0.000	0.200	0.197	0.000	0.491
	$(2, 1)$	0.000	0.035	0.191	0.000	0.452
	$(2.5, 1)$	0.000	0.027	0.198	0.000	0.427
	$(2.5+, 1)$	0.000	0.032	0.026	0.000	0.414
NDC-classes	$(0, 0)$	0.614	0.741	0.962	0.252	1.585
	$(1, 0)$	0.000	0.388	0.368	0.248	1.322
	$(2, 0)$	0.000	0.046	0.208	0.248	0.799
	$(2.5, 0)$	0.000	0.045	0.201	0.248	0.712
	$(2.5+, 0)$	0.000	0.043	0.035	0.248	0.467
	$(0, 1)$	0.597	0.751	0.951	0.000	1.612
	$(1, 1)$	0.000	0.389	0.328	0.000	1.417
	$(2, 1)$	0.000	0.022	0.158	0.000	0.749
	$(2.5, 1)$	0.000	0.019	0.156	0.000	0.722
	$(2.5+, 1)$	0.000	0.021	0.023	0.000	0.609
primary-school	$(0, 0)$	0.380	0.429	0.848	0.303	0.856
	$(1, 0)$	0.000	0.089	0.121	0.307	0.707
	$(2, 0)$	0.000	0.006	0.111	0.307	0.374
	$(2.5, 0)$	0.000	0.005	0.109	0.307	0.371
	$(2.5+, 0)$	0.000	0.007	0.008	0.307	0.313
	$(0, 1)$	0.368	0.451	0.868	0.000	0.535
	$(1, 1)$	0.000	0.089	0.126	0.000	0.434
	$(2, 1)$	0.000	0.014	0.166	0.000	0.246
	$(2.5, 1)$	0.000	0.015	0.180	0.000	0.218
	$(2.5+, 1)$	0.000	0.014	0.010	0.000	0.276

for the original hypergraph and the generated hypergraphs. For the other properties, we calculated the normalized L^1 distance between the properties for the original hypergraph and the generated hypergraphs.

We make the following observations for the three empirical hypergraphs. First, the distances for $P(k)$ is equal to 0 for $d_v = 2.5+$ and $d_e \in \{0, 1\}$, as expected. Second, the distances for $k_{nn}(k)$ are quite small values in the models

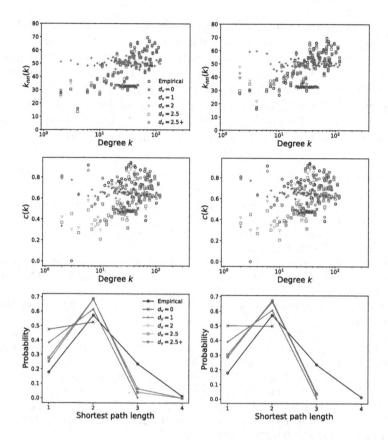

Fig. 1. Structural properties of the email-Enron hypergraph and the hypergraphs generated by the reference models. The figures in the left column show the results for $d_e = 0$. The figures in the right column show the results for $d_e = 1$.

with $d_v \in \{2.5, 2.5+\}$ and $d_e \in \{0, 1\}$, which indicates that the models approximately preserve $k_{nn}(k)$. Third, the distance for $c(k)$ is much smaller in the model with $d_v = 2.5+$ than that in the model with $d_v = 2.5$ for any $d_e \in \{0, 1\}$. Fourth, the model with $(d_v, d_e) = (2.5+, 0)$ and $d_e = 0$ has a somewhat distance for $P(s)$ but that with $(d_v, d_e) = (2.5+, 1)$ has no distance for $P(s)$, as expected. Finally, the model with $d_v = 2.5+$ often preserves more accurately $P(l)$ than that with $d_v = 2.5$ for any $d_e = \{0, 1\}$, whereas the distance is still not small in both models. Figure 1 shows $k_{nn}(k)$, $c(k)$, and $P(l)$ for the original email-Enron hypergraph and hypergraphs generated by the hyper dK-series.

5   Conclusion

We extended the hyper dK-series to generate a randomized hypergraph that preserves the pairwise joint degree distribution and the two-mode clustering coefficient. We applied the extended hyper dK-series to three empirical hypergraphs. We numerically showed that the model with $d_v = 2.5+$ preserves exactly the degree distribution of the node, approximately the pairwise joint degree distribution of the node, and approximately the degree-dependent two-mode clustering coefficient. We also found that the model with $d_v = 2.5+$ often preserves more accurately the distribution of the shortest-path length than the existing model with $d_v = 2.5$. Future work includes the application of the extended hyper dK-series to simulations of dynamical processes in hypergraphs [1].

References

1. Battiston, F., et al.: The physics of higher-order interactions in complex systems. Nat. Phys. **17**, 1093–1098 (2021)
2. Battiston, F., et al.: Networks beyond pairwise interactions: structure and dynamics. Phys. Rep. **874**, 1–92 (2020)
3. Benson, A.R., Abebe, R., Schaub, M.T., Jadbabaie, A., Kleinberg, J.: Simplicial closure and higher-order link prediction. Proc. Natl. Acad. Sci. **115**, E11221–E11230 (2018)
4. Boccaletti, S., Latora, V., Moreno, Y., Chavez, M., Hwang, D.U.: Complex networks: structure and dynamics. Phys. Rep. **424**, 175–308 (2006)
5. Cimini, G., Squartini, T., Saracco, F., Garlaschelli, D., Gabrielli, A., Caldarelli, G.: The statistical physics of real-world networks. Nat. Rev. Phys. **1**, 58–71 (2019)
6. Gjoka, M., Kurant, M., Markopoulou, A.: 2.5K-graphs: From sampling to generation. In: 2013 Proceedings IEEE INFOCOM, pp. 1968–1976 (2013)
7. Klimt, B., Yang, Y.: The Enron corpus: a new dataset for email classification research. In: Boulicaut, J.-F., Esposito, F., Giannotti, F., Pedreschi, D. (eds.) ECML 2004. LNCS (LNAI), vol. 3201, pp. 217–226. Springer, Heidelberg (2004). https://doi.org/10.1007/978-3-540-30115-8_22
8. Mahadevan, P., Krioukov, D., Fall, K., Vahdat, A.: Systematic topology analysis and generation using degree correlations. In: Proceedings of the 2006 Conference on Applications, Technologies, Architectures, and Protocols for Computer Communications, pp. 135–146 (2006)
9. Miyashita, R., Nakajima, K., Fukuda, M., Shudo, K.: Randomizing hypergraphs preserving two-mode clustering coefficient. In: 2023 IEEE International Conference on Big Data and Smart Computing (BigComp), pp. 316–317 (2023)
10. Nakajima, K., Shudo, K., Masuda, N.: Randomizing hypergraphs preserving degree correlation and local clustering. IEEE Trans. Netw. Sci. Eng. **9**, 1139–1153 (2022)
11. Newman, M.E.J.: The structure of scientific collaboration networks. Proc. Natl. Acad. Sci. **98**, 404–409 (2001)
12. Newman, M.E.J., Strogatz, S.H., Watts, D.J.: Random graphs with arbitrary degree distributions and their applications. Phys. Rev. E **64**, 026118 (2001)
13. Opsahl, T.: Triadic closure in two-mode networks: redefining the global and local clustering coefficients. Soc. Netw. **35**, 159–167 (2013)
14. Orsini, C., et al.: Quantifying randomness in real networks. Nat. Commun. **6**, 8627 (2015)
15. Patania, A., Petri, G., Vaccarino, F.: The shape of collaborations. EPJ Data Sci. **6**, 18 (2017)
16. Stehlá, J., et al.: High-resolution measurements of face-to-face contact patterns in a primary school. PLoS ONE **6**, e23176 (2011)

A Non-overlapping Community Detection Approach Based on α-Structural Similarity

Motaz Ben Hassine[1,2(✉)], Saïd Jabbour[1], Mourad Kmimech[3],
Badran Raddaoui[4,5], and Mohamed Graiet[6]

[1] CRIL, University of Artois and CNRS, Lens, France
{benhassine,jabbour}@cril.fr
[2] University of Monastir, UR-OASIS-ENIT, Monastir, Tunisia
motaz.benhassine@fsm.u-monastir.tn
[3] ESILV, Courbevoie, France
mourad.kmimech@devinci.fr
[4] SAMOVAR, Télécom SudParis, Institut Polytechnique de Paris, Palaiseau, France
badran.raddaoui@telecom-sudparis.eu
[5] Institute for Philosophy II, Ruhr University Bochum, Bochum, Germany
[6] LS2N Nantes, Nantes, France
mohamed.graiet@imt-atlantique.fr

Abstract. Community detection in social networks is a widely studied topic in Artificial Intelligence and graph analysis. It can be useful to discover hidden relations between users, the target audience in digital marketing, and the recommender system, amongst others. In this context, some of the existing proposals for finding communities in networks are agglomerative methods. These methods used similarities or link prediction between nodes to discover the communities in graphs. The different similarity metrics used in these proposals focused mainly on common neighbors between similar nodes. However, such definitions are missing in the sense that they do not take into account the connection between common neighbors. In this paper, we propose a new similarity measure, named α-Structural Similarity, that focuses not only on common neighbors of nodes but also on their connections. Afterwards, in the light of α-Structural Similarity, we extend the Hierarchical Clustering algorithm to identify disjoint communities in networks. Finally, we conduct extensive experiments on synthetic networks and various well-known real-world networks to confirm the efficiency of our approach.

Keywords: Local similarity · Community detection · Social network · Agglomerative approaches

1 Introduction

Over the years, graphs have been widely used to model various real-world applications where vertices represent objects and edges represent relationships

© The Author(s), under exclusive license to Springer Nature Switzerland AG 2023
R. Wrembel et al. (Eds.): DaWaK 2023, LNCS 14148, pp. 197–211, 2023.
https://doi.org/10.1007/978-3-031-39831-5_19

between these objects. Social network analysis is one such application where the automatic discovery of communities has been a major challenge in recent years [7]. Community detection involves identifying a set of nodes that are strongly connected within but weakly connected outside a community [15]. Community detection algorithms can be categorized into *overlapping* and *non-overlapping* approaches, with the latter being of particular interest in this paper. Long ago, various non-overlapping approaches have been studied. More precisely, Enright et al. [5] introduced a novel approach called TRIBE-MCL. The authors used the sequence protein similarity to detect the sequence protein families. Furthermore, the well-known approach, coined LPA (Label propagation algorithm), was proposed by Raghavan et al. [16] in which the nodes having the same label form the same community. In addition, Rodrigo et al. [1] introduced a novel optimized measure for detecting communities called surprise. In addition, Traag et al. [21] introduced a novel method based on a new measure called significance for detecting clusters. Moreover, Traag et al. [22] proposed a novel approach that improved the Louvain algorithm [2]. The authors found that 25% of communities are poorly connected, and then they presented a novel algorithm named LEIDEN to overcome this issue. Moreover, they enhanced the running time. Despite exhibiting strong performance, these existing proposals are still limited in terms of community quality, due to the wide variety of the real-world social network structures. Notice also that other non-overlapping approaches exist, which are based on local similarities and modularity maximization. Precisely, Yi-CHENG CHEN et al. [4] developed an approach named Hierarchical Clustering (HC, for short) used in the context of the influence maximization problem. Their algorithm starts by considering each vertex as an initial community. Then, the authors merged each pair of communities having the highest similarity values whose merging gives the greatest increase in modularity. Afterwards, they applied a local Structural Similarity (in short 2S) having the same definition of the Salton similarity [19] but utilizing differently defined neighborhood sets. Despite the good results in terms of community quality demonstrated by the use of the 2S in the HC approach, there are cases where the definition of the 2S may not be sufficient. This raises the question of whether the interaction between common neighbors enriched with the definition of the 2S would ultimately improve the quality of the detected communities. In this context, we propose a new similarity measure, named α-Structural Similarity (α-2S), that focuses not only on common neighbors of nodes but also on their connections. Ultimately, considering α-2S, we extended the HC algorithm to identify disjoint communities in networks. In this paper, we introduce some formal notations in Sect. 2. Then, in Sect. 3, we deal with a HC based on α-2S which we call α-HC. Section 4 presents our experiments on both synthetic and real-world datasets. Finally, Sect. 5 concludes the paper with hints for future work.

2 Preliminaries

In this paper, we consider a simple undirected graph $G = (V, E)$, where V is the set of vertices, and E is the set of edges. The set of **neighbors** of a node

$u \in V$ is defined as $N(u) = \{v \mid (u,v) \in E\}$. The **degree** of $u \in V$ is then $|N(u)|$. For a given node $u \in V$, we write $adj(u)$ for the set of neighbors of u including u itself, i.e., $adj(u) = N(u) \cup \{u\}$. Given a set of nodes $X \subseteq V$, a **subgraph** induced by X, denoted as $G_X = (X, E_X)$, is a graph over X s.t. $E_X = \{(u,v) \in E \mid u, v \in X\}$. Further, let $w : E \rightarrow \mathbb{R}_{>0}$ be a function that maps each edge from E to a non negative real value which $\in \,]0..1]$. We write E^w_{max} for the set of edges sets with a maximum weight w, i.e., $E^w_{max} = \{\{u,v\}$ s.t. $(u,v) \in E \mid \nexists \, (u',v') \in E$ s.t. $w(u',v') > w(u,v)\}$. A graph $G = (V,E)$ is called a **clique** iff. $\forall \, u \in V, |adj(u)| = |V|$. A graph $G = (V,E)$ can be splitted into numerous subgroups called **communities**, denoted as $C_G = \{c_1, c_2, \ldots, c_m\}$. Let $P(V) = 2^V$ the power set of V. The **Merge** function is defined as $Merge :$ $2^{P(V)} \times 2^{P(V)} \rightarrow 2^{P(V)}; (C_G, E^w_{max}) \mapsto Merge(C_G, E^w_{max})$ returns a merged set of subsets i.e., $Merge(C_G, E^w_{max}) = \{c_i \cup \{u,v\}, \{u,v\} \in E^w_{max}, c_i \in C_G, \mid \forall \, 1 \leq i \leq m$ s.t. $c_i \cap \{u,v\} \neq \emptyset\}$. Let d_{max} be the **maximum degree** of G, i.e., $d_{max} = \max_{|N(u)|}\{u \in V\}$. Besides, let d_{av} be the **average number of neighbors** of all the vertices in X, i.e., $d_{av} = \frac{1}{|X|} \sum_{u \in X} |N(u)|$.

The similarity measure $2S$ [4] is a local function that uses the immediate neighborhood between vertices as defined below.

Definition 1 (Structural Similarity). Let $G = (V,E)$ be an undirected graph and $(u,v) \in E$, then the 2S of u and v, denoted by $s_2(u,v)$, is defined as:

$$s_2(u,v) = \frac{|adj(u) \cap adj(v)|}{\sqrt{|adj(u)| \times |adj(v)|}} \tag{1}$$

For the quality of the set of communities, the modularity is formally defined as follows:

Definition 2 (Modularity [4]). Let $G = (V,E)$ be an undirected graph with $C_G = \{c_1, c_2, \ldots, c_m\}$ is the set of communities of G and s be a general similarity measure. The modularity is defined as:

$$Q(C_G) = \sum_{i=1}^{m} \left[\frac{IS_i}{TS} - \left(\frac{DS_i}{TS} \right)^2 \right] \tag{2}$$

where $IS_i = \sum_{u,\,v \,\in\, c_i} s(u,v)$, $DS_i = \sum_{u \,\in\, c_i,\, v \,\in\, V} s(u,v)$, and $TS = \sum_{u,\,v \,\in\, V} s(u,v)$.

3   A Hierarchical Clustering Approach Based on α-Structural Similarity

We propose an extended version of the 2S. Indeed, we added another term in Eq. 1, which will denote the rate of interaction between common neighbors. we added the concept of connection between common neighbors. Therefore, the identification of communities will be more significant. To be more precise, let

$G = (V, E)$ be a graph and given $(u, v) \in E$. We consider the ratio between the number of connections between common neighbors (i.e., $adj(u) \cap adj(v)$) and the minimum between the number of edges of the subgraph induced by $adj(u)$ and the one induced by $adj(v)$. Then, the new version of the similarity metric, which will be called α-2S is formally defined as follows:

$$s_2^\alpha(u, v) = (1 - \alpha) \frac{|adj(u) \cap adj(v)|}{\sqrt{|adj(u)| \times |adj(v)|}} + \alpha \frac{|E_{adj(u) \cap adj(v)}|}{\min(|E_{adj(u)}|, |E_{adj(v)}|)} \quad (3)$$

where α is a parameter in $[0..1]$. It should be noted that $s_2^\alpha(u, v) \in \,]0..1]$.

The parameter α in Eq. 3 ensures the trade-off between the notion of common neighborhood and the rate of their interactions. Thus, it is interesting to determine the value of α. It should be noted that when $\alpha = 0$, the α-2S is identical to the 2S. We illustrate the behaviour of our α-2S through the following example. We set $\alpha = 0.8$.

Example 1. Let us consider the undirected graph depicted in Fig. 1.

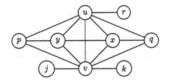

Then, we have that:

$$s_2^{0.8}(u, v) = (1 - 0.8) \frac{|adj(u) \cap adj(v)|}{\sqrt{|adj(u)| \times |adj(v)|}} + 0.8 \frac{|E_{adj(u) \cap adj(v)}|}{min(|E_{adj(u)}|, |E_{adj(v)}|)}$$

$$= (1 - 0.8) \times \frac{6}{\sqrt{7 \times 8}} + 0.8 \times \frac{12}{min(13, 14)}$$

$$= (1 - 0.8) \times \frac{6}{\sqrt{56}} + 0.8 \times \frac{12}{13} = 0.89$$

Fig. 1. A simple undirected graph with $\alpha = 0.8$.

In what follows, to show the effectiveness of our new similarity metric α-2S, let us consider the case of two disjoint cliques C_1 and C_2. A set of links is then added over C_1 and C_2 to form a new clique C_3 overlapping with C_1 and C_2. We will show how the two communities formed by the initial cliques C_1 and C_2 remain identifiable when varying the set of links between C_1 and C_2. The obtained graph will be coined k-linked-cliques graph.

Definition 3. Let $G = (V, E)$ be a graph and k, n two integers s.t. $1 \leq k \leq n$ and $n \geq 4$. Then, G is called a **k-linked-cliques graph** iff. G is formed by three cliques $C_1 = (V_1, E_1)$, $C_2 = (V_2, E_2)$, and $C_3 = (V_3, E_3)$ where:

- $V = V_1 \uplus V_2$ s.t. $|V_1| = |V_2| = n$, and $E_1 \cap E_2 = \emptyset$
- $V_3 \subseteq V_1 \cup V_2$, $|V_1 \cap V_3| = k$, and $|V_2 \cap V_3| = 2$

In the sequel, we consider $V_2 \cap V_3 = \{u_0, v_0\}$. Below we are interested in computing the values of k making the similarity inside C_1 and C_2 higher than the one of the edges of C_3 using both $2S$ and α-$2S$ metrics.

Proposition 1. *Let G be a k-linked-cliques graph. $\forall$ $(u_1, v_1) \in (E_1 \cup E_2)$ and $(u_2, v_2) \in E_3 \setminus (E_1 \cup E_2)$, we have $s_2(u_2, v_2) < s_2(u_1, v_1)$ iff. $k < n - 1$.*

Proof. (Refer to Appendix A)   □

Proposition 2. *Let G be a k-linked-cliques graph. $\forall$ $(u_1, v_1) \in (E_1 \cup E_2)$ and $(u_2, v_2) \in E_3 \setminus (E_1 \cup E_2)$, we have $s_2^\alpha(u_2, v_2) < s_2^\alpha(u_1, v_1)$ iff. $k \le n - 1$ and $\alpha > 0.06$.*

Proof. (Refer to Appendix B)   □

Proposition 1 states that for a k-linked-cliques graph, the 2S of edges linking C_1 and C_2 is lower than the ones of edges within the two cliques C_1 and C_2 for 1 until $n - 2$. While in the Proposition 2, α-2S is from 1 until $n - 1$ which makes it better than the 2S on a k-linked-cliques graph.

Example 2. Let's consider the k-linked-cliques graph depicted in Fig. 2. This figure illustrates an example of k-linked-clique graph for $k = 1$ to $k = 4$. As shown, the similarity values of the linking edges (colored in purple) are lower than the rest of the edges in the graph from $k = 1$ to $k = 2$ while according to 0.8-2S, it is from $k = 1$ to $k = 3$.

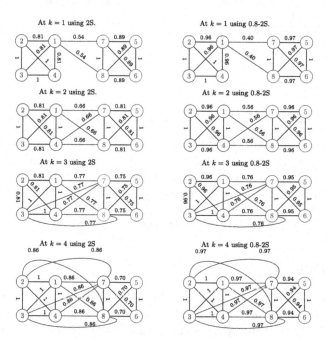

Fig. 2. An example of k-linked-cliques with $n = 4$ and $\alpha = 0.8$.(Color figure online)

Based on α-2S, our approach follows the one HC where 2S is substituted with α-2S. Algorithm 1 named α-HC describe our approach. First, we computed the similarity for each edge in G. Second, we initialized the set of communities, where each vertex is considered as community. Then, we calculated the corresponding modularity. Third, at each iteration, we merged each pair of nodes having the strongest similarity. Fourth, the modularity is recalculated on the current merged set. If we have a modularity gain, then the process continue. Otherwise, the previous result is considered as the best clustering set.

Algorithm 1: α-HC

Input: $G(V, E), \alpha$
Output : C_G
begin
    for $(u, v) \in E$ **do**
      | $w(u, v) \leftarrow s_2^{\alpha}(u, v)$
    end
    $C_G \leftarrow \emptyset$
    for $u \in V$ **do**
      | $C_G \leftarrow C_G \cup \{\{u\}\}$
    end
    $PreviousModularity \leftarrow Q(C_G)$
    $CurrentModularity \leftarrow PreviousModularity$
    $C \leftarrow C_G$
    while $CurrentModularity \geq PreviousModularity$ **do**
        $C_G \leftarrow C$
        $C \leftarrow Merge(C, E_{max}^w)$
        $PreviousModularity \leftarrow CurrentModularity$
        $CurrentModularity \leftarrow Q(C)$
    end
    return C_G
end

Computational Complexity. Usually, clustering algorithms based on modularity maximization require $O((|E| + |V|)|V|)$ [12]. Similarity computation should be considered. Indeed, the 2S requires $O(|V|d_{max}^3)$ [14]. The extraction of edges of an induced subgraph requires $O(|X|d_{av})$ [13]. Then, α-2S requires $O(|V|d_{max}^3 + 4|E||X|d_{av})$ and therefore α-HC requires $O(|V|d_{max}^3 + 4|E||X|d_{av} + (|E| + |V|)|V|)$. The complexity is polynomial.

4   Experiments

To validate our proposal, we propose an implementation [23]. We performed two kinds of experiments. First, α-HC and HC were tested on artificial networks called LFR networks [9] by changing a parameter called mixing parameter μ from 0.1 to 0.9. The parameter μ allows to control the mixture between communities. When

μ is growing the identification of the communities becomes harder. Our goal is to identify how community quality is correlated to μ. In our setting, for each value of μ, the value of α is varied from 0.1 to 1 to identify the best value of α providing the best quality. The quality of the founded communities were measured using the well-known F1-score [18] and NMI [6] metrics. In our second experiment, α is fixed, and then we compare our approach with other disjoint community detection approaches on 8 well-known real-world datasets. We illustrated all the datasets in Tables 1 and 2. We denote by AD the average degree, GT the ground truth communities, and MinCS the minimum community size.

Table 1. Real-world datasets

Datasets	Nodes/Edges	GT	Source
Karate	34/78	2	[24]
Dolphin	62/159	2	[11]
Books	105/441	3	[8]
Citeseer	3264/4536	6	[3]
Email-Eu-Core	1005/25571	42	[10]
Cora	23166/89157	70	[20]
Amazon	334863/925872	75149	[10]
YouTube	1134890/2987624	8385	[10]

Table 2. LFR networks

μ	AD	MinCS	Nodes	Source
[0.1..0.9]	5	50	1000	[17]
[0.1..0.9]	5	50	5000	[17]
[0.1..0.9]	5	50	10000	[17]
[0.1..0.9]	5	50	50000	[17]

In the first phase of the experiment, α-HC is compared to HC according to NMI and F1-score by considering various LFR networks. The Figs. 3, 4, 5, 6, 7, 8, 9 and 10 illustrates the obtained results. The histograms reveal that, for $\mu \geq 0.3$, there is always at least an $\alpha \neq 0$ (more precisely, $\alpha \geq 0.6$) for which α-HC outperforms HC in terms of NMI and F1-score. These findings suggest that α-HC is more reliable than HC for detecting mixed communities, as confirmed by the results of Propositions 1 and 2.

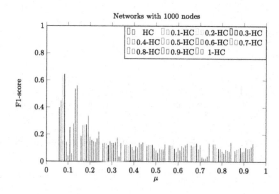

Fig. 3. HC vs. α-HC on LFR networks with 1000 nodes based on F1-score.

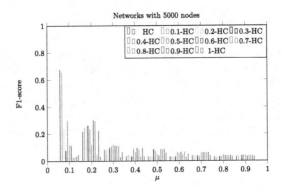

Fig. 4. HC vs. α-HC on LFR networks with 5000 nodes based on F1-score.

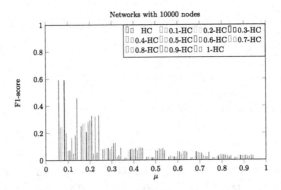

Fig. 5. HC vs. α-HC on LFR networks with 10000 nodes based on F1-score.

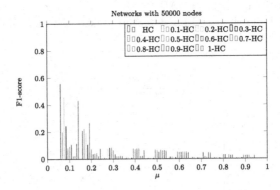

Fig. 6. HC vs. α-HC on LFR networks with 50000 nodes based on F1-score.

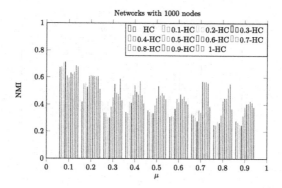

Fig. 7. HC vs. α-HC on LFR networks with 1000 nodes based on NMI.

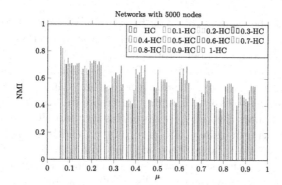

Fig. 8. HC vs. α-HC on LFR networks with 5000 nodes based on NMI.

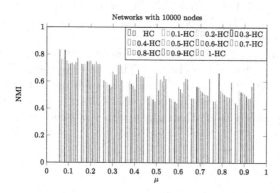

Fig. 9. HC vs. α-HC on LFR networks with 10000 nodes based on NMI.

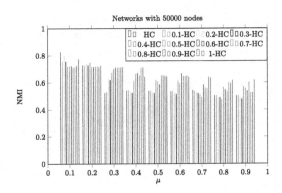

Fig. 10. HC vs. α-HC on LFR networks with 50000 nodes based on NMI.

In the rest of the experiment, α is fixed to 1. We made two comparisons. First, 1-HC with HC, then 1-HC with other some non-overlapping approaches mentioned in Sect. 1 like (Label Propagation Algorithm (LPA) [16], LEIDEN algorithm [22], Surprise Communities (SC) [1], Significance Communities Approach (SCA) [21] and Markov Clustering algorithm (MC) [5]) on 8 well-known real-world datasets. Tables 3, 4, 5, and 6 illustrates the results.

Tables 3 and 4 show that 1-HC performs better than HC. Indeed, 1-HC outperforms HC allowing a gain of 4% according to the NMI. For the F1-score, an improvement of 4% is also obtained. The results shown in Table 5 prove that 1-HC overpasses other state-of-the-art approaches considered in this paper. Indeed, 1-HC exceeds on average 13%, 14%, 14%, 9%, and 23% LPA, LEIDEN, SC, SCA, and MC respectively according to NMI. In Table 6, 1-HC overpasses the above mentioned approaches. In fact, 1-HC exceeds on average 4%, 6%, 10%, 8%, and 13% LPA, LEIDEN, SC, SCA and MC respectively according to F1-score. Therefore, 1-HC shows good results for communities compared to the above-mentioned algorithms.

Despite the potentially fruitful results observed with 1-HC, there exist some datasets where its performance is limited. This can be explained by the variety of the structures of the networks. Furthermore, $\alpha = 1$ is not universally applicable. In fact it may be not the best optimal result. Then, it is important to search the most appropriate α value that aligns with the structure of the considered network.

Table 3. 1-HC vs. HC based on NMI.

Comparison based on NMI		
Datasets	HC	1-HC
Karate	0.579	**0.777**
Email-Eu-Core	**0.705**	0.660
Citeseer	**0.098**	0.069
Dolphin	0.429	**0.509**
Books	**0.49**	0.474
Amazon	0.635	**0.752**
Cora	**0.623**	0.613
YouTube	0.127	**0.199**
Average	0.46	**0.50**

Table 4. 1-HC vs. HC based on F1.

Comparison based on F1-score		
Datasets	HC	1-HC
Karate	0.669	**0.919**
Email-Eu-Core	0.164	**0.280**
Citeseer	**0.147**	0.099
Dolphin	0.470	**0.660**
Books	**0.606**	0.345
Amazon	**0.425**	0.415
Cora	0.085	**0.102**
YouTube	0.178	**0.287**
Average	0.34	**0.38**

Table 5. 1-HC vs. others based on NMI.

Comparison based on NMI						
Datasets	1-HC	LPA	LEIDEN	SC	SCA	MC
Karate	**0.777**	0.207	0.202	0.220	0.462	0.164
Email-Eu-Core	**0.660**	0.180	0.593	0.648	0.671	0.428
Citeseer	0.069	0.087	**0.128**	0.102	0.093	0.080
Dolphin	**0.509**	0.436	0.098	0.120	0.164	0.090
Books	0.474	0.534	**0.573**	0.441	0.441	0.526
Amazon	**0.752**	0.579	0.206	0.539	0.586	0.601
Cora	**0.613**	0.551	0.472	0.552	0.584	0.337
YouTube	0.199	0.393	**0.616**	0.307	0.286	0.008
Average	**0.50**	0.37	0.36	0.36	0.41	0.27

Table 6. 1-HC vs. others based on F1.

Comparison based on F1-score						
Datasets	1-HC	LPA	LEIDEN	SC	SCA	MC
Karate	**0.919**	0.630	0.560	0.490	0.490	0.735
Email-Eu-Core	0.280	0.065	0.217	0.075	**0.387**	0.159
Citeseer	0.099	0.074	**0.286**	0.081	0.073	0.046
Dolphin	**0.660**	0.585	0.480	0.360	0.256	0.320
Books	0.345	0.656	**0.776**	0.590	0.435	0.696
Amazon	**0.415**	0.397	0.028	0.318	0.370	0.084
Cora	0.102	0.221	0.238	**0.257**	0.255	0.032
YouTube	**0.287**	0.112	0.005	0.087	0.164	0.007
Average	**0.38**	0.34	0.32	0.28	0.30	0.25

5   Conclusion and Future Work

In this paper, we extended the HC method called α-HC based on α-2S to find disjoint communities. While using 2S in HC takes only into account the neighborhood, our approach improves such formula by taking into account the number of interactions between common neighbors. We proved theoretically that for a k-linked-cliques graph, identifying the two cliques using α-2S is better than using 2S. Experimentation evaluation showed that our approach surpasses the above-mentioned methods. In a future work, we plan to use our approach in the context of the Influence Maximization problem to find the seed nodes. Another direction for future work is to develop an adaptive approach can predict the value of α which provides the best quality based on machine learning solutions.

A   Appendix a

We have $s_2(u_2, v_2) = \dfrac{k+2}{\sqrt{(n+2)(n+k)}}$

if $(\mathbf{u_1}, \mathbf{v_1}) \in \mathbf{E_2}$, we can distinguish two cases:

- $|\{\mathbf{u_1}, \mathbf{v_1}\} \cap \{\mathbf{u_0}, \mathbf{v_0}\}| \neq \mathbf{1}$. In this case, we have $s_2(u_1, v_1) = 1$. Then, $\dfrac{k+2}{\sqrt{(n+2)(n+k)}} < 1 \Leftrightarrow (k+2)^2 < (n+2)(n+k) \Leftrightarrow k < (n-2)+$

$\sqrt{(2-n)^2 + 4(n^2 + 2n - 4)}$. k.t. $(n-2) + \sqrt{(2-n)^2 + 4(n^2 + 2n - 4)} > n \implies s_2(u_2, v_2) < s_2(u_1, v_1) \ \forall \ 1 \le k \le n$.

- $|\{u_1, v_1\} \cap \{u_0, v_0\}| = 1$. In this case, we have $s_2(u_1, v_1) = \dfrac{n}{\sqrt{n(n+k)}}$. Then,

$$\frac{k+2}{\sqrt{(n+2)(n+k)}} < \frac{n}{\sqrt{n(n+k)}} \Leftrightarrow \frac{(k+2)^2}{(n+2)(n+k)} < \frac{n}{(n+k)} \Leftrightarrow (k+2)^2 < n(n+2) \Leftrightarrow$$
$k+2 < \sqrt{n(n+2)} \Leftrightarrow k < \sqrt{n(n+2)} - 2$. k.t. $\sqrt{n(n+2)} - 2 < n-1 \ \forall \ n \ge 4 \implies s_2(u_2, v_2) < s_2(u_1, v_1)$ iff. $k < n-1$.

if $(u_1, v_1) \in E_1$, there are also two cases:

- $|\{u_1, v_1\} \cap \{u_2, v_2\}| = 0$ or $(u_1, v_1) \in E_1 \cap E_3$. In this case, $s_2(u_1, v_1) = 1$. Then, $s_2(u_2, v_2) < s_2(u_1, v_1) \ \forall \ 1 \le k \le n$. (proved).
- $|\{u_1, v_1\} \cap \{u_2, v_2\}| = 1$. In this case, $s_2(u_1, v_1) = \dfrac{n}{\sqrt{n(n+2)}}$. Then,

$$\frac{k+2}{\sqrt{(n+2)(n+k)}} < \frac{n}{\sqrt{n(n+2)}} \Leftrightarrow (k+2)^2 < n(n+k) \Leftrightarrow k < \frac{(n-4) + \sqrt{n(5n-8)}}{2} \Leftrightarrow$$
$k \le n-1 < \dfrac{(n-4) + \sqrt{n(5n-8)}}{2} \Leftrightarrow s_2(u_2, v_2) < s_2(u_1, v_1)$ iff. $k \le n-1$.

B Appendix B

$s_2^\alpha(u_2, v_2) = (1-\alpha)\dfrac{k+2}{\sqrt{(n+2)(n+k)}} + \alpha\dfrac{(k+2)(k+1)}{min(n(n-1)+4k+2,\ n(n-1)+(k+2)(k+1)-2)}$

It should be noted that $min(n(n-1)+6,\ n(n-1)+4) = n(n-1)+4$ iff $k=1$ and $min(n(n-1)+4k+2,\ n(n-1)+(k+2)(k+1)-2) = n(n-1)+4k+2$ iff $k \ge 2$

if $(u_1, v_1) \in E_2$, we have the same cases as mentioned in the proof of 2S:

- $|\{u_1, v_1\} \cap \{u_0, v_0\}| \ne 1$. In this case, we have $s_2^\alpha(u_1, v_1) = 1$. Then,
  - if $k = 1$: k.t. $\dfrac{3}{\sqrt{(n+2)(n+1)}} < 1 \ \forall \ n \ge 4 \Leftrightarrow (1-\alpha)\dfrac{3}{\sqrt{(n+2)(n+1)}} \le (1-\alpha) \ \forall \ \alpha \in [0..1]$ and k.t. $\dfrac{6}{n(n-1)+4} < 1 \ \forall \ n \ge 4 \Leftrightarrow \alpha\dfrac{6}{n(n-1)+4} < \alpha \ \forall \ \alpha \in]0..1] \Leftrightarrow (1-\alpha)\dfrac{3}{\sqrt{(n+2)(n+1)}} + \alpha\dfrac{6}{n(n-1)+4} < 1 \ \forall \ \alpha \in]0..1]$
  - if $k \ge 2$: k.t. $\dfrac{k+2}{\sqrt{(n+2)(n+k)}} < 1 \ \forall \ 1 \le k \le n$ (proved) $\Leftrightarrow (1-\alpha)\dfrac{k+2}{\sqrt{(n+2)(n+k)}} \le (1-\alpha) \ \forall \ 1 \le k \le n, \ \forall \ \alpha \in [0..1]$ and wtp $\dfrac{(k+2)(k+1)}{n(n-1)+4k+2} < 1$ iff. $k < n$. $\dfrac{(k+2)(k+1)}{n(n-1)+4k+2} < 1 \Leftrightarrow (k+2)(k+1) < n(n-1)+4k+2 \Leftrightarrow k^2 + 3k + 2 - 4k < n^2 - n + 2 \Leftrightarrow k^2 - k < n^2 - n \Leftrightarrow k < n$. $\Leftrightarrow \alpha\dfrac{(k+2)(k+1)}{n(n-1)+4k+2} < \alpha$ iff. $k < n, \ \forall \ \alpha \in]0..1] \Leftrightarrow (1-\alpha)\dfrac{k+2}{\sqrt{(n+2)(n+k)}} + \alpha\dfrac{(k+2)(k+1)}{n(n-1)+4k+2} < 1$ iff $k < n, \ \forall \ \alpha \in]0..1]$.
- $|\{u_1, v_1\} \cap \{u_0, v_0\}| = 1$. We have $s_2^\alpha(u_1, v_1) = (1-\alpha)\dfrac{n}{\sqrt{n(n+k)}} + \alpha$. Then,
  - if $k = 1$: k.t. $(1-\alpha)\dfrac{3}{\sqrt{(n+2)(n+1)}} \le (1-\alpha)\dfrac{n}{\sqrt{n(n+1)}} \ \forall \ n \ge 4, \ \forall \ \alpha \in [0..1]$ and $\alpha\dfrac{6}{n(n-1)+4} < \alpha \ \forall \ n \ge 4, \ \forall \ \alpha \in]0..1] \Leftrightarrow (1-\alpha)\dfrac{3}{\sqrt{(n+2)(n+1)}} + \alpha\dfrac{6}{n(n-1)+4} < (1-\alpha)\dfrac{n}{\sqrt{n(n+1)}} + \alpha \ \forall \ \alpha \in]0..1]$.

- if $2 \leq k < n - 1$: k.t. $\frac{k+2}{\sqrt{(n+2)(n+k)}} - \frac{n}{\sqrt{n(n+k)}} < 0$ iff. $k < n-1$ (proved)

  and $\frac{(k+2)(k+1)}{n(n-1)+4k+2} - 1 < 0$ iff. $k < n$. (proved) $\Leftrightarrow (1-\alpha)[\frac{k+2}{\sqrt{(n+2)(n+k)}} -$

  $\frac{n}{\sqrt{n(n+k)}}] \leq 0$ iff. $k < n-1$, $\forall \alpha \in [0..1]$ and $\alpha[\frac{(k+2)(k+1)}{n(n-1)+4k+2}-1] < 0$ iff. $k <$

  n, $\forall \alpha \in]0..1] \Leftrightarrow (1-\alpha)[\frac{k+2}{\sqrt{(n+2)(n+k)}} - \frac{n}{\sqrt{n(n+k)}}]+\alpha[\frac{(k+2)(k+1)}{n(n-1)+4k+2}-1] < 0$

  iff. $k < n - 1$, $\forall \alpha \in]0..1] \Leftrightarrow (1 - \alpha)\frac{k+2}{\sqrt{(n+2)(n+k)}} + \alpha\frac{(k+2)(k+1)}{n(n-1)+4k+2} <$

  $(1 - \alpha)\frac{n}{\sqrt{n(n+k)}} + \alpha$ iff. $k < n - 1$, $\forall \alpha \in]0..1]$.

- if $k = n - 1$: let be $a = (1 - \alpha)\frac{n+1}{\sqrt{(n+2)(2n-1)}} + \alpha\frac{(n+1)n}{(n+4)(n-1)+2}$ and

  $b = (1-\alpha)\frac{n}{\sqrt{n(2n-1)}}+\alpha \Leftrightarrow a-b = (1-\alpha)\frac{n+1}{\sqrt{(n+2)(2n-1)}}+\alpha\frac{(n+1)n}{(n+4)(n-1)+2} -$

  $(1 - \alpha)\frac{n}{\sqrt{n(2n-1)}} - \alpha = [\frac{n+1}{\sqrt{(n+2)(2n-1)}} - \frac{n}{\sqrt{n(2n-1)}}] - \alpha[\frac{n+1}{\sqrt{(n+2)(2n-1)}} -$

  $\frac{n}{\sqrt{n(2n-1)}} + 1 - \frac{(n+1)n}{(n+4)(n-1)+2}]$

  $\Leftrightarrow a - b < 0$ iff. $\alpha > \dfrac{\frac{n+1}{\sqrt{(n+2)(2n-1)}} - \frac{n}{\sqrt{n(2n-1)}}}{\frac{n+1}{\sqrt{(n+2)(2n-1)}} - \frac{n}{\sqrt{n(2n-1)}}+1-\frac{(n+1)n}{(n+4)(n-1)+2}}$. Let's consider

  $f(n) = \dfrac{\frac{n+1}{\sqrt{(n+2)(2n-1)}} - \frac{n}{\sqrt{n(2n-1)}}}{\frac{n+1}{\sqrt{(n+2)(2n-1)}} - \frac{n}{\sqrt{n(2n-1)}}+1-\frac{(n+1)n}{(n+4)(n-1)+2}}$ a continuous decreasing

  function on $[4, +\infty[$. Calculating the limits : $\lim\limits_{n \to +\infty} f(n) = 0$ and

  $\lim\limits_{n \to 4} f(n) \approx 0.06 \Leftrightarrow 0 \leq f(n) \leq 0.06$. Then, $a < b$ iff. $\alpha > 0.06$.

- if $k = n$: let be $a = (1-\alpha)\sqrt{\frac{n+2}{2n}}+\alpha$ and $b = (1-\alpha)\frac{1}{\sqrt{2}}+\alpha$. Let's consider

  $g(n) = \sqrt{\frac{n+2}{2n}}$ a continuous and strictly positive decreasing function on

  $[4, +\infty[$. Calculating the limits : $\lim\limits_{n \to +\infty} g(n) = \frac{1}{\sqrt{2}}$ and $\lim\limits_{n \to 4} g(n) = \sqrt{\frac{3}{4}} \Leftrightarrow$

  $\frac{1}{\sqrt{2}} \leq g(n) \leq \sqrt{\frac{3}{4}} \Leftrightarrow g(n) \geq \frac{1}{\sqrt{2}} \Leftrightarrow (1 - \alpha)\, g(n) + \alpha \geq (1 - \alpha)\frac{1}{\sqrt{2}} + \alpha \Leftrightarrow$

  $a \geq b \Leftrightarrow$ if $k = n, s_2^\alpha(u_2, v_2) \geq s_2^\alpha(u_1, v_1)\ \forall \alpha \in [0..1]$.

  $\implies s_2^\alpha(\mathbf{u_2}, \mathbf{v_2}) < s_2^\alpha(\mathbf{u_1}, \mathbf{v_1})$ iff. $\mathbf{k \leq n - 1}$ and $\alpha > 0.06$.

if $(\mathbf{u_1}, \mathbf{v_1}) \in \mathbf{E_1}$, there are also two cases, which are the same in the proof of 2S:

– $|\{\mathbf{u_1}, \mathbf{v_1}\} \cap \{\mathbf{u_2}, \mathbf{v_2}\}| = \mathbf{0}$ or $(\mathbf{u_1}, \mathbf{v_1}) \in \mathbf{E_1} \cap \mathbf{E_3}$. In this case, $s_2^\alpha(u_1, v_1) = 1$. Then, $s_2^\alpha(u_2, v_2) < s_2^\alpha(u_1, v_1)$ iff $k < n$ (proved).

– $|\{\mathbf{u_1}, \mathbf{v_1}\} \cap \{\mathbf{u_2}, \mathbf{v_2}\}| = \mathbf{1}$. In this case, $s_2^\alpha(u_1, v_1) = (1-\alpha)\frac{n}{\sqrt{n(n+2)}}+\alpha$. Then,

  - if $k = 1$: k.t. $\frac{3}{\sqrt{(n+2)(n+1)}} < \frac{n}{\sqrt{n(n+2)}} \Leftrightarrow (1 - \alpha)\frac{3}{\sqrt{(n+2)(n+1)}} \leq (1 -$

    $\alpha)\frac{n}{\sqrt{n(n+2)}}\ \forall \alpha \in [0..1]$, $\forall n \geq 4$. k.t. $\frac{6}{n(n-1)+4} < 1 \Leftrightarrow \alpha\frac{6}{n(n-1)+4} <$

    $\alpha\ \forall \alpha \in]0..1] \Leftrightarrow (1 - \alpha)\frac{3}{\sqrt{(n+2)(n+1)}} + \alpha\frac{6}{n(n-1)+4} <. (1 - \alpha)\frac{n}{\sqrt{n(n+2)}} +$

    $\alpha\ \forall \alpha \in]0..1]$, $\forall n \geq 4$.

  - if $k \geq 2$: k.t. $\frac{k+2}{\sqrt{(n+2)(n+k)}} < \frac{n}{\sqrt{n(n+2)}}$ iff $k \leq n - 1$ (proved) $\Leftrightarrow (1 -$

    $\alpha)\frac{k+2}{\sqrt{(n+2)(n+k)}} \leq (1 - \alpha)\frac{n}{\sqrt{n(n+2)}}$ iff $k \leq n - 1$, $\forall \alpha \in [0..1]$. k.t.

$\frac{(k+2)(k+1)}{n(n-1)+4k+2} < 1$ iff $k < n$. (proved) $\Leftrightarrow \alpha \frac{(k+2)(k+1)}{n(n-1)+4k+2} < \alpha \ \forall \ \alpha \in]0..1]$

$\Leftrightarrow (1-\alpha)\frac{k+2}{\sqrt{(n+2)(n+k)}} + \alpha \frac{(k+2)(k+1)}{n(n-1)+4k+2} < (1-\alpha)\frac{n}{\sqrt{n(n+2)}} + \alpha$ iff $k \leq$

$n-1 \ \forall \ \alpha \in]0..1] \Leftrightarrow s_2^\alpha(u_2, v_2) < s_2^\alpha(u_1, v_1)$ iff. $k \leq n-1$.

References

1. Aldecoa, R., Marín, I.: Deciphering network community structure by surprise. PLoS ONE **6**, e24195 (2011)
2. Blondel, V.D., Guillaume, J.L., Lambiotte, R., Lefebvre, E.: Fast unfolding of communities in large networks. J. Stat. Mech: Theory Exp. **2008**, P10008 (2008)
3. Bollacker, K.D., Lawrence, S., Giles, C.L.: CiteSeer: an autonomous web agent for automatic retrieval and identification of interesting publications. In: Proceedings of the Second International Conference on Autonomous Agents, pp. 116–123 (1998)
4. Chen, Y.C., Zhu, W.Y., Peng, W.C., Lee, W.C., Lee, S.Y.: CIM: community-based influence maximization in social networks. .ACM Trans. Intell. Syst. Technol. (TIST) **5**(2), 1–31 (2014)
5. Enright, A.J., Van Dongen, S., Ouzounis, C.A.: An efficient algorithm for large-scale detection of protein families. Nucleic Acids Res. **30**(7), 1575–1584 (2002)
6. Fortunato, S., Lancichinetti, A.: Community detection algorithms: a comparative analysis: invited presentation, extended abstract. In: 4th International ICST Conference on Performance Evaluation Methodologies and Tools (2010)
7. Ganley, D., Lampe, C.: The ties that bind: social network principles in online communities. Decis. Support Syst. **47**(3), 266–274 (2009)
8. Krebs, V.: Books about us politics. Unpublished (2004). http://www.orgnet.com
9. Lancichinetti, A., Fortunato, S., Radicchi, F.: Benchmark graphs for testing community detection algorithms. Phys. Rev. E **78**(4) (2008). https://doi.org/10.1103/physreve.78.046110
10. Leskovec, J., Krevl, A.: SNAP datasets: stanford large network dataset collection at http://snap.stanford.edu/data (2014)
11. Lusseau, D., Schneider, K., Boisseau, O.J., Haase, P., Slooten, E., Dawson, S.M.: The bottlenose dolphin community of doubtful sound features a large proportion of long-lasting associations. Behav. Ecol. Sociobiol. **54**(4), 396–405 (2003)
12. Luís, R.: Towards data science: modularity maximization greedy algorithm. https://towardsdatascience.com/modularity-maximization-5cfa6495b286. Accessed 30 May 2020
13. Martin, J.: Theoretical computer science: Fast extraction of the edges of an induced subgraph. https://cstheory.stackexchange.com/questions/33440/fast-extraction-of-the-edges-of-an-induced-subgraph. Accessed 26 December 2015
14. Martínez, V., Berzal, F., Cubero, J.C.: A survey of link prediction in complex networks. ACM Comput. Surv. **49**(4), 1–33 (2016). https://doi.org/10.1145/3012704
15. Newman, M.E.: Fast algorithm for detecting community structure in networks. Phys. Rev. E **69**(6), 066133 (2004)
16. Raghavan, U.N., Albert, R., Kumara, S.: Near linear time algorithm to detect community structures in large-scale networks. Phys. Rev. E **76**, 036106 (2007)
17. Rossetti, G., Milli, L., Cazabet, R.: CDLIB: a python library to extract, compare and evaluate communities from complex networks. Appl. Netw. Sci. **4**, 1–26 (2019)

18. Rossetti, G., Pappalardo, L., Rinzivillo, S.: A novel approach to evaluate community detection algorithms on ground truth. In: Cherifi, H., Gonçalves, B., Menezes, R., Sinatra, R. (eds.) Complex Networks VII. SCI, vol. 644, pp. 133–144. Springer, Cham (2016). https://doi.org/10.1007/978-3-319-30569-1_10
19. Salton, G., McGill, M.J.: Introduction to Modern Information Retrieval (1983)
20. Šubelj, L., Bajec, M.: Model of complex networks based on citation dynamics. In: Proceedings of the 22nd International Conference on World Wide Web (2013)
21. Traag, V.A., Krings, G., Van Dooren, P.: Significant scales in community structure. Sci. Rep. 3(1), 1–10 (2013)
22. Traag, V.A., Waltman, L., Van Eck, N.J.: From Louvain to Leiden: guaranteeing well-connected communities. Sci. Rep. 9(1), 1–12 (2019)
23. Unknown: CDP (2022). https://github.com/2x254/CDP
24. Zachary, W.W.: An information flow model for conflict and fission in small groups. J. Anthropol. Res. 33(4), 452–473 (1977)

Improving Stochastic Gradient Descent Initializing with Data Summarization

Robin Varghese$^{(\boxtimes)}$ and Carlos Ordonez

Department of Computer Science, University of Houston, Houston, TX 77204, USA
rsvargh2@cougarnet.uh.edu

Abstract. Linear Regression (LR) is the prototypical statistical model, which can be applied on a wide range of predictive problems. Ordinary Least Squares (OLS), is the standard technique for estimating the parameters of the LR model. However, such computation can be slow and resource-hungry for large data sets with high dimensionality, due to heavy matrix operations. More importantly, OLS may be impractical for large data sets as the entire data set is required to be loaded into main memory, often exceeding RAM capacity. These limitations emphasize the need for optimization techniques to compute LR. Two state of the art algorithms used to compute LR are: Stochastic Gradient Descent (SGD) and Data Summarization (DS), combined with a matrix factorization. A few decades ago DS was the main technique to accelerate data mining computations, followed by SGD. Nowadays, SGD has become the workhorse behind most ML algorithms and deep neural networks. Merging both techniques, we propose to initialize SGD with a solution computed via DS on the initial batch of points, leaving SGD computation on the remaining points "as is". An experimental evaluation with several data sets shows our improved SGD algorithm reaches higher quality solutions (lower MSE error, higher R^2) and it converges faster (less iterations, reduced data usage, less computation time). We believe our simple SGD change can benefit many more ML models beyond LR.

1 Introduction

Linear Regression is the basis for linear models. LR is highly regarded for its simplicity and ease of implementation, which makes it accessible to users with varying levels of technical expertise. This includes researchers, analysts, and practitioners in various fields, such as finance, healthcare, social sciences, and engineering. The intuitive nature of LR also makes it useful for educators and students seeking to understand fundamental concepts in statistics and data analysis. The need to accelerate the computation of LR arises because it can be computationally expensive, particularly when dealing with large data sets. This is because LR involves solving a system of equations to find the optimal values of the coefficients, that minimize the sum of the squared errors. Furthermore as the size of the data set increases, the number of calculations required to solve these equations also increases, leading to long computation times and slow performance.

© The Author(s), under exclusive license to Springer Nature Switzerland AG 2023
R. Wrembel et al. (Eds.): DaWaK 2023, LNCS 14148, pp. 212–223, 2023.
https://doi.org/10.1007/978-3-031-39831-5_20

Gradient descent (GD) has been making significant strides in the field of machine learning (ML) also for its simplicity, but more importantly its ability to be applied to a wide variety of problems [15]. It can be said to be the definitive optimization algorithm in ML. Logistic regression builds on the concepts of LR and is a step towards creating non-linear models. Neural networks, used in both statistical models, have their quality and speed affected by GD's initialization. Many aspects and properties of GD for LR have been extensively researched including, optimal learning rates, learning rate schedulers, cost functions and data shuffling [4,10,14]. Furthermore from [16], we see how initialization can have significant impact on model performance. Compared to the traditional base case initializations of 0 and 1, we can see from [3] how random initialization can result in better model performance. In this paper we propose using data summarization as weight initialization. Additionally there are many variations of GD, most commonly Stochastic Gradient Descent (SGD) and mini-batch SGD. GD is computed on the entire data set, SGD on a single point, and mini-batch SGD on a batch of data. Mini-batch SGD is often favored because it is a balance between GD and SGD. This paper implements mini-batch SGD, but is referred to throughout the paper as simply SGD.

Gamma (Γ) is but one form of DS and can be used to integrate sufficient statistics (SS) of an input data set into a single matrix. Then, the computation of many ML models and tasks such as Naive Bayes, Principal Component Analysis (PCA), Linear Discriminant Analysis and Linear Regression can be accelerated [11]. The authors of [1], compute Γ in Python and have shown to be as fast as popular Python ML libraries including scikit-learn (sklearn) [13].

Our contributions and experimental findings show that, using data summarization for initialization in gradient descent can result in:

1. Faster convergence with less data
2. Lower model error
3. Higher quality (R^2)

The paper is organized as follows: Sect. 2 presents background information, key concepts, definitions, and notations that are mentioned throughout the paper. Section 3 details the system and theoretical aspects of the underlying related algorithms and our contribution. Section 4 provides an overview of the experimental setup, implementation details, results, and analysis. In Sect. 5, we explore related works in gradient descent optimizations and initialization techniques. Finally, in Sect. 6 we give concluding remarks and discuss the potential for future research.

2 Definitions

2.1 Input Data Set

Each input data set is defined as X. Here, X is a $d \times n$ matrix, equivalent to a set of n column vectors corresponding to each observation in the data set, having d

dimensions or attributes. In the case of LR, augment X with a row vector of n ones along the X_0 dimension producing the $(d+1) \times n$ matrix $\mathbf{X}$. $\mathbf{Y}$ is a n row vector corresponding to the output for each n observation.

2.2 LR Model

Given a $(d+1) \times n$ input matrix $\mathbf{X}$, and a $(d \times 1)$ column vector $\hat{\beta}$ of coefficients, compute $\hat{\mathbf{Y}}$. $\hat{\mathbf{Y}}$ is a n row vector, corresponding to the predicted outputs of each n observation, produced by the LR model.

$$\hat{\mathbf{Y}} = \hat{\beta}^T \mathbf{X} + \epsilon \tag{1}$$

To achieve a close approximation of $\mathbf{Y}$ the true predicted outputs, fit the LR model Eq. 1. The model can be fit using many methods most commonly, Ordinary Least Squares (OLS). This method finds the $\hat{\beta}$ that minimizes the Residual Sum of Squares (RSS), resulting in $\hat{\mathbf{Y}} \approx \mathbf{Y}$ [5].

$$RSS = (\mathbf{Y} - \hat{\beta}^T \mathbf{X})(\mathbf{Y} - \hat{\beta}^T \mathbf{X})^T \tag{2}$$

To minimize RSS, differentiate w.r.t. $\hat{\beta}$ and set equal to 0, deriving the unique solution:

$$\hat{\beta} = (\mathbf{X}\mathbf{X}^T)^{-1}\mathbf{X}\mathbf{Y}^T \tag{3}$$

Given the trained coefficients $\hat{\beta}$ from fitting the model in Eq. 3, the model can be utilized to make predictions on new or future observations.

3 System and Algorithms

3.1 Gamma Summarization (Γ)

For computing gamma (Γ), augment $\mathbf{X}$ with $\mathbf{Y}$. In general, this $(d+2) \times n$ matrix is defined as $\mathbf{Z}$, but in practice is $(d+2) \times c$, where c is equivalent to chunk size. It should be emphasized that c is considered a hyper-parameter and can vary to achieve optimal performances. However for our experiments c is held constant to give a fair comparison between the algorithms, Γ and SGD. The terms $\mathbf{X}\mathbf{X}^T$ and $\mathbf{X}\mathbf{Y}^T$ in Eq. 3 and other sufficient statistics (SS), are contained in the single matrix Γ. Therefore the computation of Γ becomes a two phase algorithm as follows, Phase 1: Obtain SS stored in Γ; Phase 2: Solve $\hat{\beta}$ exploiting SS.

$$\mathbf{Z} = \begin{bmatrix} 1 & 1 & \dots & 1 \\ x_{11} & x_{12} & \dots & x_{1c} \\ x_{d1} & x_{d2} & \dots & x_{dc} \\ y_{11} & y_{12} & \dots & y_{1c} \end{bmatrix} \tag{4}$$

Phase 1. Begin by reading one chunk of the input data set of size $d \times c$. Augment this chunk with a row vector of c ones and $\mathbf{Y}$ (dependant variable) also a c row vector. This will produce $\mathbf{Z}$, a $(d+2) \times c$ dense matrix. Compute $\mathbf{ZZ}^T$ to produce the partial gamma Γ_i. It should be noted Γ is computed on the entire data set, but because only the first chunk is used, we utilize partial gamma Γ_i to initialize SGD.

$$\Gamma_i = \mathbf{ZZ}^T = \begin{bmatrix} n & L^T & \mathbf{1}^T \cdot Y^T \\ L & Q & XY^T \\ Y \cdot \mathbf{1} & YX^T & YY^T \end{bmatrix} \tag{5}$$

The sufficient statistics (SS) L, n and Q are defined as follows: $n = |X|$, $L = \sum_{i=1}^n x_i$, and $Q = XX^T = \sum_{i=1}^n x_i \cdot x_i^T$, n is total number of points in the data set, L is the linear sum of x_i and Q is the sum of vector outer products of x_i. It should be noted that Phase 1 takes majority of the computation time, between the two phases.

Phase 2. Begin Phase 2 by exploiting the sufficient statistics integrated into the single matrix Γ_i to further compute a ML model or task. In this case, the task is to compute the regression coefficients $\hat{\beta}$ also seen in Eq. 3. The sufficient statistic Q is exploited by substituting it into the OLS analytical solution (Eq. 3) as shown in Eq. 6. It should be emphasized that the initial chunk is no longer needed and is summarized within the significantly smaller matrix Γ_i. This allows $\hat{\beta}$ to be solved in main memory in $O(d^3)$.

$$\hat{\beta} = (\mathbf{XX}^T)^{-1}\mathbf{XY}^T = Q^{-1}(\mathbf{XY}^T) \tag{6}$$

3.2   Mini-batch SGD

The closed-form solution of LR as shown in Eq. 3, is computationally expensive. An alternative would be to implement gradient descent, where MSE is iteratively optimized until reaching a local minimum Eq. 7, 8. In mini-batch SGD, Mean Squared Error (MSE) is typically used over RSS because MSE averages the squared errors, providing normalization that makes training more scalable. As the number of data points increases, there are more residuals to square and sum, leading to large RSS values regardless if the errors are relatively small. This means that even if a model is making relatively good predictions, the RSS value can still be large simply due to the large number of data points. Mathematically, MSE is equivalent to: MSE = RSS/c

$$f = MSE = \frac{1}{c}(\mathbf{Y} - \hat{\beta}^T\mathbf{X})(\mathbf{Y} - \hat{\beta}^T\mathbf{X})^T \tag{7}$$

After each chunk, obtain the gradient ∇f by taking the partial derivative w.r.t $\hat{\beta}$ of Eq. 7, resulting in Eq. 8.

$$\nabla f = \frac{1}{c}\sum_{i=1}^c -2x_i(y_i - (\hat{\beta}x_i + \hat{\beta}_0)) \tag{8}$$

$$\hat{\beta} = \hat{\beta} - \alpha \nabla f \qquad (9)$$

By utilizing the gradient for the processed chunk and scaled by the learning rate α, update $\hat{\beta}$ with its new value. After each update, $\hat{\beta}$ iteratively minimizes the models loss until reaching convergence.

3.3 Mini-batch SGD Initialization Using Gamma

Our contribution improves SGD by incorporating Γ as initialization. Although this may seem like a small and simple alteration, it has yielded significant and impactful results. Depending on the initialization method, the number of updates required to reach convergence can vary greatly. Therefore, we propose computing Γ on a chunk and using the computed $\hat{\beta}$ to initialize Mini-Batch SGD. In general, the algorithm will take an input data set X of d features, n observations, and the corresponding outputs for each observation $\mathbf{Y}$. As output, the trained $\hat{\beta}$'s, R^2, and MSE are returned. The outline of our proposed approach, as seen in Fig. 1, is as follows:

1. Read one chunk of data
2. Compute Γ_i (Phase 1)
3. Exploit SS to compute $\hat{\beta}$ (Phase 2)
4. Initialize Mini-Batch SGD weights using the previously computed $\hat{\beta}$ from step 3
5. Begin model training using the initial and remaining chunks

It is important to note that Γ is computed only on the first chunk. Then Mini-Batch SGD begins training as standard, continuously and sequentially reading in chunks until convergence. Additionally, Γ can be computed sequentially and also more favorably, in parallel [11]. This highlights another aspect where Γ can help aid in Mini-Batch SGD.

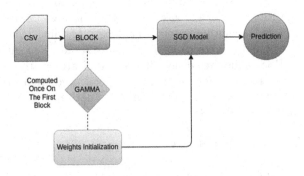

Fig. 1. Gamma Initialization System Design.

4 Experiments

4.1 Experimental Setup

The YearPredictionMSD data set was obtained from the publicly available, UCI machine learning repository [7]. The creators of the YearPredictionMSD data set [2], created it to serve as a benchmark data set for regression tasks in ML, particularly for the problem of predicting the release year of a song, given a set of musical features. The data set was released as part of the Million Song Dataset project, which aimed to provide a comprehensive data set for music analysis and recommendation systems. The data set consists of $n = 515,345$ songs and $d = 90$ features. Each song is represented by a 90-dimensional feature vector, which includes information such as tempo, timbre, and loudness. The goal of the regression task is to predict the year in which the song was released, given the 90-dimensional feature vector. Additionally as stated in the original data set, the following train/test split should be respected to avoid the 'producer effect' (making sure no song from a given artist ends up in both the train and test set):

1. train: first 463,715 examples
2. test: last 51,630 examples

The California Housing data set can be obtained directly from sklearn (`sklearn.datasets.fetch_california_housing`). The data set is also a widely used LR benchmark in machine learning. It contains information on the median house prices, the number of households, the median income, and other factors in various neighborhoods across California. It was collected from the 1990 U.S. Census, and consists of $n = 20,640$ census block groups and $d = 10$ features [12]. A census block group is the smallest geographical unit the Census Bureau uses to collect data. The target variable for the model is the median house value (in units of 100,000 dollars). Since no train/test split is given by the authors, we use sklearns `sklearn.model_selection.train_test_split` method:

1. train: 70%
2. test: 30%

Table 1. Data sets.

Data set	Description	n	d
YearPredictionMSD	Song Year Prediction	515,345	90
California Housing	Median House Value	20640	9

Using k-fold cross-validation may not be necessary for comparing different initialization methods of SGD because cross-validation is typically used to assess how well a model performs on new, unseen data. However, when testing initialization methods, the main goal is to compare their effects on the optimization process of SGD, rather than evaluating the model's performance on unseen data.

The system used for the experiments is a Pentium(R) Quadcore CPU running at 1.60 GHZ, 8 GB RAM, 1 TB storage and with Linux Ubuntu as the operating system. Pre-processing and standardizing the data is important in gradient descent to prevent under/overflows occurring resulting in inaccurate results. Standardization is a common pre-processing step used in machine learning to transform data so that it has a mean of 0 and a standard deviation of 1. We utilize sklearns StandardScaler() on the data set to achieve this. To read chunks of the .csv, we use the Pandas data frame library [9]. Pandas is a powerful tool for data analysis and manipulation in Python. It provides a fast and efficient way to handle data in a variety of formats, including CSV, Excel, SQL databases, and more.

We use sklearn's SGDRegressor as an honest SGD implementation. Sklearn's SGD, internally has many optimizations and automatic hyper-parameter tuning. Most importantly, the initial learning rate, learning rate scheduler, and early stopping. For fair comparisons between initialization methods, we keep all hyper-parameters constant between trials and default to what sklearn provides. This includes using the default learning rate and default inverse scaling scheduler, in addition to early stopping. Only the initialization method is changed before training begins. Early stopping is a regularization technique commonly used in SGD to prevent overfitting and improving generalization. Overfitting occurs when the model learns the noise in the training data and fails to generalize well to new, unseen data. In our experiments, when the model does not achieve a lower error (MSE) after e steps, we stop training. We tested using 5, 10, and 20 steps however we only include 10 and 20 in this paper as there were little to no change in accuracy. This highlights one of the main benefits of SGD over OLS. That is, being able to approximately reach the optimal accuracy without requiring training on the entire data set. Once the weights are computed by Γ, we are easily able to initialize the sklearn model using the `reg.coef_` attribute. For 0 and 1 initialization, we set the `reg.coef_` attribute respectively. For random initialization, we simply initialize the sklearn SGD object and begin the training loop. In the case of Γ initialization, we use the first initial chunk to compute Γ, and then set `reg.coef_` using the produced β's. Additionally, we used 1% of the total data set as chunk size for both the initial Gamma chunk and SGD chunks. In the case of the YearPredictionMSD data set, chunk size was 5153, 1% of 515,345. For the California Housing data set, chunk size was 144 1% of 14,448. Due to the randomness factor in SGD, the results are taken as the average over 100 trials in addition to the experiment being ran multiple times with identical averages each time.

4.2   Experimental Results

The YearPredictionMSD data set is a challenging regression problem, as the relationship between the musical features and the release year is complex and non-linear. This can be verified by the low R^2 value (0.39) produced by computing OLS on the entire data set. The coefficient of determination (R^2), is a commonly used metric in LR because it measures the proportion of variance in

Table 2. R^2 Results For Initialization (YearPredictionMSD: $R^2 = 0.39$, California Housing: $R^2 = 0.59$), $e = $ early stopping steps.

Data set	e	Gamma SGD	Random SGD	0 SGD	1 SGD
YearPredictionMSD	10	**0.26**	0.23	0.23	0.23
California Housing	10	**0.60**	0.58	0.58	0.56
YearPredictionMSD	20	**0.27**	0.24	0.23	0.23
California Housing	20	**0.60**	0.58	0.58	0.56

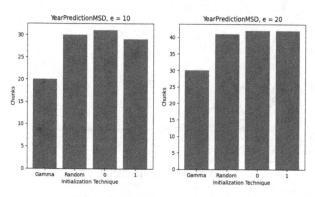

(a) YearPredictionMSD

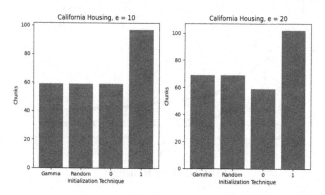

(b) California Housing

Fig. 2. Number of chunks used in training.

the dependent variable explained by the independent variables included in the model. An R^2 score of 1.0 indicates that the model perfectly fits the data. On the contrary, the California Housing data set has a relatively higher R^2 value (.59), indicating much more linearity in comparison.

Table 3. Speed Comparisons: OLS vs. SGD with Gamma, Random, 0, and 1 Initializations.

Data set	OLS	Gamma SGD	Random SGD	0 SGD	1 SGD
YearPredictionMSD	26.517	**4.354**	5.583	5.694	5.516
California Housing	**0.942**	1.152	1.174	1.111	1.258

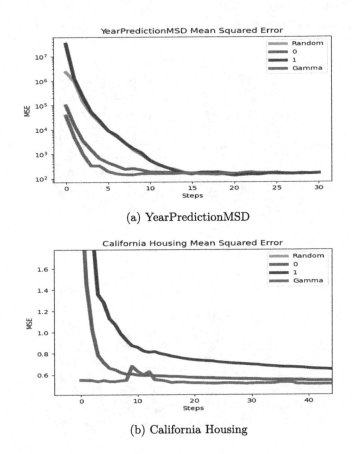

(a) YearPredictionMSD

(b) California Housing

Fig. 3. Model MSE during training.

Discussion. Our analysis of the performance of different models on the YearPredictionMSD data set, as shown in Table 2 and Fig. 2(a), demonstrates that Gamma outperforms other models in terms of R^2 value despite using significantly less data. Whereas randomization the second best performing model used 30 chunks, Gamma used only 20 showing a 1.5× (30/20) increase in training efficiency. This can be attributed to Gamma's better ability to handle non-linear and noisy data. As previously mentioned, the YearPredictionMSD data set is challenging due to its low optimal (OLS) R^2, which makes it difficult for linear models to learn effectively. Therefore, the relative improvements in R^2 and data usage achieved by Gamma is noteworthy.

Conversely our results on the significantly more linear data set California Housing, across the models are identical. While Gamma does achieve a higher R^2, from Fig. 2(b) it is clear there is no longer a major improvement in training efficiency. This highlights that while Gamma provides advantages for non-linear data sets, it also does not perform worse than other methods on linear data sets.

We also verify SGD's efficiency over OLS in Table 3. On the YearPrediction-MSD data set, we can see OLS taking a substantial 26.5 s to compute. Using Gamma initialization, SGD is able to converge in 4.3 s, a speed up of approximately 6× (26.5/4.3). In order to identify potential findings, trends, or avoiding local minima and to ensure that each initialization technique received a fair evaluation, we conducted initial experiments with a maximum of 5 steps for early stopping. However, the resulting accuracy's were below acceptable levels with all models reaching below $R^2 = 0.18$. We then increased the maximum number of steps to 10, which led to a significant improvement in accuracy, approaching optimality. However, our experimental findings show that increasing the number of steps further to 20 did not result in any noticeable performance improvements, as the performance of the initialization techniques remained similar. We also did not include the plot for each models MSE when $e = 20$ steps for this reason, as the trend is same. From Fig. 3, the intuition behind Gamma initialization becomes clear. SGD is able to begin its descent from a smaller peak and eventually reach convergence.

5 Related Work

In this section, we overview previous work in SGD optimizations and weight initialization.

SGD Optimizations: Although gradient descent algorithms are widely used, it is still often viewed as a black-box optimizer. As a result, research in gradient descent has produced many improvements in the algorithm.

Optimal hyper-parameter tuning is vital during training and can lead to significant improvements [16]. Specifically, the authors show the importance of momentum and initialization in performance. Momentum is a fundamental method in learning rate scheduling. Intuitively, momentum will accelerate the progress towards the steepest descent thereby, accelerating convergence in comparison to a constant learning rate. Additionally ADAGRAD and ADADELTA are first and second order methods respectively that improve upon momentum and are commonly used [8,17]. ADAGRAD has been shown to be well-suited to sparse data sets as the learning rate adapts towards the frequency of parameter updates. Its extension ADADELTA, addresses the problem of ADAGRAD's aggressive decreasing of learning rate.

In addition to hyper-parameter tuning, research has also produced many variants in the overall gradient descent algorithm. Depending on the training environment, modifications to the algorithm such as batch size, density, or sparsity can be leveraged to improve performance [15].

Initialization: Weight initialization is a vital step before training of a neural network begins. During training, the weights are repeatedly updated until the models loss or error converges to a minimum value. Therefore, weight initialization directly affects the convergence or training time of a model [6]. For its simplicity and versatility, randomization is a well known technique that can be found all across machine learning [3].

6   Conclusions

SGD is a building block for complex machine learning algorithms such as neural networks. Improving initialization of SGD can lead to faster and more accurate predictions. In this work, we propose a method for using DS to accelerate the computation of SGD. Our results show that this method can lead to faster convergence with less data, higher R^2 values, and lower error (MSE) in LR models. We provide experimental results on the YearPredictionMSD data set, a challenging regression problem with a complex and non-linear relationship between the musical features and the release year. Our analysis demonstrates that Γ initialization outperforms other models in terms of R^2 value, particularly for non-linear and noisy data. Additionally, our results on the more linear California Housing data set show that Γ initialization is equally effective as other methods. Our proposed change can be easily integrated into common Python SGD implementations. Furthermore, we provide an example in sklearn.

In future work, we plan to improve the quality of sparse SGD for LR implementations, Logistic Regression, and explore Γ as a hyper-parameter. Sparse SGD offers memory and computational efficiency by focusing only on non-zero features, leading to faster computation times and reduced memory consumption compared to dense SGD. In complement to LR, logistic regression is often viewed as the first step towards a non-linear model. This may broaden the possibilities for our research to delve into the extensive realm of classification applications. In this study, we propose the use of DS on a single data chunk to initialize SGD and subsequently continue with the standard model training. However, further research might investigate how the number of chunks or the amount of data used during the initialization stage with DS compares to the number of chunks used during the SGD model training.

References

1. Al-Amin, S.T., Ordonez, C.: Incremental and accurate computation of machine learning models with smart data summarization. J. Intell. Inf. Syst. (JIIS) **59**, 149–172 (2022). https://doi.org/10.1007/s10844-021-00690-5
2. Bertin-Mahieux, T., Ellis, D.P.W., Whitman, B., Lamere, P.: The Million Song Dataset (2011). https://labrosa.ee.columbia.edu/millionsong/

3. Chen, Y., Chi, Y., Fan, J., Ma, C.: Gradient descent with random initialization: fast global convergence for nonconvex phase retrieval. Math. Program. **176**, 5–37 (2019). https://doi.org/10.1007/s10107-019-01363-6

4. Hu, T., Wu, Q., Zhou, D.X.: Convergence of gradient descent for minimum error entropy principle in linear regression. IEEE Trans. Sig. Process. **64**(24), 6571–6579 (2016)

5. James, G., Witten, D., Hastie, T., Tibshirani, R.: An Introduction to Statistical Learning, vol. 112. Springer, New York (2013). https://doi.org/10.1007/978-1-4614-7138-7

6. Kumar, S.K.: On weight initialization in deep neural networks. arXiv preprint arXiv:1704.08863 (2017)

7. Lichman, M.: UCI machine learning repository (2013). http://archive.ics.uci.edu/ml

8. Lydia, A., Francis, S.: Adagrad-an optimizer for stochastic gradient descent. Int. J. Inf. Comput. Sci. **6**(5), 566–568 (2019)

9. McKinney, W.: Data structures for statistical computing in Python. In: van der Walt, S., Millman, J. (eds.) Proceedings of the 9th Python in Science Conference, pp. 56–61 (2010). https://doi.org/10.25080/Majora-92bf1922-00a

10. Meng, Q., Chen, W., Wang, Y., Ma, Z.M., Liu, T.Y.: Convergence analysis of distributed stochastic gradient descent with shuffling. Neurocomputing **337**, 46–57 (2019). https://doi.org/10.1016/j.neucom.2019.01.037. https://www.sciencedirect.com/science/article/pii/S0925231219300578

11. Ordonez, C., Zhang, Y., Cabrera, W.: The Gamma matrix to summarize dense and sparse data sets for big data analytics. IEEE Trans. Knowl. Data Eng. (TKDE) **28**(7), 1906–1918 (2016)

12. Pace, R.K., Barry, R.: Sparse spatial autoregressions. Stat. Probab. Lett. **33**(3), 291–297 (1997)

13. Pedregosa, F., Varoquaux, G., Gramfort, A., Michel, V.: Scikit-learn: machine learning in Python. J. Mach. Learn. Res. **12**, 2825–2830 (2011)

14. Picheny, V., Dutordoir, V., Artemev, A., Durrande, N.: Automatic tuning of stochastic gradient descent with Bayesian optimisation. In: Hutter, F., Kersting, K., Lijffijt, J., Valera, I. (eds.) ECML PKDD 2020. LNCS (LNAI), vol. 12459, pp. 431–446. Springer, Cham (2021). https://doi.org/10.1007/978-3-030-67664-3_26

15. Ruder, S.: An overview of gradient descent optimization algorithms. arXiv preprint arXiv:1609.04747 (2016)

16. Sutskever, I., Martens, J., Dahl, G., Hinton, G.: On the importance of initialization and momentum in deep learning. In: International Conference on Machine Learning, pp. 1139–1147. PMLR (2013)

17. Zeiler, M.D.: ADADELTA: an adaptive learning rate method. arXiv preprint arXiv:1212.5701 (2012)

Feature Analysis of Regional Behavioral Facilitation Information Based on Source Location and Target People in Disaster

Kosuke Wakasugi[1](✉), Futo Yamamoto[1], Yu Suzuki[2], and Akiyo Nadamoto[1]

[1] Konan University, Okamoto 8–9–1, Higashinada–ku, Kobe, Japan
m2324004@s.konan-u.ac.jp, nadamoto@konan-u.ac.jp
[2] Gihu University, 1–1, Yanagido, Gifu–shi, Gifu 501-1193, Japan
ysuzuki@gifu-u.ac.jp

Abstract. In the event of a disaster, there is information related to social networking services (SNSs) that is specific to a region and which encourages readers to take action. For this study, we designate this information "regional behavioral facilitation information". This information is extremely important for disaster victims because it is widely dispersed on SNSs. To convey that necessary information to disaster victims quickly and accurately, we generate and analyze a model that classifies the regional behavioral facilitation information by source location (disaster or non-disaster area) and target people (disaster victims or non-victims). We target typhoons as a disaster type, and they have different characteristics. We find that a model learned from a single typhoon data is insufficient for classification. For this study, we propose a model that can classify typhoons appropriately with features that differ from those of the training data. Specifically, we combine three typhoon data with different characteristics and create a classification model for each combination. Then, we compare the classification accuracies of different models.

Keywords: Disaster · Social Network Systems · Twitter · Classification

1 Introduction

In recent years, many disasters such as typhoons, hurricanes, earthquakes, and heavy rains have occurred. Disaster events generate large amounts of disaster-related information on social networking services (SNSs). Among such information, promoting (or discouraging) behavior is particularly influential to the readers. Therefore, we have analyzed "Behavioral facilitation tweets (hereafter BF tweets)" on Twitter, which promote or discourage the reader's behavior, with typhoons as the target events. Results demonstrate that information that includes words indicating the disaster area in the BF tweets has more influence on the readers. We have proposed a method for automatically extracting these tweets, which include words that indicate the disaster area, and analyzed

R. Wrembel et al. (Eds.): DaWaK 2023, LNCS 14148, pp. 224–232, 2023.
https://doi.org/10.1007/978-3-031-39831-5_21

the extracted tweets. We call such BF tweets that include local information "Regional behavioral facilitation tweets(hear after RBF tweets)". Furthermore, since RBF tweets have various types of information, sending the necessary information to disaster victims quickly and accurately is difficult. We have proposed a method to classify RBF tweets by source location (disaster area, non-disaster area) and target people (disaster victims, non-victims), and have subsequently conducted feature analyses based on the combination of these four categories [8]. This method makes it possible to convey information to disaster victims quickly and accurately.

Although there are many kinds of disasters, this research specifically examined typhoons, because, with global warming, typhoons, and hurricanes are becoming more frequent. Moreover, their associated damage is increasing yearly. Our former proposed classification model targets a single typhoon. However, typhoons have various characteristics, such as wind typhoons with strong winds and rain typhoons with heavy rains. The associated information can be expected to differ depending on the typhoon's characteristics and scale. The most important information can be expected to be different. Therefore, we consider that our proposed RBF tweet classification model for a single typhoon cannot extract typhoons with different features with sufficient accuracy. Our proposed model is not sufficiently accurate to accommodate typhoons with various characteristics. A model must be able to extract RBF tweets quickly and accurately when a typhoon strikes an area. As described herein, we propose a method to extract RBF tweets by generating a model that combines multiple typhoons with various characteristics, thereby accommodating large-scale typhoons, small-scale typhoons, rain typhoons, and wind typhoons. Specifically, we target the following three typhoons: From data of three typhoons, we generate a classification model that uses one typhoon data as training data and a classification model that combines the two typhoon data as training data. Then we conduct an experiment to compare the accuracy of each classification model. Finally, we compare the extraction accuracies of the respective models using test data of each typhoon and data of different typhoons. Findings obtained from this research will enable the rapid transmission of necessary and accurate RBF information to those who need it when typhoons strike.

2   Related Work

Many studies have been conducted to extract disaster information from Twitter messages. Paul et al. [6] collect hurricane-related tweets from 2012 through 2018 and use machine learning to categorize tweets about power outages and communication outages. Yasin et al. [4] use machine learning to categorize disaster-related tweets for floods of six types. By doing this classification, they identify the necessary information to support the user. Gao et al. [3] propose the extraction of disaster-related tweets from disaster tweets with different characteristics using a multimodal adversarial neural network. These studies show high accuracy in classification using data from multiple disasters as training data. These studies show features and methods which are similar to those of the present study. However, the present study differs from earlier ones in that it incorporates consideration of the combination of each feature to create a model that can properly classify typhoons.

Some studies identify where tweets are sent. Ahmed et al. [1] create a hierarchical model of tweet topics. They identify the location of tweets based on the relation between topic and location. Tian et al. [7] filter out tweets that are irrelevant to geographic attributes. They propose a user-location inference method based on a heterogeneous graph of relationships among users whose locations are known and those whose locations are not known, and between users and location words. Dutt et al. [2] propose a real-time location extraction method from tweets posted during an emergency by using hashtags and by creating a suffix list containing words after place names. Whereas these studies identify the locations of the affected areas by extracting relationships among users or place names, our study differs in that it identifies whether the tweets are created from the affected or non-affected area based on the context.

3   Basic Concept of RBF Tweet Classification

3.1   Extraction of BF Tweets

We have proposed a method for automatically extracting BF tweets in the event of a large-scale disaster. For such times, we compare five methods: rule-based, LSTM, Bi-LSTM, BERT, and RoBERTa [5]. Findings from our experiments indicate that our model based on RoBERTa provides the best accuracy [8]. As described in this paper, we use RoBERTa to extract BF tweets. We also use PyTorch[1], a machine learning library, to implement RoBERTa. We use the RoBERTa Japanese pre-trained model[2] for the pre-training model. After inputting them into the RoBERTa Japanese pre-trained model, we use the embedding representation obtained from the final layer of RoBERTa. We determined the hyperparameters for use with RoBERTa by grid search and set the hidden layer to 12, the vector size to 768, the batch size to 32, the number of epochs to 5, the learning rate to 0.001, and the dropout rate to 0.1. We use Typhoon 15 of 2019 to create the BF tweet extraction model. The data used are 12,215 BF tweets and non-BF tweets, for a total of 24,430 tweets.

3.2   RBF Tweet Extraction and Classification

Similarly to an earlier study [8], we define an RBF tweet as a tweet that includes at least one word that can identify the disaster area in a BF tweet. We regard a tweet that contains one or more region-specific words as an RBF tweet. By classification for RBF tweets according to their origin (source location) and by their destination (target person), we can present the information to those who

---

[1] https://pytorch.org/.
[2] https://huggingface.co/nlp-waseda/roberta-base-japanese.

need it. Therefore, using the method of our earlier research [8], we classify the extracted BRF tweets by source location (disaster area and non-disaster area) and target people (disaster victims and non-victims).

Classification of Source Location

We classify the source locations of the extracted tweets into disaster and non-disaster areas. We use RoBERTa for source location classification. First, the system performs morphological analysis of tweets excluding URLs and Twitter user names using Juman++. Next, the system inputs them into the RoBERTa, Japanese pre-trained model, and uses word embedding obtained from the final layer of RoBERTa as the word feature vector. Word embedding becomes the input to the fully connected layer. The system calculates fine-tuning to infer whether the tweet is associated with a disaster area or a non-disaster area. We determine the parameters of RoBERTa using grid search. For RoBERTa, we use 12 hidden layers, vector size of 768, batch size of 32 or 16, the number of epochs of 5, the learning rate of 0.00002 or 0.00003, and the dropout rate of 0.1. The optimizer is Adam.

Classification of Target People

As with the source location, we classify the target people of the extracted tweets into disaster victims and non-victims. We also use RoBERTa for classification of the target people. The process and parameters are the same as those for source location classification.

4   Analysis of RBF Tweets

We have proposed a model using a single typhoon as training data. Then we analyzed the model using the same typhoon. Results demonstrate that the proposed model can classify the source location and target people. As described in this paper, we propose a model that can accommodate typhoons of various types. The proposed model can respond quickly and accurately to new typhoons.

4.1   Training and Test Data

Although typhoons are of various types, we generate models for each using typhoons of three types, as described below. (A) is a windy and rainy typhoon with large-scale, localized damage (mainly in Chiba prefecture) (Typhoon No. 15 in 2019). (B) is a wind typhoon with medium-scale, widespread damage across Japan (Typhoon No. 14 in 2022). (C) is a rainy typhoon with small-size, localized damage (Shizuoka Prefecture)(Typhoon No. 15 in 2022). The numbers of BF tweets for the typhoons are (A) 12,215, (B) 67,378, and (C) 51,599. The number of RBF tweets for the typhoons are (A) 4,025, (B) 9,932, and (C) 6,994. The data acquisition procedures for RBF tweet classification are presented below.

1. Extract each typhoon tweet using the typhoon's name as a query.

Table 1. Number of Disaster data.

No	Disaster	source location			target people		
		disaster area	non-disaster area	sum	disaster victims	non-victims	sum
1	A	1,240	1,240	2,480	985	985	1,970
2	B	970	970	1,940	1,600	1,600	3,200
3	C	1,740	1,740	3,480	1,825	1,825	3,650

2. Extract BF tweets using our proposed method from the tweets extracted in (1).
3. From the BF tweets extracted in (2) using our proposed method, extract RBF tweets that include words which identify disaster areas. Here, the words which specify the disaster area at the time of each typhoon are determined as the target area of the main disaster area.
4. Label the source location of the extracted RBF tweets as "disaster area", "non-disaster area" or "unknown" using crowdsourcing. Five subjects judged each RBF tweet. The RBF tweets which are labeled as either affected or unaffected by the disaster by at least three subjects judge using the same labeling were regarded as correct data.
5. We also use crowdsourcing to label tweets of three types: disaster victim, non-victim, and unknown. The tweets labeled as "disaster victims" and "non-victims" are used as correct data. The numbers of data used for experimentation are shown in Table 1.

4.2   Research Question

We use the following research queries to investigate the composite model.

RQ1: Can a model using one typhoon data as training data classify other typhoon data?

We use data of one typhoon as training data and data of another typhoon for which data have not been used as training data as test data. We consider that mixing multiple typhoon data is not necessary if data of one typhoon are useful to train other typhoon data. Specifically, we use the data of each of (A), (B), and (C) as training data, and calculate fine-tuning to classify the source location (disaster area, non-disaster area) and the target people (disaster victims, non-victims), respectively, to generate three source location classification models and three target people classification models. Then we calculate the accuracy of each model using another typhoon data as test data.

RQ2: Can a model with two typhoon data as training data classify other typhoon data?

We combine two typhoon data as training data, such as (A) and (B), (B) and (C), and (C) and (A), and perform fine-tuning of the source location and target people, respectively. Thereby, three source location classification models and three target people classification models are generated. Then, we calculate the accuracy of each model using the training data and different typhoon data as test data. If RQ2 is successful, then the model can be expected to be able to cope with new typhoons. Specifically,

4.3   Results and Discussion of Research Questions

We have conducted an experiment using our proposed models. We performed five-cross validation using the model trained on 80% of the training data and the remaining 20% of the test data for each Research Question.

Results and Discussion of RQ1

We evaluated whether the proposed model can learn the same data. Table 2(I) shows that the model is able to classify the same data. Next, we conducted an experiment using different typhoon data as test data, to ascertain whether our proposed model can classify them or not. Table 2 (II) presents shows the results for source location, and the results for the target people. From the results obtained for No. 1 and No. 2 of data for typhoon (A), the model trained only with the training data for typhoon (A) shows a good precision of 0.897 and 0.844 for the disaster area. The good results obtained for precision for rain typhoons (B) and (C) are attributable to the fact that typhoon (A) is a rain and wind typhoon. However, the recall of disaster-area for typhoon (A) is low because typhoon (B) and typhoon (C) were unaffected by wind, which is a characteristic of typhoon (A), resulting in low recall. This result is the same as a result found for the target people. Furthermore, the respective precision results obtained for non-disaster area and non-victims are low because many tweets related to typhoons were sent to non-disaster area from non-disaster area.

The model trained on typhoon (B) was found to have good recall of more than 0.9, whereas the precision value for Nos. 3 and 4 was 0.68 for the disaster area. Recall of tweets from non-disaster area was also low because the disaster area of typhoon (B) was a wide area, and most of the tweets from typhoon (A) and typhoon (C), where the disaster area was a small area, were extracted for the disaster area. Consequently, many inappropriate tweets were included in the extracted data, The precision was low. Moreover, the precision values found for the non-disaster area of typhoon (A) and typhoon (C), where the non-disaster areas are wide areas, was poor. For typhoon (C), the results for typhoon (B) are slightly worse than those reported for typhoon (A). The reason is that typhoon (C) and typhoon (A) are small areas. As described above, the training data of a single typhoon might or might not be sufficient to classify typhoons based on the learned typhoon feature inputs. Based on these results, we inferred that we can classify the BF tweets by the model trained by mixing several typhoons.

Results and Discussion of RQ2

Table 2 (III) presents classification results for the other disasters using the model with the respective mixed data. The results are good, except for No. 3. In the case of No. 3, where typhoon (A) and typhoon (C) are the training data and typhoon

Table 2. The result of each Research Question.

(I) Same typhoon

No	disaster		disaster area			non-disaster area			AUC
	training data	test data	Precision	Recall	F-measure	Precision	Recall	F-measure	
1	A	A	0.891	0.827	0.856	0.838	0.899	0.868	0.937
2	B	B	0.841	0.763	0.800	0.783	0.857	0.818	0.896
3	C	C	0.829	0.862	0.845	0.856	0.822	0.839	0.926
	Average		0.854	0.817	0.834	0.826	0.859	0.855	0.920
No	disaster		disaster victims			non-victims			AUC
	training data	test data	Precision	Recall	F-measure	Precision	Recall	F-measure	
4	A	A	0.880	0.934	0.906	0.930	0.873	0.901	0.961
5	B	B	0.756	0.784	0.770	0.776	0.747	0.761	0.843
6	C	C	0.825	0.929	0.874	0.918	0.803	0.857	0.943
	Average		0.820	0.881	0.850	0.875	0.808	0.840	0.916

(II) Single typhoon

No	disaster		disaster area			non-disaster area			AUC
w	training data	test data	Precision	Recall	F-measure	Precision	Recall	F-measure	
1	A	B	0.897	0.252	0.393	0.565	0.971	0.714	0.785
2	A	C	0.844	0.290	0.432	0.571	0.947	0.713	0.835
3	B	A	0.677	0.927	0.783	0.885	0.557	0.684	0.859
4	B	C	0.679	0.957	0.794	0.928	0.547	0.688	0.871
5	C	A	0.845	0.796	0.820	0.807	0.854	0.830	0.889
6	C	B	0.935	0.503	0.654	0.660	0.965	0.784	0.876
	Average		0.813	0.621	0.646	0.736	0.807	0.730	0.853
No	disaster		disaster victims			non-victims			AUC
	train data	test data	Precision	Recall	F-measure	Precision	Recall	F-measure	
1	A	B	0.681	0.374	0.483	0.568	0.825	0.673	0.700
2	A	C	0.900	0.576	0.702	0.688	0.936	0.793	0.896
3	B	A	0.730	0.847	0.784	0.818	0.687	0.747	0.863
4	B	C	0.797	0.816	0.807	0.812	0.792	0.802	0.874
5	C	A	0.820	0.864	0.842	0.857	0.810	0.833	0.925
6	C	B	0.812	0.657	0.726	0.712	0.848	0.774	0.842
	Average		0.790	0.689	0.724	0.743	0.811	0.770	0.850

(III) Multi typhoons

No	disaster		disaster area			non-disaster area			AUC
	train data	test data	Precision	Recall	F-measure	Precision	Recall	F-measure	
1	A & B	C	0.762	0.874	0.814	0.853	0.728	0.785	0.880
2	B & C	A	0.740	0.918	0.820	0.892	0.678	0.770	0.886
3	C & A	B	0.938	0.434	0.593	0.632	0.971	0.766	0.859
	Average		0.813	0.742	0.742	0.792	0.792	0.774	0.875
No	disaster		disaster victims			non-victims			AUC
	train data	test data	Precision	Recall	F-measure	Precision	Recall	F-measure	
1	A & B	C	0.889	0.858	0.873	0.862	0.893	0.878	0.944
2	B & C	A	0.846	0.890	0.867	0.884	0.838	0.860	0.942
3	C & A	B	0.821	0.674	0.741	0.723	0.853	0.783	0.850
	Average		0.852	0.807	0.827	0.823	0.861	0.840	0.912

(B) is the test data, the recall is particularly low when the source location is the disaster area. We regard typhoon (A) and typhoon (C) as localized typhoons, and typhoon (B) as a wide-area typhoon because typhoon (A) and typhoon (C) include more detailed information related to the land than typhoon (B) does. Therefore, the precision is better. Moreover, the recall is lower. The results of precision are regarded as sufficient for use with new typhoon data. The results of other models are also good. The classification of new typhoons is possible by mixing different typhoons. Adapting a model trained for one typhoon to other typhoons is difficult. However, a model using training data that include a mixture of data from these typhoons can be useful to classify other typhoons.

5   Conclusion

In this paper, we proposed a classification model that classifies the source location and target people using multiple typhoon data as training data, with the aim of quickly and accurately providing RBF tweets to disaster victims. After we generated a classification model using the RBF tweets of one typhoon's training data and combined two typhoons' training data, we compared their respective classification accuracies. Results suggest that combining training data for an experiment yields better overall classification performance than training on a single set of data. In addition, combining data with data related to disasters of different sizes can efficiently produce good classification model. In the near future, we expect to propose a method, using our proposed model, to extract useful information for disaster victims from classification results.

Acknowledgements. This work was partially supported by Research Institute of Konan University, and by JSPS KAKENHI 20K12085 and 19H04218.

References

1. Ahmed, A., Hong, L., Smola, A.J.: Hierarchical geographical modeling of user locations from social media posts. In: Proceedings of the 22nd International Conference on World Wide Web, pp. 25–36. Association for Computing Machinery (2013)
2. Dutt, R., Hiware, K., Ghosh, A., Bhaskaran, R.: SAVITR: a system for real-time location extraction from microblogs during emergencies. In: Companion Proceedings of the The Web Conference 2018 (WWW 18), pp. 1643–1649 (2018)
3. Gao, W., Li, L., Zhu, X., Wang, Y.: Detecting disaster-related tweets via multimodal adversarial neural network. IEEE Multimed. **27**(4), 28–37 (2020)
4. Kabir, M.Y., Madria, S.: A deep learning approach for tweet classification and rescue scheduling for effective disaster management. In: In Proceedings of the 27th ACM SIGSPATIAL International Conference on Advances in Geographic Information Systems, pp. 269–278 (2019)
5. Liu, Y., et al.: RoBERTa: a robustly optimized BERT pretraining approach. ArXiv abs/1907.11692 (2019)
6. Paul, U., Ermakov, A., Nekrasov, M., Adarsh, V., Belding, E.: # outage: detecting power and communication outages from social networks. In: Proceedings of The Web Conference 2020, pp. 1819–1829 (2020)

7. Tian, H., Zhang, M., Luo, X., Liu, F., Qiao, Y.: Twitter user location inference based on representation learning and label propagation. In: Proceedings of The Web Conference 2020 (WWW 2020), pp. 2648–2654 (2020)
8. Yamamoto, F., Suzuki, Y., Nadamoto, A.: Extraction and analysis of regionally specific behavioral facilitation information in the event of a large-scale disaster. In: IEEE/WIC/ACM International Conference on Web Intelligence and Intelligent Agent Technology, pp. 538–543 (2021)

Exploring Dialog Act Recognition in Open Domain Conversational Agents

Maliha Sultana[✉] and Osmar R. Zaïane

Computing Science, University of Alberta, Edmonton, AB, Canada
{sultana2,zaiane}@ualberta.ca

Abstract. Recognizing dialog acts of users is an essential component in building successful conversational agents. In this work, we propose a dialog act (DA) classifier for two of our open domain dialog systems. For this, we first build a hierarchical taxonomy of 8 DAs suitable for classifying user utterances in open-domain setting. Next, we curate a high-quality, multi-domain dataset with over 24k user dialogs and annotate it with our 8 DAs. Next, we fine-tune our pretrained BERT-based DA classifier on this dataset. Through extensive experimentation, we show that our proposed model not only outperforms the baseline SVM classifier by achieving state-of-the-art accuracy but also generalizes extremely well on previously unseen data.

Keywords: Dialog Acts · Speech Act Recognition · Natural Language Processing

1 Introduction

Human beings are inherently social. Through frequent conversations, we convey our intentions, thoughts and opinions to our peers. Naturally, we grow accustomed to the everyday sentences we utter and the dialog acts we perform. In natural language understanding, a dialog act (DA) is an utterance in the context of a conversational dialog that serves a precise function in the dialog (or sometimes more than one) [1]. It can be a question, a statement or a request for action. Effective communication relies on recognizing the different DAs and responding accordingly. For example: someone asking a question expects an answer as a response whereas someone giving an order expects its execution or acknowledgment of its execution.

Dialog systems have long been researched in the field of AI dating back to 1966 with the advent of Eliza, a chatbot [2]. Although intended to be a mere caricature of human conversation, users were soon treating ELIZA like a companion-confiding their most intimate thoughts. Nowadays, with the advancement in AI, chatbots are being used as virtual assistants in different fields to enhance productivity and reduce service costs. Recent studies have found that users often consider chatbots as friendly companions and not just mere assistants. In fact, over 40% of user requests received by customer service chatbots on social media

R. Wrembel et al. (Eds.): DaWaK 2023, LNCS 14148, pp. 233–247, 2023.
https://doi.org/10.1007/978-3-031-39831-5_22

have been observed to be emotional than informative [3]. How much trust a chat-bot gains from its users depends on how human-like the chatbot is, i.e, how well it can handle natural language. As a result, recognizing the DA of users to generate better response has become an integral component in chatbots. Dialog systems usually include a taxonomy of dialog types or tags that are used to classify the different functions DAs can play. Depending on the task or domain in question, user intents vary and so do the proposed DA tag-sets. For example, to facilitate the development of dialog systems for mental-health counselling, Malhotra et al. [4] proposed a dataset called HOPE which consists of 12.9K patient-therapist utterances annotated with 12 dialog-act labels related to therapy. They also pro-posed a transformer based DA classifier which achieves state-of-the-art (SOTA) performance on HOPE.

With the aim of improving open-domain conversational agents, our work focuses on DA classification of users. We speculate that a dialog system can gen-erate better responses through proper identification of user dialog acts. For this, we first identified the relevant dialog acts for our existing chatbots-MIRA [5] and ANA [6]. We then curated a corresponding high-quality, multi-domain dataset of ~24k utterances belonging to one of our 8 proposed DAs- Statement, Factual Question, Yes/No Question, Direct Order, Indirect Order, Greeting, Feedback, Apology. Structuring this as a multi-class classification problem, we propose a pretrained BERT-base model as our DA classifier. Upon fine-tuning it on our curated dataset, the model achieves SOTA accuracy; outperforming the baseline SVM classifier by 3%. Our proposed DA classifier is also robust and generalizes well on never-before-seen dataset. In summary, the key contributions of our work are as follows:

1. We propose a hierarchical taxonomy consisting of 8 DAs suitable for open-domain conversational agents
2. We curate a high-quality, large-scale dataset of ~24k user utterances from multiple domains and rich data sources
3. We propose a fine-tuned BERT-based model for DA classification which not only achieves SOTA performance on our dataset but also generalizes well on unseen data

The remainder of this paper is structured as follows: We present a summary of the related works in Sect. 2. We describe our proposed DA taxonomy in Sect. 3 and provide details on our data collection process. In Sect. 4, we present the architecture of our DA classifier and summarize the results of our comprehensive evaluation. Finally, we conclude our work in Sect. 5.

2   Related Works

Building conversational AI is a long-standing challenge in NLP. Human con-versations are inherently complex and ambiguous. Training a dialog system that understands the semantic and syntactic nuances and generates natural and engaging response is difficult to achieve. However, recent works have shown the promise of combining dialog acts for neural response generation [7]. DAs can help conversational agents by providing a representation of the underlying meaning of a user's utterance.

In order to drive the research on building better dialog systems, a number of conversational corpora have been released in the past. The Switchboard Dialog Act Corpus (SwDA) [8] and the ICSI Meeting Recorder Dialog Act (MRDA) Corpus [9] are widely used to train dialog systems in open-domain setting. They consist of human-human utterances that are hand-labelled with over 40 dialog acts like Statement-non-opinion, Statement-opinion, Appreciation, Yes-No-Question, Wh-Question, Open-Question, Apology and so on. Authors like Colombo et al. [10] leveraged a sequence-to-sequence model and achieved an accuracy of 85% on SWDA, and SOTA accuracy of 91.6% on MRDA. Likewise, Li et al. [11] proposed a dual-attention hierarchical RNN with a CRF as their DA classifier. The model reached an accuracy of 92.2% on MRDA and SOTA accuracy of 82.3% on SWDA. On the other hand, Raheja et al. [12] proposed a DA classifier which can learn richer, more effective utterance representations with the help of self-attention and achieve an accuracy of 82.9% on SWDA and 91.1% on MRDA. More recently, to explicitly model the interaction between DA recognition and sentiment classification, Qin et al. [13] utilized co-interactive relation networks. Their classifier produced significant results for both tasks and even achieved performance boost after incorporating BERT [14]. Likewise, Saha et al. [15] jointly learnt dialog-act classification and emotion recognition tasks in a multi-modal setup.

Researchers have also looked into building DA classifiers for specific domains. To develop better learning environments and virtual mentors, Gautam et al. [16] proposed 8 unique dialog-act labels to classify their dataset consisting of student-mentor conversations in Nephrotex, a virtual internship. They also explored several machine learning methods to categorize the DAs and achieved promising results. Quinn et al. [6] looked into improving their chatbot, ANA, by proposing 3 DAs: Declarative, Interrogative, and Imperative because they fit into ANA's definition of a potential user utterance. They used an SVM model as their DA classifier and achieved 72% accuracy on the dataset. Zhang et al. [17] proposed classifying Tweets into 5 user acts- Statement, Question, Suggestion, Comment and Miscellaneous. Using a set of word-based and character-based features, their model achieved an average F1 score of nearly 0.70 on their dataset. Noticing how differently humans interact with other humans vs with machine, Yu et al. [18] proposed a DA annotation scheme called MIDAS based on human-machine conversations in open domain setting. The authors also collected and annotated 24k segmented sentences using MIDAS and deployed transfer learning to train a multi-label DA prediction model on it which achieved an F1-Score of 0.79.

Like the previous works, our paper aims towards building open-ended conversational agents that respond naturally by accurate detection of user DA. In particular, we focus on building a DA classifier that is applicable for our pre-existing text-based chatbots- ANA and MIRA. Apart from answering questions and sending reminders, Automated Nursing Agent or ANA aims to have a fluent and personalized conversation with the elderlies [6]. On the other hand, MIRA is a Mental Health Virtual Assistant which provides mental health resources to health care workers and their families [5]. It also has a module called 'Chatty MIRA' which allows users to have open-ended conversations with the chatbot. Given the difference in domain and task intents, our goal is to propose a common DA schema and its corresponding classifier. The next section gives a detailed explanation on how our DA tag-set was chosen.

3   Proposed Dialog Act Taxonomy

Although we had initially planned on curating a larger dataset using Quinn et al.'s [6] proposed dialog acts (Imperative, Declarative, Interrogative), we soon realised that their DA tag-set was too general and failed to capture multiple cases that require different response from the chatbot. For example: 'Can penguins fly?' and 'What is the name of our galaxy?' are both questions. However, the first one expects a yes/no answer whereas the second one expects a factual answer. To generate better responses, our chatbots need to learn how to distinguish between the two. After looking into the related works and having iterative discussions with two of our psychology students, we selected the following 8 DAs that adequately capture the intentions of our users. For a better understanding, Table 1 provides examples of each of the DAs categorized into a hierarchy.

1. Apology: Includes sentences through which the user expresses apology.
2. Greeting: Includes sentences through which the user greets the chatbot either at the beginning or towards the end of a session.
3. Informative: Includes queries asked by the user with the intention of gaining some information. Depending on the type of response, questions can broadly be of 2 types:
   (a) Yes/No: Includes close-ended queries that can be sufficiently answered with a simple yes or no.
   (b) Factual: Includes open-ended queries that seek fact-based answers. A majority of these questions are WH-questions but utterances like 'Name the highest rated therapist in my area' are also included here due to the similarity in user intent.
4. Directive: Includes orders given by the user to the chatbot for accomplishing a task. This again can be of two types:
   (a) Direct Order: Includes straightforward orders that are easy to detect, understand and carry out.
   (b) Indirect Order: Includes utterances that indirectly expect or request some type of action. These are a bit difficult to comprehend and might require the chatbot to first make an assumption and then prompt for a confirmation before execution. For example: 'I need help with managing anxiety' or 'Can you help me manage my anxiety?' can be interpreted as 'Show me resources for managing anxiety'.

Table 1. Selected dialog acts with examples.

Dialog Act	Sub Category	Example
Apology		Sorry about that!
		My bad
Greeting		Hey, how are you?
		Bye, see you soon!
Informative	Yes/No	Is it possible to treat ADHD?
	Factual	What year did Bangladesh achieve their independence?
		Name the best therapist in my area
Directive	Direct Order	Show me the list of hospitals nearby
	Indirect Order	I need help with managing anxiety
		Can you turn on the music please?
Statement		I am being bullied at school lately
		I like spending time with my family
Feedback		This is exactly what I was looking for! Thanks
		This is not what I wanted. You suck!

5. Statement: Includes user utterances that do not request for an action or information. Rather, these are dialogs through which the user casually converses with the chatbot. By analyzing the emotion behind these utterances, the chatbot can either choose to give a sympathetic response or ask follow-up questions.
6. Feedback: Includes feedback from the user once the chatbot accomplishes a task e.g. carries out an order or answers a question. Feedback can be positive (when the chatbot is successful) or negative (when the chatbot is unsuccessful). By detecting the sentiment behind it, the chatbot can either thank the user or apologize and/or attempt the task again.

3.1 Data Sources

Once the DA taxonomy was decided upon, we moved onto curating the corresponding dataset. To make our chatbots can easily recognize user intents, we intended to include user utterances our conversational agents are likely to encounter. Moreover, since our goal is to build an open-domain DA classifier applicable to both of our chatbots, we want our training dataset to be versatile. For this, we included examples not only from mental health (for MIRA) and popular chatbot commands (for ANA), but also from common domains like banking, air lines, product reviews and so on. We expect the mix of multiple data sources to add variation in sentence structure and make the dataset more diverse. For this, we first looked at some of the popular datasets that has a few dialog-act tags that are similar to ours. As for the rest, we scraped various websites and forums using simple rules. It is to be noted that, during data collection,

Table 2. Overview of our proposed training and test dataset.

Dialog Act	Sub Category	Train Examples	Test Examples	% Distribution
Informative	Yes/No	3385	847	17.31
	Factual	3697	924	18.89
Directive	Direct Orders	3125	781	15.97
	Indirect Orders	5400	1349	27.60
Statement		3250	812	16.61
Greeting		239	60	1.22
Feedback		392	98	2
Apology		73	18	0.4

we decided to include only those examples that followed our definition of each of the DAs. Moreover, to avoid dominance of a particular domain or type of sentence, we decided not to include too many examples from a single source. Below, we give a brief overview of the data sources we had used for each DA:

1. Informative: We used popular question-answering datasets and mental health FAQ websites.
   (a) Yes/No Question: BoolQ [19], SNIPS [20]
   (b) Factual Question: SNIPS [20], SQUAD [21]
2. Directive: We used task-completion dialog intent datasets to collect Direct and Indirect Orders. Simple extraction rules were used to distinguish between the two. Datasets include Taskmaster [22], SNIPS [20], ATIS [23] and ACID [24] to name a few.
3. Statement: We mostly used the dataset shared by a mental health forum called 'Counsel-Chat' which consists of anonymous user posts related to mental health. We also included some examples from Wiki-Article, IMDB Movie Review and Amazon Product Review datasets.
4. Feedback, Apology and Greeting: We scraped a few basic English learning websites to extract positive and negative appraisals, apologies and greetings.

Table 2 shows how the examples are distributed per class. Among the 8 dialog acts, Apology, Greeting and Feedback are our minority classes. Due to the lack of variation in the ways users greet, apologize and provide feedback in real life, these 3 classes have a small number of examples in comparison. In total, our dataset has 24450 examples. We split it into two in order to create a train and a test dataset. The split was done in a way to include 25% of the examples of each class into the test dataset in order to offset the class imbalance. Given that we used high-quality datasets as source, our curated dataset is also refined, with little to no mislabelling.

4   Proposed Dialog Act Classifier

Now, we will discuss in depth the architecture of our proposed DA classifier and report the results obtained through extensive experimentation. We further analyze the results and provide our inference.

4.1   Experimental Setup

Given the success rate of BERT in achieving SOTA result in multiple NLP tasks [25], our proposed DA classifier is a pretrained BERT-based model. BERT, which stands for Bidirectional Encoder Representations from Transformers is based on the transformer architecture that uses bidirectional training to have a deeper sense of language context. Moreover, because BERT was trained on a huge corpus, it can easily be fine-tuned on a new dataset and achieve great results. For the task of DA classification, we first convert our labels into categorical data. Next, we load the pretrained 'bert-base-cased' model from Tensorflow and fine-tune it on our training dataset. Next, we use the corresponding tokenizer with the maximum length set to 70. The BERT layers accept 3 input arrays but since 'tokenTypeIds' is necessary only for the question-answering model, we work with 2 input layers-'inputIds' and 'attentionMask'. We use also 'Global-MaxPooling1D' and then a dense layer to build the CNN layers using hidden states of BERT. These CNN layers yield the output. We use the Adam optimizer with a learning rate of 5e-05, a decay of 0.01 and 'CategoricalCrossentropy' as loss. Once training is completed in 3 epochs, we calculate its accuracy on the test data.

To compare the performance of our DA classifier against a baseline, we use an SVM model. Support Vector Machine or SVM is a popular supervised learning algorithm that works by creating a decision boundary that best segregates an n-dimensional space into distinct classes. SVM is suitable for text categorization task as they can efficiently handle high dimensional input space, few irrelevant features and sparse document vectors [26]. Moreover, since text categorization problems are mostly linearly separable, SVM is the perfect candidate. They are very light-weight and are much faster to train than large language models. Given their benefits, a number of authors have successfully used SVMs for text classification tasks [6,27–29]. We use LinearSVC as our baseline which is similar to SVC with the parameter kernel='linear', but provides more flexibility in the choice of penalties and loss functions in Scikit-learn [30]. As for converting the text files into numerical feature vectors, we use the Bag-of-Words (BoW) technique (CountVectorizer) and later run the TF-IDF technique (TfidfTransformer) over the features generated by BoW. Lastly, we train the baseline on our proposed dataset and evaluate it on our test dataset.

4.2   Performance Evaluation

From Table 3, it is evident that both the baseline and our DA classifier perform well on our proposed dataset. SVM yields an accuracy of 96% and BERT outperforms SVM by 3% by achieving an accuracy of 99%. One of the reasons for such high accuracy rates might be because of the stark differences in the structure of sentences for each of the dialog acts. For further analysis of the wrongly predicted examples, we take a look at the confusion matrices in Fig. 1 where the y-axis shows the true labels of the examples and the x-axis shows the predicted labels.

For majority classes (~3771 examples/label) like Directive Direct Order (DD), Question Factual (QF) and Directive Indirect Order (DI), SVM achieves individual accuracies of 96%, 98% and 99% with only a very few instances of misclassification. The accuracy, however, is comparatively low for other majority classes like Statement (S) and Question Yes/No (QYN) (92%). Upon further analysis, we see that 8% of Feedback (F) are misclassified as Statement. For example: 'This works well' and 'I'm glad you are my friend' are all Feedback but are wrongly predicted as Statement. This makes sense given the similarity in sentence structure for both of these classes. In case of Yes/No Question, the low accuracy rate comes from misclassifying a large number of Statement (4%) and Feedback (2%). For example: 'My issue is that there is always drama' and 'It is good' were wrongly predicted as Yes/No Question. Possible reason for this might be the presence of the helping verb 'is' towards the beginning of the sentence which is similar to the structure of a Yes/No Question ('Is it cold in here?'). The baseline model fails to learn the difference in these cases. As for the minority classes, although SVM scores a high accuracy for Apology (A) class, it struggles to detect Greeting (88%) and Feedback (87%) in comparison. Possible reason for this might be because the training data is not enough for the baseline to learn specific patterns to recognize them. Future work might look into using rules to detect these minority classes instead and compare the performance.

Now, we take a look at our fine-tuned pretrained BERT-based model. Unlike SVM, it does a very good job at achieving 99% accuracy for all seven of the eight classes. The Feedback class, however, has a slightly low accuracy rate (96%) for misclassifying some of the examples as Statement. Given the similarities shared by the examples of these two classes, it makes sense why the misclassification happened. On the bright side, it is important to mention that unlike the baseline, our BERT-based model does not struggle with accurately predicting Yes/No Question, Greeting or Statement. Thus, our proposed DA classifier not only outperforms the baseline model but also achieves SOTA result on our high-quality dataset.

4.3   Generalizability of Model

The generalizability (or robustness) of a model is a measure of its successful application to datasets other than the one used for training and testing. To compare and evaluate the generalizability of the baseline and our proposed DA

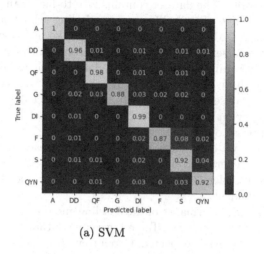

(a) SVM

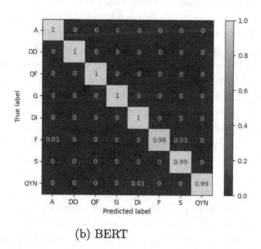

(b) BERT

Fig. 1. Confusion matrices of the baseline (SVM) and our proposed classifier (BERT) on our proposed dataset.

Table 3. Comparison of performance between the baseline (SVM) and our proposed classifier (BERT) on our proposed dataset.

Model	Accuracy	Precision	Recall	F1-Score
SVM	0.96	0.96	0.94	0.95
BERT	0.99	0.99	0.99	0.99

classifier, we decided to create a new dataset called 'generalized dataset'. The plan was to find a data source that was never used for curating our original train and test data and then manually label it with our proposed taxonomy of 8 DAs. We chose the DialogSum dataset [31] for this purpose. It is a large-scale dialog summarization dataset consisting of ~13k dialogs from 3 public dialog corpora, namely Dailydialog [32], DREAM [33] and MuTual [34], as well as an English speaking practice website. The dataset contains face-to-face high quality spoken dialogs from a wide range of daily-life topics including schooling, work, medication, travel and so on. Most of the conversations take place between friends, colleagues, and between service providers and customers. This, however, is an issue for us because we trained our model on a dataset that has conversations a user is more likely to have with a chatbot- not a person. We mitigated this by only including dialogs a user is more likely to have with a chatbot. For example: sentences like 'Zach, what's that on your arm?', 'Here, let me help you with your coat and we'll be on our way' were avoided. Moreover, given how large the DialogSum dataset is, we only chose a few samples for each DA manually. In the end, the curated generalized dataset consisted of 8 Apology, 9 Greeting, 9 Feedback, 30 Indirect Order, 36 Direct Order, 43 Factual Question, 45 Yes/No Question and 47 Statement.

Now we evaluate the performance of the baseline and our proposed DA classifier on the generalized dataset. From Table 4, we can see that the performance of both the models drops which is expected. Given that the generalized dataset has more human-human conversations whereas our proposed dataset i.e the dataset the models were trained on has more human-machine conversations, this makes sense. However, what is impressive is that, although the accuracy of the baseline drops drastically by 10% on the generalized dataset (from 96% to 86%), our proposed DA classifier holds up really well. Even on the never-before-seen dataset, it achieves an accuracy of 96%- a mere 3% drop from its performance on our test dataset which is remarkable. This proves that our proposed DA classifier is both generalizable and robust on unseen data.

For further analysis, we take a look at the examples that were mislabelled. Figure 2 shows the confusion matrices for both the models. We can see that the baseline struggles the most with identifying Direct Orders. For example: sentences like 'Please wrap it for me and I'll take it', 'Go back to sleep then but only five more minutes' etc. are mislabelled as Statement and Yes/No Question. Since the model was trained on some very common chatbot commands like playing a song, booking a flight or reserving a seat- it has a hard time predicting these unconventional commands as Direct Order. The baseline also struggles with classifying a number of Yes/No Question correctly. For example: sentences like 'Excuse me, do you speak English?' and 'Have you turned on the air-conditioner?' are mislabelled as Statement and Indirect Order. This might be because our training dataset includes Yes/No Question that are usually factual and not casual (i.e. sentences like 'Is Canada in the United States of America?' instead of 'Do you like to play the piano?'). Moreover, the phrase 'excuse me' in our dataset is mostly associated with the class Apology which might be another reason for the wrong prediction. On the flip side, the accuracy rate for the class Indirect Order is very high (97%). Overall, the accuracy rate of all the classes is above 80% which is not ideal but reasonable.

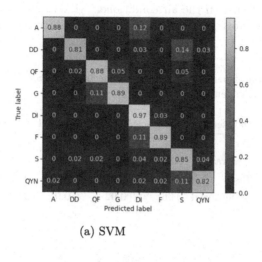

(a) SVM

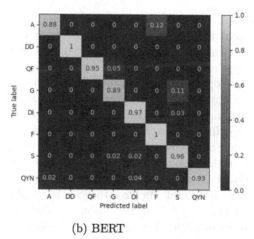

(b) BERT

Fig. 2. Confusion matrices of the baseline (SVM) and our proposed classifier (BERT) on generalized dataset.

As for our proposed DA classifier, we notice the least accuracy in the minority classes Apology (88%) and Greeting (89%). This happens for mislabelling two sentences 'I hope you can forgive me' and 'Hi, my name is Susan' as Indirect Order and Statement. Probable reason for this is the lack of enough training data for these two classes. As a result, the model is not able to learn all sorts of variations properly. On the bright side, the remaining classes all have an impressive accuracy rate (over 90%). Despite having an accuracy of 93%, the Yes/No Question class struggles a bit with sentences like 'Have you turned on the air-conditioner', 'Can I exchange it?'. This might be because these sentences somewhat resemble the structure of Indirect Order examples present in the training data ('Turn on the air-conditioner', 'Exchange it'). All in all, the experiments clearly prove that our proposed DA classifier generalizes far better than the baseline.

Table 4. Comparison of performance between the baseline (SVM) and our proposed classifier (BERT) on the generalized dataset.

Model	Dataset	Accuracy	Precision	Recall	F1-Score
SVM	Test	0.96	0.96	0.94	0.95
	Generalized	0.86	0.85	0.87	0.86
BERT	Test	0.99	0.99	0.99	0.99
	Generalized	0.96	0.92	0.95	0.93

5   Conclusion

Classifying the intent of a user dialog in a conversation, also known as dialog act, is a key component in building conversational agents. By identifying the different DAs, chatbots can respond more coherently and assist users in accomplishing their tasks more effectively. In this work, we propose a BERT-based DA classifier for two of our open domain conversational agents- ANA and MIRA. For this, we first investigated the current literature and through iterative discussions, proposed a taxonomy of 8 DAs that are suitable for our chatbot users. We then curated a high-quality, large-scale dataset consisting of ~24k user utterances from multiple domains. Upon fine-tuning our proposed classifier on this dataset, it outperforms the baseline SVM model by achieving SOTA accuracy. Through further evaluations, we prove the generalizability and robustness of our proposed model on unseen dataset. As for future work, we plan on investigating the effectiveness of structuring DA classification as a multi-label instead of a multi-class classification problem. We also want to look into the feasibility of including more dialog acts into our taxonomy.

References

1. Popescu-Belis, A.: Abstracting a dialog act tagset for meeting processing. In: Proceedings of the Fourth International Conference on Language Resources and Evaluation (LREC 2004), Lisbon, Portugal. European Language Resources Association (ELRA) (2004). http://www.lrec-conf.org/proceedings/lrec2004/pdf/268.pdf
2. Wei, C., Yu, Z., Fong, S.: How to build a chatbot: chatbot framework and its capabilities. In: Proceedings of the 2018 10th International Conference on Machine Learning and Computing, pp. 369–373 (2018)

3. Xu, A., Liu, Z., Guo, Y., Sinha, V., Akkiraju, R.: A new chatbot for customer service on social media. In: Proceedings of the 2017 CHI Conference on Human Factors in Computing Systems, pp. 3506–3510 (2017)
4. Malhotra, G., Waheed, A., Srivastava, A., Akhtar, M.S., Chakraborty, T.: Speaker and time-aware joint contextual learning for dialogue-act classification in counselling conversations. In: Proceedings of the Fifteenth ACM International Conference on Web Search and Data Mining, pp. 735–745 (2022)
5. Noble, J.M., et al.: Developing, implementing, and evaluating an artificial intelligence-guided mental health resource navigation chatbot for health care workers and their families during and following the covid-19 pandemic: Protocol for a cross-sectional study. JMIR Res. Protoc. 11(7), 33717 (2022)
6. Quinn, K., Zaiane, O.: Identifying questions & requests in conversation. In: Proceedings of the 2014 International C* Conference on Computer Science & Software Engineering, pp. 1–6 (2014)
7. Welivita, A., Pu, P.: A taxonomy of empathetic response intents in human social conversations. In: Proceedings of the 28th International Conference on Computational Linguistics, Barcelona, Spain, pp. 4886–4899. International Committee on Computational Linguistics (2020). https://doi.org/10.18653/v1/2020.coling-main.429. https://aclanthology.org/2020.coling-main.429
8. Godfrey, J.J., Holliman, E.: Switchboard-1 release 2. Linguistic Data Consortium, Philadelphia, vol. 926, p. 927 (1997)
9. Dhillon, R., Bhagat, S., Carvey, H., Shriberg, E.: Meeting recorder project: dialog act labeling guide. Technical report, International Computer Science Inst Berkeley CA (2004)
10. Colombo, P., Chapuis, E., Manica, M., Vignon, E., Varni, G., Clavel, C.: Guiding attention in sequence-to-sequence models for dialogue act prediction. In: Proceedings of the AAAI Conference on Artificial Intelligence, vol. 34, pp. 7594–7601 (2020)
11. Li, R., Lin, C., Collinson, M., Li, X., Chen, G.: A dual-attention hierarchical recurrent neural network for dialogue act classification. In: Proceedings of the 23rd Conference on Computational Natural Language Learning (CoNLL), pp. 383–392 (2019)
12. Raheja, V., Tetreault, J.: Dialogue act classification with context-aware self-attention. In: Proceedings of the 2019 Conference of the North American Chapter of the Association for Computational Linguistics: Human Language Technologies, Volume 1 (Long and Short Papers), pp. 3727–3733 (2019)
13. Qin, L., Che, W., Li, Y., Ni, M., Liu, T.: DCR-Net: a deep co-interactive relation network for joint dialog act recognition and sentiment classification. In: Proceedings of the AAAI Conference on Artificial Intelligence, vol. 34, pp. 8665–8672 (2020)
14. Devlin, J., Chang, M.-W., Lee, K., Toutanova, K.: BERT: pre-training of deep bidirectional transformers for language understanding. In: Proceedings of the 2019 Conference of the North American Chapter of the Association for Computational Linguistics: Human Language Technologies, Volume 1 (Long and Short Papers), Minneapolis, Minnesota, pp. 4171–4186. Association for Computational Linguistics (2019). https://doi.org/10.18653/v1/N19-1423. https://aclanthology.org/N19-1423
15. Saha, T., Gupta, D., Saha, S., Bhattacharyya, P.: Emotion aided dialogue act classification for task-independent conversations in a multi-modal framework. Cogn. Comput. 1–13 (2020)

16. Gautam, D., Maharjan, N., Graesser, A.C., Rus, V.: Automated speech act categorization of chat utterances in virtual internships. In: EDM (2018)

17. Zhang, R., Gao, D., Li, W.: What are tweeters doing: Recognizing speech acts in twitter. In: Analyzing Microtext (2011)

18. Yu, D., Yu, Z.: Midas: A dialog act annotation scheme for open domain humanmachine spoken conversations. In: Proceedings of the 16th Conference of the European Chapter of the Association for Computational Linguistics: Main Volume, pp. 1103–1120 (2021)

19. Clark, C., Lee, K., Chang, M.-W., Kwiatkowski, T., Collins, M., Toutanova, K.: Boolq: exploring the surprising difficulty of natural yes/no questions. In: Proceedings of the 2019 Conference of the North American Chapter of the Association for Computational Linguistics: Human Language Technologies, Volume 1 (Long and Short Papers), pp. 2924–2936 (2019)

20. Coucke, A., et al.: Snips voice platform: an embedded spoken language understanding system for private-by-design voice interfaces. arXiv preprint arXiv:1805.10190 (2018)

21. Rajpurkar, P., Zhang, J., Lopyrev, K., Liang, P.: Squad: 100,000+ questions for machine comprehension of text. In: Proceedings of the 2016 Conference on Empirical Methods in Natural Language Processing, pp. 2383–2392 (2016)

22. Byrne, B., et al.: Taskmaster-1: toward a realistic and diverse dialog dataset. In: Proceedings of the 2019 Conference on Empirical Methods in Natural Language Processing and the 9th International Joint Conference on Natural Language Processing (EMNLP-IJCNLP), pp. 4516–4525 (2019)

23. Hemphill, C.T., Godfrey, J.J., Doddington, G.R.: The ATIS spoken language systems pilot corpus. In: Speech and Natural Language: Proceedings of a Workshop Held at Hidden Valley, Pennsylvania, 24–27 June 1990 (1990)

24. Acharya, S., Fung, G.: Using optimal embeddings to learn new intents with few examples: an application in the insurance domain (2020)

25. González-Carvajal, S., Garrido-Merchán, E.C.: Comparing bert against traditional machine learning text classification. arXiv preprint arXiv:2005.13012 (2020)

26. Joachims, T.: Text categorization with support vector machines: learning with many relevant features. In: Nédellec, C., Rouveirol, C. (eds.) ECML 1998. LNCS, vol. 1398, pp. 137–142. Springer, Heidelberg (1998). https://doi.org/10.1007/BFb0026683

27. Luo, X.: Efficient English text classification using selected machine learning techniques. Alex. Eng. J. **60**(3), 3401–3409 (2021). https://doi.org/10.1016/j.aej.2021.02.009

28. Morales-Hernández, R.C., Becerra-Alonso, D., Vivas, E.R., Gutiérrez, J.: Comparison between SVM and distilbert for multi-label text classification of scientific papers aligned with sustainable development goals. In: Pichardo Lagunas, O., Martínez-Miranda, J., Martínez Seis, B. (eds.) MICAI 2022. LNCS, vol. 13613, pp. 57–67. Springer, Cham (2022). https://doi.org/10.1007/978-3-031-19496-2_5

29. Kambar, M.E.Z.N., Nahed, P., Cacho, J.R.F., Lee, G., Cummings, J., Taghva, K.: Clinical text classification of Alzheimer's drugs' mechanism of action. In: Yang, X.-S., Sherratt, S., Dey, N., Joshi, A. (eds.) Proceedings of Sixth International Congress on Information and Communication Technology. LNNS, vol. 235, pp. 513–521. Springer, Singapore (2022). https://doi.org/10.1007/978-981-16-2377-6_48

30. Pedregosa, F., et al.: Scikit-learn: machine learning in python. J. Mach. Learn. Res. **12**, 2825–2830 (2011)

31. Chen, Y., Liu, Y., Chen, L., Zhang, Y.: DialogSum: a real-life scenario dialogue summarization dataset. In: Findings of the Association for Computational Linguistics: ACL-IJCNLP 2021, pp. 5062–5074. Association for Computational Linguistics, Online (2021). https://doi.org/10.18653/v1/2021.findings-acl.449. https://aclanthology.org/2021.findings-acl.449

32. Li, Y., Su, H., Shen, X., Li, W., Cao, Z., Niu, S.: DailyDialog: a manually labelled multi-turn dialogue dataset. In: Proceedings of the Eighth International Joint Conference on Natural Language Processing (Volume 1: Long Papers), Taipei, Taiwan, pp. 986–995. Asian Federation of Natural Language Processing (2017). https://aclanthology.org/I17-1099

33. Sun, K., Yu, D., Chen, J., Yu, D., Choi, Y., Cardie, C.: DREAM: a challenge data set and models for dialogue-based reading comprehension. Trans. Assoc. Comput. Linguist. **7**, 217–231 (2019). https://doi.org/10.1162/tacl_a_00264

34. Cui, L., Wu, Y., Liu, S., Zhang, Y., Zhou, M.: MuTual: a dataset for multi-turn dialogue reasoning. In: Proceedings of the 58th Annual Meeting of the Association for Computational Linguistics, pp. 1406–1416. Association for Computational Linguistics, Online (2020). https://doi.org/10.18653/v1/2020.acl-main.130. https://aclanthology.org/2020.acl-main.130

UniCausal: Unified Benchmark and Repository for Causal Text Mining

Fiona Anting Tan[1](✉) ⓘ, Xinyu Zuo[2], and See-Kiong Ng[1]

[1] Institute of Data Science, National University of Singapore, Singapore, Singapore
`tan.f@u.nus.edu, seekiong@nus.edu.sg`
[2] Tencent Technology, Beijing, China
`xylonzuo@tencent.com`
`https://github.com/tanfiona/UniCausal`

Abstract. Current causal text mining datasets vary in objectives, data coverage, and annotation schemes. These inconsistent efforts prevent modeling capabilities and fair comparisons of model performance. Furthermore, few datasets include cause-effect span annotations, which are needed for end-to-end causal relation extraction. To address these issues, we propose UniCausal, a unified benchmark for causal text mining across three tasks: (I) Causal Sequence Classification, (II) Cause-Effect Span Detection and (III) Causal Pair Classification. We consolidated and aligned annotations of six high quality, mainly human-annotated, corpora, resulting in a total of 58,720, 12,144 and 69,165 examples for each task respectively. Since the definition of causality can be subjective, our framework was designed to allow researchers to work on some or all datasets and tasks. To create an initial benchmark, we fine-tuned BERT pre-trained language models to each task, achieving 70.10% Binary F1, 52.42% Macro F1, and 84.68% Binary F1 scores respectively.

Keywords: datasets · causal text mining · causal relation extraction

1 Introduction

Causal text mining relates to the extraction of causal information from text. Given an input sequence, we are interested to know if and where causal information occurs. The extracted causal information can serve as a knowledge base [9,15,16] for researchers in tasks such as summarization or prediction [24]. Since causality is an important part of human cognition, causal text mining also has important natural language understanding applications [6,8,28]. Figure 1 illustrates three causal text mining tasks (Sequence Classification, Span Detection and Pair Classification) and their expected output.

Currently, large and diverse causal text mining corpora are limited [1]. Furthermore, annotation guidelines vary across datasets [33]. These issues hinder both modeling capabilities and fair comparisons between models. Additionally, working on independent tasks and datasets runs the risk of training task-specialized and dataset-specific models that are not generalizable.

R. Wrembel et al. (Eds.): DaWaK 2023, LNCS 14148, pp. 248–262, 2023.
https://doi.org/10.1007/978-3-031-39831-5_23

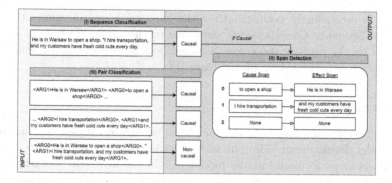

Fig. 1. A two-sentence example that contains causal relations. (I) Sequence Classification aims to return a *Causal* label. (II) Span Detection identifies the text related to the Cause-Effect spans. This example contains three annotated relation pairs, where two are labeled *Causal* and one is labeled *Non-causal*. These are target labels of the (III) Pair Classification task.

As such, we propose the construction of UniCausal, a unified benchmark and repository for causal text mining, which has the following contributions:

- To the best of our knowledge, we are the first to produce a unified benchmark and resource for causal text mining. Apart from consolidating six datasets for three tasks, we also employed competitive natural language processing (NLP) models to obtain baseline scores, reported in this paper.
- Our framework provides a seamless way for researchers to design individual or joint models, while benchmarking their performance against clearly defined test sets across some or all the processed corpora.
- Our codes and processed data is available online. Our trained baseline model checkpoints are uploaded to Huggingface Hub.[1]

The rest of the paper is organized as follows: Sect. 2 performs a literature review. Section 3 describes how we created the consolidated corpus and outlines our baseline models. Section 4 discusses our findings and Sect. 5 concludes.

2   Related Work

We are interested to support end-to-end causal relation extraction. More specifically, given an input sequence, a model should be designed to detect whether the sequence contains causal relations, and if so, where its causal arguments are.

2.1   Tasks

In the earlier days, researchers extracted causal relations and assessed the validity of the extracted relations directly [8,12]. In the recent years, a two-stepped

---

[1] Our repository is at https://github.com/tanfiona/UniCausal. Links to our trained baseline models are available in the repository.

approach is increasingly popular: After the successful identification of causal sequences (Sequence Classification), detection of the cause-effect spans can be conducted on the positive sequences (Cause-Effect Span Detection). This step-wise approach is practiced by shared tasks [17,18,29,30].

2.2 Datasets

In causal text mining research, the lack of adequate annotated training data limits model development [33]. The lack of standardized datasets also hinders comparisons in performance across models [1]. In the next few paragraphs, we describe six common causal text mining datasets in terms of their original intentions and limitations.

AltLex [11][2] investigates causal relations with alternative lexicalizations (AltLex) connectives in single sentences. AltLex connectives are an open-class of markers with varied linguistic constructions [32], like *"so (close) that"* and *"This may help explain why"*. The limitations of the AltLex dataset is that it (1) is small in size, (2) ignores explicit and implicit signals in causal relations, (3) ignores inter-sentence causal relations, and (4) assumes Cause and Effect spans to be all the words before and after the signals.

BECAUSE 2.0 [7][3] contains annotations for causal language in single sentences. Cause, Effect or Connective spans were annotated based on principles of Construction Grammar. Documents and articles were selected from four data sources: Congressional Hearings from 2014 NLP Unshared Task in Poli-Informatics (CHRG), Penn Treebank (PTB), Manually Annotated Sub-Corpus (MASC), and the New York Times Annotated Corpus (NYT). The limitations of BECAUSE 2.0 is that (1) it is relatively small in size and (2) ignores inter-sentence causal relations.

CausalTimeBank (CTB) [19,20][4] annotated only explicit causal relations in the TempEval-3 based on rule-based algorithm. EventStoryLine (ESL) [4][5] annotated both explicit and implicit causal relations in the Event Coreference Bank+. Both CTB and ESL includes annotations of intra- and inter-sentence causal relations between events. These two datasets are popular amongst researchers studying Event Causality Identification (ECI) [3,34,35], which aims to classify if a pair of events are causal or not given its context. However, these two datasets have limited usage outside the event text mining space because they only annotate events, and furthermore, exclude the context of the event in the argument span.

For Penn Discourse Treebank V3.0 (PDTB) [32][6] and SemEval 2010 Task 8 (SemEval) [10], causal relations were not the original focus of the dataset. Causal relations are one out of the many relations they annotated. PDTB annotated discourse relations between arguments, expressed either explicitly, implicitly or in

---

[2] https://github.com/chridey/altlex.
[3] https://github.com/duncanka/BECAUSE/tree/2.0.
[4] https://github.com/paramitamirza/Causal-TimeBank.
[5] https://github.com/tommasoc80/EventStoryLine.
[6] https://catalog.ldc.upenn.edu/LDC2019T05.

AltLex forms. The main limitation of PDTB is that causal relations expressed within clauses are not annotated. SemEval was annotated for the purpose of semantic relations classification. They accepted only noun phrases with common-noun heads as relation arguments. SemEval's limitations are: (1) the context is not included in the argument span and (2) it also ignores inter-sentence causal relations.

Table 1. Six popular causal corpora and their annotation coverage in terms of: data source, sentence lengths, linguistic construction and argument types.

Corpus	Source	Inter-sent	Linguistic	Arguments
AltLex [11]	News	No	AltLex	Words before/after signal
BECAUSE 2.0 [7]	News, Congress Hearings	No	Explicit	Phrases
CausalTimeBank (CTB) [19]	News	Yes	All	Event head word(s)
EventStoryLine V1.0 (ESL) [4]	News	Yes	All	Event head word(s)
Penn Discourse Treebank V3.0 (PDTB) [32]	News	Yes	All	Clauses
SemEval 2010 Task 8 (SemEval) [10]	Web	No	All	Noun phrases

Table 1 summarizes the differences across the six datasets. The current disarray across causal text mining datasets leads to three key missed opportunities: Firstly, for training, researchers are unable to seamlessly increase models' exposure to a wide range of examples. Secondly, for evaluation, researchers cannot make fair comparisons of model performance with one another. Thirdly, it is inconvenient for researchers to test their models' generalizability to other corpora. To address these issues, we propose UniCausal, a large consolidated resource of annotated texts for causal text mining. We relied on the above six high quality corpora and aligned each corpus' definitions, where possible, to cater to the three causal text mining tasks. With the exception of CTB, all other datasets were annotated by humans.

2.3   Other Large Causal Resources

Although some large corpora or knowledge bases (KBs) that include causal relations already exists, they are annotated in a semi-supervised manner and constructed using rule-based methods.[7] For example, CauseNet [9] detected causal relations automatically using causal dependency path patterns obtained in a boostrap fashion. Examples from such corpora are of lower quality and have less linguistic variation. Thus, although they are useful databases for common causal relations, they contribute minimally to training and fair testing of models' reasoning abilities. We perform a short study on this matter in Sect. 4.3. Different from them, our corpus encompasses examples that exhibit linguistic, syntactic, and semantic variation. By training a text mining model on our diverse corpus, researchers can potentially create an even larger causal KB by extracting more relations compared to rule-based methods.

---

[7] E.g.: Bootstrapped versions of AltLex [11] and SCITE [14]; Causal KBs: CauseNet [9], CausalNet [16] and CausalBank [15]; Semantic KBs that include causal relations: FrameNet [26] and ConceptNet [27].

3 Methodology

3.1 Creation of UniCausal

Causal Text Mining Datasets. We combine six datasets: AltLex [11], BECAUSE 2.0 [7], CTB [19,20], ESL [4], PDTB [32] and SemEval [10].

Causal Text Mining Tasks. We focus on three causal text mining tasks:

(I) Sequence Classification: Given an example, does it contain any causal relations?

(II) Span Detection: Given a causal example, which words in the example correspond to the Cause and Effect arguments? Identify up to three causal relations and their spans.

(III) Pair Classification: Given a marked argument or entity pair in an example, are the pairs causally related such that the first argument (ARG0) causes the second argument (ARG1)?

Since pairs can be *Non-causal*, we marked the arguments with ARG0 and ARG1 instead of CAUSE and EFFECT.

Data Processing. Causal text mining is a novel and challenging task. Thus, we believe that researchers should first tackle simpler inter-sentence and intra-sentence examples before progressing to complex, document-level scenarios. Therefore, we eliminate roughly 20% of the available data because they were longer than three sentences. Of which, 92% of the discarded examples came from ESL, suggesting that this operation has a limited impact, and at most, only on the Sequence Classification task. Subsequently, for examples containing causal relations, we retain only the sentences that contain the arguments. In terms of train-test splits, we follow recommendations of previous works unless the splits were not specified, then we split 10% of the dataset out randomly to form the test set. Finally, we process each dataset to fit into the format required for our three tasks described below.

Let a unique sequence of text be represented by a vector $\boldsymbol{w} = w_1, w_2, ..., w_N$ of N word tokens. Each sequence has a binary label s of either 1 or 0, representing *Causal* or *Non-causal* respectively.

(I) Sequence Classification: Worked on both *Causal* and *Non-causal* sequences. Each sequence is unique with a target label, s.

(II) Span Detection: Worked on *Causal* sequences. Each sequence is unique. We approach the task as a token classification task, where the annotated sequences were converted to BIO-format (Begin (B), Inside (I), Outside (O)) [25] for two types (Cause (C), Effect (E)). Therefore, there were five possible labels per word: B-C, I-C, B-E, I-E and O. The corresponding target token vector is $\boldsymbol{t} = t_1, t_2, ..., t_N$, where each t_n represents one of the five labels. Currently, we only focus on sequences with up to three causal relations. For sequences with multiple causal relations, we sort them based on the location of the B-C, followed by B-E if tied. For a sequence with three causal relations, w would have multiple token vectors t^v, where $v = 0, 1, 2$.

(III) Pair Classification: Worked on both *Causal* and *Non-causal* sequences. Each sequence is unique after taking into account of where special tokens ARG0 and ARG1 are located. For a sequence with N word tokens, we include $2 \cdot a$ special beginning and end tokens[8] such that the input word vector u is now of length $N + 2 \cdot a$. a represents the number of arguments in the example. w can have multiple versions of u, depending on the location of the special tokens. For example, in Fig. 1, there are three Pair Classification examples for one Sequence Classification example.

Table 2. Size of dataset. "-" indicates tasks not applicable to the corpus.

Corpus	Split	(I) Seq		(II) Span	(III) Pair	
		Non-causal	Causal	Causal	Non-causal	Causal
AltLex	Train	277	300	300	296	315
	Test	286	115	115	289	127
BECAUSE	Train	183	716	716	266	902
	Test	10	41	41	14	46
CTB	Train	1,651	234	-	3,047	270
	Test	274	42	-	444	48
ESL	Train	957	1,043	-	-	-
	Test	119	113	-	-	-
PDTB	Train	24,901	8,917	8,917	32,587	9,809
	Test	5,796	2,055	2,055	7,694	2,294
Sem-Eval	Train	6,976	999	-	6,997	1,003
	Test	2,387	328	-	2,389	328
Total		43,817	14,903	12,144	54,023	15,142

The final data sizes are reflected in Table 2. The number of Span Detection example tallies with the positive instances of Sequence Classification examples because multiple cause-effect relation spans (i.e. t^0, t^1 and t^2) were grouped into a unique example (i.e. same w), which we term as a 'grouped' example. At evaluation, performance metrics were calculated at the 'ungrouped' level so that every causal relation is evaluated against equally.

Since each data source has a different data format, our codes had to extract the text sequences and relations from different annotation types: Our codes work for BECAUSE's 'brat', CTB's 'TimeML'/'XML' and ESL's 'CAT' data formats.

---

[8] (<ARG0>, </ARG0>) marks the boundaries of a Cause span, while (<ARG1>, </ARG1>) marks the boundaries of a corresponding Effect span.

PDTB uses its own standoff annotations format. AltLex and SemEval datasets are more user-friendly, in that they are stored in 'CSV' and 'JSON' formats, and can be interpreted directly in a single file. Given the limited space, we describe our data processing steps in detail in our Supplementary Material available in our repository. The codes for processing data from each source is provided too. The final, post-processed datasets are all stored in 'CSV' for convenience. We also built a custom dataset loader based on Huggingface's `load_dataset` function, such that users only need to indicate the datasets of interest either as a list of inputs within the script (E.g. 'dataset_name=['altlex','because']'), or directly in the command line (E.g. '--dataset_name altlex because').

3.2   Baseline Model

Transformer-based pre-trained language models are the state-of-the-art in NLP. To create our initial benchmark, we used pre-trained Bidirectional Encoder Representations from Transformers (BERT) models [5]. First, sequences were tokenized into token embeddings (r_n). Special start ([CLS]) and end ([SEP]) tokens were added to the input sequence. For the Pair Classification task, we added four special tokens to the vocabulary to represent the boundaries of the two arguments. The BERT encoder was fine-tuned to our task during training.

Sequence and Pair Classification. For each classification task, we pooled the token embeddings into a sequence embedding by extracting the embedding on the [CLS] token. The sequence embeddings were then fed into a sequence classifier $g(.)$ to predict logits for the two labels: *Causal* and *Non-causal*, as in $\hat{y} = g(r_{[CLS]})$. We compared the logits with the true sequence label y to calculate Cross-Entropy (CE) Loss for learning using the following formula:

$$\mathcal{L}_{seq} = -y \cdot \log(\hat{y}) - (1 - y) \cdot \log(1 - \hat{y}) \tag{1}$$

Span Detection. We fed the token embeddings into a token classifier ($f(.)$), as in $\hat{t_n} = f(r_n)$. The classifier returned predicted logits for the five BIO-CE token labels. Again, the logits and true token vector t^v were used to calculate CE Loss based on the following formula:

$$\mathcal{L}_{token} = -\sum_{n=1}^{N} \sum_{i=1}^{C} t_{n,i} \cdot \log(\hat{t_{n,i}}) \tag{2}$$

Note that given the current simple set up, the span detection model only predicts one cause-effect relation per input sequence. While this model is acceptable as an initial baseline, we do intend to replicate or design more complex models that can identify multiple causal relations in the future.

Evaluation Metrics. During inference, for both Classification and Span Detection, we apply `argmax` on the predicted logits to obtain the predicted sequence and token labels respectively. For the two Classification tasks, we calculated the Accuracy (Acc), Precision (P), Recall (R) and F1 scores per experiment. For Span Detection, we referred to evaluation metrics from earlier Cause-Effect Span Detection [17,18,29] shared tasks, and used the Macro P, R and F1 metrics. The token classification evaluation scheme by `seqeval` [21,25] reverts the BIO-formatted labels to the original form (i.e. Cause (C) and Effect (E)) for evaluation. Our default evaluation scripts report metrics for all and each corpus. In the next section, we present the average and standard deviation scores obtained from multiple runs using five random seeds. For model hyperparameters, additional experimental details and results, please refer to our Supplementary Material.

4   Experiments

In this section, we describe experiments performed on the UniCausal corpus.

4.1   Baseline Performance

Table 3. Mean and standard deviation of performance metrics for different test sets across the three tasks, across five random seeds. All models were trained on all six datasets, where applicable. Tasks that are not applicable to the dataset are indicated by "-". Scores are reported in percentages (%).

Test Set	(I) Sequence Classification				(II) Span Detection			(III) Pair Classification			
	P	R	F1	Acc	P	R	F1	P	R	F1	Acc
All	71.13 ± 0.80	69.14 ± 1.60	70.10 ± 0.58	86.27 ± 0.15	46.33 ± 1.22	60.35 ± 0.30	52.42 ± 0.90	85.44 ± 0.96	83.93 ± 0.44	84.68 ± 0.27	93.68 ± 0.16
AltLex	50.76 ± 1.61	63.48 ± 4.60	56.37 ± 2.49	71.87 ± 1.19	27.74 ± 1.20	42.99 ± 0.85	33.72 ± 1.12	82.60 ± 1.99	87.09 ± 1.53	84.76 ± 0.66	90.43 ± 0.55
BECAUSE	92.32 ± 1.69	70.24 ± 2.04	79.77 ± 1.68	71.37 ± 2.24	32.51 ± 2.82	44.30 ± 2.33	37.47 ± 2.57	87.93 ± 1.73	94.78 ± 1.94	91.21 ± 1.18	86.00 ± 1.90
CTB	42.37 ± 2.11	66.19 ± 4.26	51.58 ± 1.82	83.48 ± 1.21	-	-	-	75.66 ± 3.61	72.50 ± 6.81	73.94 ± 4.68	95.04 ± 0.78
ESL	76.11 ± 2.04	67.43 ± 3.45	71.45 ± 1.89	73.79 ± 1.34	-	-	-	-	-	-	-
PDTB	72.59 ± 0.61	66.34 ± 1.63	69.31 ± 0.70	84.63 ± 0.17	47.77 ± 1.22	61.54 ± 0.29	53.78 ± 0.88	84.56 ± 1.17	82.04 ± 0.46	83.28 ± 0.36	92.43 ± 0.23
SemEval	73.39 ± 1.18	89.51 ± 1.59	80.64 ± 0.46	94.81 ± 0.16	-	-	-	93.38 ± 0.88	96.10 ± 0.59	94.71 ± 0.23	98.70 ± 0.07

In Table 3, we present the performance of the baseline BERT models when trained on all datasets, and tested on all and each dataset. Across all test sets, baseline models achieved 70.10% Binary F1 score for Sequence Classification, 52.42% Macro F1 score for Span Detection, and 84.68% Binary F1 score for Pair Classification.

Overall, regardless of the dataset, performance for Pair Classification is always better than Sequence Classification. F1 scores for Span Detection is poor in comparison to the Classification tasks. This finding correlates with the difficulty of each task: For Pair Classification, since the prompts that already identifies the arguments are provided, it is arguably a simpler task than Sequence

Table 4. Evaluation metrics for each dataset in the literature review compared to our benchmark (Ours). We do not cover methods that rely on the connectives as features for Classification tasks. Notations: ^ Rebalanced the dataset, § Evaluated on k-folds or different folds, ‡ Slightly different definitions for class labels. Abbreviations: Self-Attention Embeddings (SA), Logistic Regression (LR), Bidirectional GRU + Self-Attention (BIGRUATT), Feed-forward neural network (FFNN)

(I) Sequence Classification							
Corpus	Source	Features	Model	P	R	F1	Acc
AltLex	[11]	Lexical	Support Vector Machine	**61.98**	58.51	**60.19**	67.68
	Ours (All)	BERT	BERT+LR	50.76	**63.48**	56.37	**71.87**
	Ours (AltLex)	BERT	BERT+LR	50.58	53.57	51.85	71.52
BEC-AUSE	[36]§ ‡	Discourse, SA	PSAN	-	-	81.70	-
	Ours (All)	BERT	BERT+LR	**92.32**	70.24	79.77	71.37
	Ours (BECAUSE)	BERT	BERT+LR	86.20	**96.01**	**90.77**	**84.31**
CTB	[13] ^	n-gram	LR	100.00	22.22	36.36	-
		word2vec	BIGRUATT	67.04	73.89	69.98	-
		ELMO	BIGRUATT	**81.29**	70.28	75.08	-
		BERT	BERT+LR	71.17	**93.33**	**80.55**	-
		BERT	BERT+BIGRUATT	74.52	86.94	80.06	-
	Ours (All)	BERT	BERT+LR	42.37	66.19	51.58	83.48
	Ours (CTB)	BERT	BERT+LR	71.46	58.57	63.65	**91.27**
ESL	[13] ^	n-gram	LR	**100.00**	27.27	42.86	-
		word2vec	BIGRUATT	70.09	60.91	63.65	-
		ELMO	BIGRUATT	77.47	59.09	66.55	-
		BERT	BERT+LR	62.44	87.17	72.35	-
		BERT	BERT+BIGRUATT	66.15	83.64	73.09	-
	Ours (All)	BERT	BERT+LR	76.11	67.43	71.45	73.79
	Ours (ESL)	BERT	BERT+LR	75.90	**87.79**	**81.21**	**80.17**
PDTB	[23]‡	Lexical	Shallow CNN	39.80	75.29	52.04	63.00
		Lexical	FFNN	42.04	71.74	53.01	66.44
		Lexical, positional, event	FFNN	42.37	**76.45**	54.52	66.35
	[36]§ ‡	Discourse, SA	PSAN	-	-	**76.60**	-
	[31]§	BERT	BERT+LR	-	-	74.45	-
	Ours (All)	BERT	BERT+LR	72.59	66.34	69.31	84.63
	Ours (PDTB)	BERT	BERT+LR	**73.54**	67.35	70.31	**85.11**
Sem-Eval	[22]	n-gram	Random Forest	-	-	52.80	-
		n-gram	LR	-	-	81.90	-
		word2vec	LSTM	-	-	85.60	-
		word2vec	LSTM + Self-Attention	-	-	86.90	-
	[13] ^	n-gram	LR	88.67	66.83	76.22	-
		word2vec	BIGRUATT	93.96	87.59	90.64	-
		ELMO	BIGRUATT	**94.45**	91.26	**92.81**	-
		BERT	BERT+LR	86.62	97.09	91.55	-
		BERT	BERT+BIGRUATT	86.80	**96.63**	91.45	-
	Ours (All)	BERT	BERT+LR	73.39	89.51	80.64	94.81
	Ours (SemEval)	BERT	BERT+LR	87.84	91.40	89.58	**97.43**
(III) Pair Classification							
Corpus	Source	Features	Model	P	R	F1	Acc
Sem-Eval	[2]	Lexical, semantic, dependency	Bayesian Classifier	-	-	66.00	93.00
		word2vec	CNN	-	-	66.00	88.00
		GrammarTags	CNN	-	-	86.60	93.00
	Ours (All)	BERT	BERT+LR	93.38	**96.10**	94.71	98.70
	Ours (SemEval)	BERT	BERT+LR	**93.96**	95.67	**94.80**	**98.73**

Table 5. Mean and standard deviation of F1 score across different training and test set combinations across five random seeds. Scores are reported in percentages (%). For each panel, the top score per column is bolded. Paired T-test was conducted against the first row per panel, where all datasets were used for training. Statistical significance: ***< 0.001, **< 0.01, *< 0.05.

(I) Sequence Classification

Training Set	Test Set						
	All	AltLex	BECAUSE	CTB	ESL	PDTB	SemEval
All	**70.10 ± 0.58**	**56.37 ± 2.49**	79.77 ± 1.68	51.58 ± 1.82	71.45 ± 1.89	69.31 ± 0.70	80.64 ± 0.46
AltLex	32.93 ± 3.57***	51.85 ± 2.53	36.47 ± 11.18***	38.21 ± 6.20*	53.30 ± 8.37**	22.91 ± 5.79***	55.83 ± 6.68***
BECAUSE	39.15 ± 0.99***	47.02 ± 1.52**	**90.77 ± 2.22*****	25.17 ± 1.34***	63.49 ± 1.94**	42.49 ± 0.68***	23.71 ± 1.93***
CTB	33.49 ± 5.48***	55.91 ± 7.63	54.73 ± 9.40**	**63.65 ± 5.55****	33.26 ± 15.44**	25.97 ± 3.73***	51.76 ± 13.85**
ESL	39.62 ± 0.89***	46.29 ± 1.15**	90.12 ± 1.05***	30.84 ± 1.35***	**81.21 ± 2.35*****	42.55 ± 1.25***	26.15 ± 2.62***
PDTB	60.99 ± 0.76***	48.94 ± 1.88**	69.61 ± 2.16**	39.54 ± 1.88***	38.71 ± 3.15***	**70.31 ± 0.56***	19.75 ± 3.35***
SemEval	28.25 ± 0.86***	28.95 ± 1.74***	16.91 ± 3.40***	38.51 ± 3.44**	45.95 ± 3.50***	10.11 ± 1.61***	**89.58 ± 0.71*****

(II) Span Detection

Training Set	Test Set			
	All	AltLex	BECAUSE	PDTB
All	**52.42 ± 0.90**	**33.72 ± 1.12**	37.47 ± 2.57	53.78 ± 0.88
AltLex	6.20 ± 0.74***	21.45 ± 1.87***	11.51 ± 1.63***	5.47 ± 0.76***
BECAUSE	12.74 ± 0.35***	7.38 ± 2.19***	**37.79 ± 5.77**	12.60 ± 0.34***
PDTB	51.97 ± 0.48	6.73 ± 0.94***	35.84 ± 2.42	**55.02 ± 0.38***

(III) Pair Classification

Training Set	Test Set					
	All	AltLex	BECAUSE	CTB	PDTB	SemEval
All	**84.68 ± 0.27**	**84.76 ± 0.66**	91.21 ± 1.18	73.94 ± 4.68	83.28 ± 0.36	94.71 ± 0.23
AltLex	31.83 ± 3.93***	80.57 ± 2.48*	48.44 ± 20.00**	20.06 ± 7.14***	25.11 ± 8.75***	57.72 ± 14.52**
BECAUSE	36.40 ± 0.64***	47.99 ± 1.33***	90.01 ± 1.95	23.58 ± 1.52***	38.39 ± 0.37***	25.23 ± 2.02***
CTB	20.17 ± 5.78***	19.16 ± 15.64***	22.00 ± 10.92***	73.29 ± 6.14	7.02 ± 6.06***	63.69 ± 5.65***
PDTB	68.13 ± 0.88***	40.34 ± 1.52***	82.59 ± 2.17***	26.74 ± 2.42***	**83.70 ± 0.34**	33.64 ± 1.76***
SemEval	26.66 ± 1.86***	37.07 ± 6.58***	25.70 ± 11.46***	50.63 ± 1.74***	8.08 ± 3.20***	**94.80 ± 0.28**

Classification. For Span Detection, it is much more challenging than both Classification tasks because it involves accurate identification of the words that corresponds to the cause and effect, not just the mere identification that they exist. Furthermore, the baseline token classification set-up was too simplistic, and unable to handle multiple cause-effect span relations in the same sequence. For each input text, only one pair of Cause and Effect will be predicted. Thus, if multiple relation exists, only one pair can be predicted correctly at best.

In Table 4, we provide a snapshot of evaluation metrics reported by previous works on the datasets. It is challenging to make claims about model superiority from this table alone, since different papers used different train-test splits and some papers altered the dataset composition by rebalancing it. Nevertheless, for datasets like AltLex and SemEval, the development set was predefined by the dataset creators. Thus, comparisons between previous work and ours can be made concretely. For Sequence Classification with AltLex, [11]'s handcrafted lexical features fed through a Support Vector Machine achieved an F1 score of 60.19%, beating us. For Sequence Classification with SemEval, our best F1 score of 89.58% surpasses methods covered by [22] which, at best, achieved 86.90% using word2vec embeddings fed through a Long-Short Term Memory Self-Attention network. Finally, for Pair Classification with SemEval, our BERT-based model consistently surpasses Bayesian Classifier and Convolutional Neural Network methods explored by [2].

All in all, our baseline model is simple but competitive. From Table 4 alone, it is apparent that the causal text mining community lacks a consistent way to benchmark performance. Therefore, we hope that from here on, the scores presented in Table 3 will serve as an initial, universal baseline score for the causal text mining community to beat.

4.2 Impact of Datasets

In Table 5, we present the F1 scores when training and testing on different corpus. This table reflects how compatible each corpus is to one another. When testing on all the datasets, we noticed that training on all datasets returned the best performance across all tasks by a large margin. Training on any one dataset was unable to achieve similar performance. Meanwhile, the generalized model trained on all datasets did not always return the best performance for each corpus. Given the differences in definitions and linguistic coverage of each dataset, it is expected that for some datasets, specializing on its own data distributions leads to better performance. However, such specialized models are more likely to overfit and lack generalizability. Thus, good performance on one dataset but not others should be handled with caution.

4.3 Adding CauseNet to Investigate the Importance of Linguistic Variation in Examples

The modular structure of our code, in terms of data loading and evaluation, allows researchers to incorporate custom data. To illustrate, we obtained around 50,000 training and 5,000 testing causal examples from CauseNet [9] suitable for Pair Classification. After data processing, it is straightforward to load CauseNet with other datasets by including its name in the command '--dataset_name'.

We found improvements in F1 for CTB (77.08%) and AltLex (85.04%), which are rule-based and relatively template-based respectively. The model also performs perfectly on CauseNet. However, BECAUSE (89.80%), PDTB (83.20%), and SemEval (94.31%) had poorer performance. These findings indicate that training the model with a large number of rule-based causal examples only helps the model identify similar rule-based causal examples. They do not help the model learn about the semantics of causal relations. In fact, the model is worse-off for non-rule-based examples due to the lack of linguistic variation covered. This again motivates our focus to include mainly human-annotated data in Uni-Causal for both better training and fairer testing.

5   Conclusion

We propose UniCausal, a unified resource and benchmark for causal text mining. Our codes were designed to allow researchers to work on some or all datasets and tasks, while still comparing their performance fairly against us or others. Researchers can easily include new datasets too. In this paper, we provided evaluation metrics per dataset as an initial benchmark for future researchers to compete against. We hope researchers will use UniCausal to design joint models that learn from multiple causal text mining tasks and datasets. A unified model that learns from diverse objectives and knowledge sources will be more adaptable and generalizable.

Acknowledgements. This research/project is supported by the National Research Foundation, Singapore under its Industry Alignment Fund - Pre-positioning (IAF-PP) Funding Initiative. Any opinions, findings and conclusions or recommendations expressed in this material are those of the author(s) and do not reflect the views of National Research Foundation, Singapore. Additionally, we thank Jiatong Han for helping with the creation of tutorials.

References

1. Asghar, N.: Automatic extraction of causal relations from natural language texts: a comprehensive survey. CoRR **abs/1605.07895** (2016). http://arxiv.org/abs/1605.07895
2. Ayyanar, R., Koomullil, G., Ramasangu, H.: Causal relation classification using convolutional neural networks and grammar tags. In: 2019 IEEE 16th India Council International Conference (INDICON), pp. 1–3 (2019). https://doi.org/10.1109/INDICON47234.2019.9028985
3. Cao, P., et al.: Knowledge-enriched event causality identification via latent structure induction networks. In: Proceedings of the 59th Annual Meeting of the Association for Computational Linguistics and the 11th International Joint Conference on Natural Language Processing (Volume 1: Long Papers), pp. 4862–4872. Association for Computational Linguistics, Online (2021). https://doi.org/10.18653/v1/2021.acl-long.376. https://aclanthology.org/2021.acl-long.376
4. Caselli, T., Vossen, P.: The event StoryLine corpus: a new benchmark for causal and temporal relation extraction. In: Proceedings of the Events and Stories in the News Workshop, Vancouver, Canada, pp. 77–86. Association for Computational Linguistics (2017). https://doi.org/10.18653/v1/W17-2711. https://aclanthology.org/W17-2711
5. Devlin, J., Chang, M.W., Lee, K., Toutanova, K.: BERT: pre-training of deep bidirectional transformers for language understanding. In: Proceedings of the 2019 Conference of the North American Chapter of the Association for Computational Linguistics: Human Language Technologies, Volume 1 (Long and Short Papers), Minneapolis, Minnesota, pp. 4171–4186. Association for Computational Linguistics (2019). https://doi.org/10.18653/v1/N19-1423. https://aclanthology.org/N19-1423
6. Dunietz, J., Burnham, G., Bharadwaj, A., Rambow, O., Chu-Carroll, J., Ferrucci, D.: To test machine comprehension, start by defining comprehension. In: Proceedings of the 58th Annual Meeting of the Association for Computational Linguistics, pp. 7839–7859. Association for Computational Linguistics, Online (2020). https://doi.org/10.18653/v1/2020.acl-main.701. https://aclanthology.org/2020.acl-main.701
7. Dunietz, J., Levin, L., Carbonell, J.: The BECauSE corpus 2.0: annotating causality and overlapping relations. In: Proceedings of the 11th Linguistic Annotation Workshop, Valencia, Spain, pp. 95–104. Association for Computational Lin-

guistics (2017). https://doi.org/10.18653/v1/W17-0812. https://aclanthology.org/W17-0812

8. Girju, R.: Automatic detection of causal relations for question answering. In: Proceedings of the ACL 2003 Workshop on Multilingual Summarization and Question Answering, Sapporo, Japan, pp. 76–83. Association for Computational Linguistics (2003). https://doi.org/10.3115/1119312.1119322. https://aclanthology.org/W03-1210

9. Heindorf, S., Scholten, Y., Wachsmuth, H., Ngomo, A.N., Potthast, M.: Causenet: towards a causality graph extracted from the web. In: d'Aquin, M., Dietze, S., Hauff, C., Curry, E., Cudré-Mauroux, P. (eds.) CIKM 2020: The 29th ACM International Conference on Information and Knowledge Management, Virtual Event, Ireland, 19–23 October 2020, pp. 3023–3030. ACM (2020). https://doi.org/10.1145/3340531.3412763

10. Hendrickx, I., et al.: SemEval-2010 task 8: multi-way classification of semantic relations between pairs of nominals. In: Proceedings of the 5th International Workshop on Semantic Evaluation, Uppsala, Sweden, pp. 33–38. Association for Computational Linguistics (2010). https://aclanthology.org/S10-1006

11. Hidey, C., McKeown, K.: Identifying causal relations using parallel Wikipedia articles. In: Proceedings of the 54th Annual Meeting of the Association for Computational Linguistics (Volume 1: Long Papers), Berlin, Germany, pp. 1424–1433. Association for Computational Linguistics (2016). https://doi.org/10.18653/v1/P16-1135. https://aclanthology.org/P16-1135

12. Ittoo, A., Bouma, G.: Minimally-supervised learning of domain-specific causal relations using an open-domain corpus as knowledge base. Data Knowl. Eng. **88**, 142–163 (2013). https://doi.org/10.1016/j.datak.2013.08.004

13. Kyriakakis, M., Androutsopoulos, I., Saudabayev, A., Ginés i Ametllé, J.: Transfer learning for causal sentence detection. In: Proceedings of the 18th BioNLP Workshop and Shared Task, Florence, Italy, pp. 292–297. Association for Computational Linguistics (2019). https://doi.org/10.18653/v1/W19-5031. https://aclanthology.org/W19-5031

14. Li, Z., Li, Q., Zou, X., Ren, J.: Causality extraction based on self-attentive BiLSTM-CRF with transferred embeddings. Neurocomputing **423**, 207–219 (2021). https://doi.org/10.1016/j.neucom.2020.08.078

15. Li, Z., Ding, X., Liu, T., Hu, J.E., Durme, B.V.: Guided generation of cause and effect. In: Bessiere, C. (ed.) Proceedings of the Twenty-Ninth International Joint Conference on Artificial Intelligence, IJCAI 2020, pp. 3629–3636. ijcai.org (2020). https://doi.org/10.24963/ijcai.2020/502

16. Luo, Z., Sha, Y., Zhu, K.Q., Hwang, S., Wang, Z.: Commonsense causal reasoning between short texts. In: Baral, C., Delgrande, J.P., Wolter, F. (eds.) Principles of Knowledge Representation and Reasoning: Proceedings of the Fifteenth International Conference, KR 2016, Cape Town, South Africa, 25–29 April 2016, pp. 421–431. AAAI Press (2016). http://www.aaai.org/ocs/index.php/KR/KR16/paper/view/12818

17. Mariko, D., Abi-Akl, H., Labidurie, E., Durfort, S., De Mazancourt, H., El-Haj, M.: The financial document causality detection shared task (FinCausal 2020). In: Proceedings of the 1st Joint Workshop on Financial Narrative Processing and MultiLing Financial Summarisation, Barcelona, Spain, pp. 23–32. COLING (Online) (2020). https://aclanthology.org/2020.fnp-1.3

18. Mariko, D., Akl, H.A., Labidurie, E., Durfort, S., de Mazancourt, H., El-Haj, M.: The financial document causality detection shared task (FinCausal 2021). In: Proceedings of the 3rd Financial Narrative Processing Workshop, Lancaster, United

Kingdom, pp. 58–60. Association for Computational Linguistics (2021). https://aclanthology.org/2021.fnp-1.10

19. Mirza, P., Sprugnoli, R., Tonelli, S., Speranza, M.: Annotating causality in the TempEval-3 corpus. In: Proceedings of the EACL 2014 Workshop on Computational Approaches to Causality in Language (CAtoCL), Gothenburg, Sweden, pp. 10–19. Association for Computational Linguistics (2014). https://doi.org/10.3115/v1/W14-0702. https://aclanthology.org/W14-0702

20. Mirza, P., Tonelli, S.: An analysis of causality between events and its relation to temporal information. In: Proceedings of COLING 2014, the 25th International Conference on Computational Linguistics: Technical Papers, Dublin, Ireland, pp. 2097–2106. Dublin City University and Association for Computational Linguistics (2014). https://aclanthology.org/C14-1198

21. Nakayama, H.: seqeval: a python framework for sequence labeling evaluation (2018). https://github.com/chakki-works/seqeval

22. Niki, Y., Sakaji, H., Izumi, K., Matsushima, H.: Causality existence classification from multilingual texts using end-to-end LSTM models. In: Papapetrou, P., Cheng, X., He, Q. (eds.) 2019 International Conference on Data Mining Workshops, ICDM Workshops 2019, Beijing, China, 8–11 November 2019, pp. 17–23. IEEE (2019). https://doi.org/10.1109/ICDMW.2019.00011

23. Ponti, E.M., Korhonen, A.: Event-related features in feedforward neural networks contribute to identifying causal relations in discourse. In: Proceedings of the 2nd Workshop on Linking Models of Lexical, Sentential and Discourse-Level Semantics, Valencia, Spain, pp. 25–30. Association for Computational Linguistics (2017). https://doi.org/10.18653/v1/W17-0903. https://aclanthology.org/W17-0903

24. Radinsky, K., Horvitz, E.: Mining the web to predict future events. In: Leonardi, S., Panconesi, A., Ferragina, P., Gionis, A. (eds.) Sixth ACM International Conference on Web Search and Data Mining, WSDM 2013, Rome, Italy, 4–8 February 2013, pp. 255–264. ACM (2013). https://doi.org/10.1145/2433396.2433431

25. Ramshaw, L., Marcus, M.: Text chunking using transformation-based learning. In: Third Workshop on Very Large Corpora (1995). https://aclanthology.org/W95-0107

26. Ruppenhofer, J., Ellsworth, M., Schwarzer-Petruck, M., Johnson, C.R., Scheffczyk, J.: Framenet II: extended theory and practice. Technical report, International Computer Science Institute (2016)

27. Speer, R., Chin, J., Havasi, C.: Conceptnet 5.5: an open multilingual graph of general knowledge. In: Singh, S.P., Markovitch, S. (eds.) Proceedings of the Thirty-First AAAI Conference on Artificial Intelligence, 4–9 February 2017, San Francisco, California, USA, pp. 4444–4451. AAAI Press (2017). http://aaai.org/ocs/index.php/AAAI/AAAI17/paper/view/14972

28. Stasaski, K., Rathod, M., Tu, T., Xiao, Y., Hearst, M.A.: Automatically generating cause-and-effect questions from passages. In: Proceedings of the 16th Workshop on Innovative Use of NLP for Building Educational Applications, pp. 158–170. Association for Computational Linguistics, Online (2021). https://aclanthology.org/2021.bea-1.17

29. Tan, F.A., et al.: Event causality identification with causal news corpus - shared task 3, CASE 2022. In: Proceedings of the 5th Workshop on Challenges and Applications of Automated Extraction of Socio-political Events from Text (CASE), Abu Dhabi, United Arab Emirates (Hybrid), pp. 195–208. Association for Computational Linguistics (2022). https://aclanthology.org/2022.case-1.28

30. Tan, F.A., et al.: The causal news corpus: annotating causal relations in event sentences from news. In: Proceedings of the Thirteenth Language Resources and Evaluation Conference, Marseille, France, pp. 2298–2310. European Language Resources Association (2022). https://aclanthology.org/2022.lrec-1.246

31. Tan, F.A., et al.: The causal news corpus: annotating causal relations in event sentences from news. In: Proceedings of the Language Resources and Evaluation Conference, Marseille, France, pp. 2298–2310. European Language Resources Association (2022). https://aclanthology.org/2022.lrec-1.246

32. Webber, B., Prasad, R., Lee, A., Joshi, A.: The penn discourse treebank 3.0 annotation manual. University of Pennsylvania, Philadelphia (2019)

33. Yang, J., Han, S.C., Poon, J.: A survey on extraction of causal relations from natural language text. Knowl. Inf. Syst. (2022). https://doi.org/10.1007/s10115-022-01665-w

34. Zuo, X., et al.: Improving event causality identification via self-supervised representation learning on external causal statement. In: Findings of the Association for Computational Linguistics: ACL-IJCNLP 2021, pp. 2162–2172. Association for Computational Linguistics, Online (2021). https://doi.org/10.18653/v1/2021.findings-acl.190. https://aclanthology.org/2021.findings-acl.190

35. Zuo, X., et al.: Improving event causality identification via self-supervised representation learning on external causal statement. In: Findings of the Association for Computational Linguistics: ACL-IJCNLP 2021, pp. 2162–2172. Association for Computational Linguistics, Online (2021). https://doi.org/10.18653/v1/2021.findings-acl.190. https://aclanthology.org/2021.findings-acl.190

36. Zuo, X., Chen, Y., Liu, K., Zhao, J.: KnowDis: knowledge enhanced data augmentation for event causality detection via distant supervision. In: Proceedings of the 28th International Conference on Computational Linguistics, Barcelona, Spain, pp. 1544–1550. International Committee on Computational Linguistics (Online) (2020). https://doi.org/10.18653/v1/2020.coling-main.135. https://aclanthology.org/2020.coling-main.135

Deep Learning

Accounting for Imputation Uncertainty During Neural Network Training

Thomas Ranvier$^{(\boxtimes)}$ (ORCID), Haytham Elghazel, Emmanuel Coquery,
and Khalid Benabdeslem

Univ Lyon, UCBL, CNRS, INSA Lyon, LIRIS, UMR5205, 43 bd du 11 Novembre
1918, 69622 Villeurbanne, France
{thomas.ranvier,haytham.elghazel,emmanuel.coquery,
khalid.benabdeslem}@univ-lyon1.fr

Abstract. In this paper we are interested in dealing with missing values in a machine learning context, and more especially when training a neural network. We focus on improving neural network training by reducing the potential biases that can occur during the training phase on artificially imputed datasets. We do so by taking into account the between-variance that can be observed between multiple imputations. We propose two new imputation frameworks, *S-HOT* and *M-HOT*, that can be used to train neural networks on completed data in a less biased way, leading to models able of more generalization, and so, to better inference results. We perform extensive comparative experiments and statistically assess the results on both benchmark and real-world datasets. We show that our frameworks compete against and even outperform existing imputation frameworks, while being both useful in different settings. We make our entire code publicly accessible to facilitate reproduction of our experimental results.

Keywords: Data Imputation · Imputation Uncertainty ·
Between-variance · Optimization · Neural Networks

1 Introduction

Two types of variances occur when multiply imputing an incomplete dataset: the within-variance and the between-variance. The within-variance corresponds to the variance within each imputed dataset. The between-variance corresponds to the variance in-between every imputed dataset. In real scenarios, it is usually not possible to know the within-variance, since most imputation methods estimate a fixed value in place of missing ones without outputting any probability measure. However, the between-variance can easily be computed between a set of completed datasets. In the following, we refer to the between-variance as imputation uncertainty.

In this paper, we make the difference between imputation methods, which aim to impute missing values in a dataset, and imputation frameworks such as

R. Wrembel et al. (Eds.): DaWaK 2023, LNCS 14148, pp. 265–280, 2023.
https://doi.org/10.1007/978-3-031-39831-5_24

Single-Imputation or Multiple-Imputation, which rely on any imputation method to deal with missing data. Our contributions are imputation frameworks, they aim to train neural networks while taking into account imputation uncertainty and can rely on any imputation method. Note that we are not interested in comparing imputation methods performance.

When dealing with incomplete data, a naive and overly used approach is Single-Imputation [11]. That is, to arbitrarily choose an imputation method, use it to impute the dataset, and treat the completed dataset as the new real dataset to perform any kind of future analysis. When training a neural network or similarly strong learners on completed data, those models are usually able to generalize enough to limit the bias occurring due to imputation uncertainty. This leads to good enough results so that one does not look for better ways to deal with missing values. It has even been shown that, when using strong inference models, almost any imputation asymptotically leads to optimal prediction [6]. This is probably one of the main reasons why accounting for imputation uncertainty has not been widely researched. However, those models might reach good results in such situations but they are biased by imputation uncertainty, we show that accounting for this uncertainty during their training phase helps to reach even better prediction results.

In this paper, we propose two imputation frameworks, *S-HOT* and *M-HOT*, that aim to train neural networks on imputed datasets while accounting for imputation uncertainty to reduce the natural bias occurring when training on completed datasets. Those frameworks are to be used in different situations, *S-HOT* is adapted to train a unique and large neural network, *M-HOT* can be used to train multiple learners in an ensemble way and reach extremely good prediction results at the expense of a higher computational cost. We conduct extensive experiments to compare our two frameworks with the existing Single-Imputation and Multiple-Imputation frameworks. We then perform statistical analysis to assess the obtained results on both benchmark and real-world medical datasets. We show that our proposed frameworks compete against, and even outperform existing imputation frameworks. This paper is a first step towards finding better ways to deal with missing values imputation in machine learning. We hope that it spikes the interest of other machine learning researchers throughout the world on this important and largely overlooked matter in the machine learning literature.

The complete results, supplementary material and source code used to conduct our experiments are available at the following GitHub repository[1].

In the rest of the paper, we first present related works on imputation frameworks and methods. We then present and describe our propositions in Sect. 3. Section 4 shows our experiments and obtained results. Finally, we conclude with a summary of our contributions.

---

[1] https://github.com/ThomasRanvier/Accounting_for_Imputation_Uncertainty_Dur
ing_Neural_Network_Training.

2   Related Works

Single-Imputation (SI) is probably the most commonly used framework for handling missing values in practice [11]. SI has the advantage of being extremely simple and straightforward: an imputation method is chosen and applied to the incomplete dataset, which yields a completed dataset where missing values have been assigned with new values, which can then be used and exploited as any complete dataset. It is a huge benefit to obtain a completed dataset since it is then possible to integrate SI in any existing pipeline or software to make them usable in the presence of missing values [5]. However, SI presents several important drawbacks. Once SI has been performed, it treats the imputed dataset as the new complete dataset, this is a problem since all methods that exploit the completed dataset will treat missing values as if they were known [11]. The extra variability due to the unknown missing values cannot be taken into account by those methods, thus, their inferences will be too sharp and overstate precision [9]. SI results in biased inference models which lack generalization capacity, but it outputs a fixed imputed dataset, which makes it extremely convenient and easy to use in any scenario.

Multiple-Imputation (MI) has originally been introduced by Rubin [11]. It consists in replacing each missing value with at least two or more substitution values representing a distribution of possibilities, which represents the uncertainty about the right value to impute [15]. Applying this framework results in several imputed datasets that are then exploited exactly as if the imputed data were the real data [11]. In a machine learning context, one learner is trained on each imputed dataset and the results from all the learners are then pooled in an ensemble manner. An advantage of MI over SI is that each missing value is represented by a sample of possible imputation values, which results in inferences that reflect the uncertainty level associated with each missing value [15]. Therefore, the results pooled from the ensemble of learners will be less biased compared to those of each learner taken independently. The use of the powerful ensemble paradigm leads to significantly better results when using the MI framework compared to SI. An obvious disadvantage is the computational cost of such a framework, the ensemble training of the learners multiplies the required amount of calculation time. While being quite an old framework, MI is not used in most cases, scientists and users of imputation methods usually still rely on Single-Imputation. This can be partially explained by the computational cost of MI which is high because of the ensemble paradigm.

Many methods exist that can be used to deal with missing values by replacing missing values with plausible ones. A very simple method that can be used to deal with missing values is mean substitution, where missing values are replaced with the mean value of the corresponding feature. This method has the advantage of being easy to implement and use, while retaining all non-missing information. More advanced imputation methods can be used to obtain better inference results on the imputed data. A popular method is the SOFTIMPUTE algorithm, introduced by Mazumder et al. [7], which works in an iterative way, at each step missing values are replaced using a single value decomposition. In 2012,

Stekhoven and Bühlmann introduced the MISSFOREST algorithm, an iterative imputation method based on random forests [12]. They have shown that MISS-FOREST can successfully handle missing values, particularly in datasets including mixed-types of variables. In 2018, Gondara and Wang introduced MIDA: Multiple-Imputation Using Denoising Autoencoders (DAEs) [3]. This method is based on autoencoders, a neural network model used to reconstruct its own input from a reduced latent representation. More recently, Yoon et al. introduced GAIN: Generative Adversarial Imputation Nets [14], which is based on adversarial models to impute missing values. In 2020, Muzellec et al. introduced SINKHORN OT, an optimal transport based method for data imputation [8]. All those imputation methods can be used differently depending on the imputation framework one wishes to apply. We use those popular methods in our experiments to demonstrate the usage and performances of each used imputation framework.

Some imputation methods deal with imputation uncertainty directly within their own algorithm. The MICE algorithm, for Multivariate Imputation by Chained Equations [1], deals with imputation uncertainty by iteratively imputing the dataset within its own algorithm. Missing values are initially imputed by the mean of the corresponding feature, MICE then iteratively trains multiple linear regressions in an ensemble way to impute each missing value by using all other features to fit the regressions until convergence. This is a way of performing *MI* within the algorithm of MICE, in this sense, MICE outputs a final imputed dataset that takes into account the between-imputation uncertainty. MICE can be considered both as an imputation method and an imputation framework as it performs itself Multiple-Imputation. In our experiments we treat MICE as an imputation framework to which we compare our frameworks results. Very recently, Hameed et al. introduced an adversarial imputation method that uses uncertainty-aware predictors [4]. They propose a neural network-based architecture trained using an adversarial strategy to estimate the uncertainty of imputed data. The method estimates the uncertainty of each imputed value and uses this level back as an input in an iterative way to minimize it, leading to better imputed values. We were not able to obtain the source code of the paper from Hameed et al. and could not replicate their method.

3   Contributions

MI is a first step towards taking account of between-imputation uncertainty. It naturally takes into account the uncertainty of imputed values through its ensemble nature, but each inference model is still biased by being trained on an arbitrarily fixed completed dataset [11].

We propose two frameworks that take this between-imputation uncertainty into account and show that neural networks trained using those frameworks have better generalization capacity. Those frameworks are based on the computation of the between-imputation uncertainty, which corresponds to the standard deviation between imputed values of all completed datasets. This uncertainty is then used as a scale to add stochasticity to the imputation of missing values directly on batch extraction during training. It leads to a kind of noise regularization that takes into account the imputation uncertainty, which improves generalization capacity, and thus, prediction results on unseen data.

3.1   Single-Hotpatching

Our Single-Hotpatching (*S-HOT*) framework is similar to *MI*, but has the advantage of training only one strong model. We named this framework Single-Hotpatching since missing values are dynamically imputed on batch extraction in a "hotpatching" manner.

It takes a huge amount of time and resources to train one large neural network, training several such models in an ensemble manner is not always a viable option. *S-HOT* aims to train only one model while relying on multiple imputations such as *MI*. It trains a model that is less biased than if it was trained using *SI* without needing as much computational time as *MI*. We experimentally show that *S-HOT* obtains significantly better results than *SI* with identical training times. Thus, it is interesting to use *S-HOT* in all situations where one aims to train a unique large model on imputed data.

When we train a model on a fixed imputed dataset in which imputed values are most certainly non-optimal, the model repeatedly learns on wrong and imprecise data, leading to a biased model lacking generalization. Instead, *S-HOT* performs multiple imputations and computes the between-imputation standard deviation associated with each imputed value, which we call the uncertainty level. This uncertainty level is used to draw random values from a normal distribution parameterized using the mean and standard deviation computed between the imputations. By training the model with batches in which missing values are dynamically drawn on the imputation distributions, we ensure that the model learns on a multitude of plausible imputations. The trained model has seen a span of possible values in place of each missing value during its training, leading to a less biased model with higher generalization capacity. Figure 1 shows the training and test phases of a neural network when using *S-HOT*. Once the model is trained, prediction results are obtained by randomly patching missing values in the test set using the same process for p iterations, leading to p predictions. The mean of the p model output probabilities is used as the final prediction, which is more robust than a unique iteration of this process.

Mathematically, we note $X \in \mathbb{R}^{n \times d}$ the original incomplete dataset, where each value X_{ij} is either observed or missing, with n and d the number of elements and features in X. We perform m imputations, leading to m different completed datasets $\tilde{X}^{1 \cdots m}$, with $\tilde{X}^k \in \mathbb{R}^{n \times d}$ the k-th completed dataset. Therefore, a missing value X_{ij} is imputed with m different values $\tilde{X}_{ij}^{1 \cdots m}$. Then, we compute the means μ and standard deviations σ of each value of the m completed datasets, with $\mu_{ij} = \frac{1}{m} \sum_{k=1}^{m} \tilde{X}_{ij}^k$ (1) and $\sigma_{ij} = \sqrt{\frac{1}{m} \sum_{k=1}^{m} (\tilde{X}_{ij}^k - \mu_{ij})^2}$ (2) . We note that values in $\tilde{X}^{1 \cdots m}$ that are observed in X have a mean of $\mu_{ij} = X_{ij}$ and a standard deviation of $\sigma_{ij} = 0$, only missing values in X have a value $\sigma_{ij} > 0$.

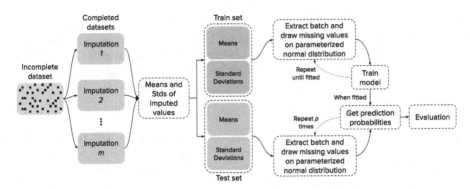

Fig. 1. Single-Hotpatching. We perform m imputations and compute the means and standard deviations of imputed values. Those are split between the train and test sets. During training, every time a batch is extracted, missing values are drawn from a normal distribution parameterized using previously computed means and stds. Once the model is trained, we extract the prediction probabilities by applying the same process p times for each test instance, which results in p predictions. The final prediction is computed as the mean of the p prediction probabilities, p is set as a few dozens to obtain robust prediction results.

Then, we train a neural network by feeding it batches that are computed from μ and σ. To extract a batch $B \in \mathbb{R}^{b \times d}$, with b the number of elements in the batch, we draw each value of the batch from a normal distribution parameterized with mean μ_{ij} and standard deviation $\alpha \cdot \sigma_{ij}$, such as $B_{ij} \sim \mathcal{N}(\mu_{ij}, \alpha \cdot \sigma_{ij})$. Where α is a scale hyper-parameter that can be set to 1 in most cases and might need to be set lower depending on the average uncertainty level. If the used imputation method outputs imputed values with a wide uncertainty the α scale should be empirically set lower than 1 to limit the stochastic impact induced by the approach. We never found a situation that benefited from increasing the α value. When a data point is presented to the neural network, observed values are set to X_{ij}, and missing values are set to a random value that follows the normal distribution of the m imputations. Thus, the neural network is not repeatedly trained on arbitrarily fixed (and very probably non-optimal) imputations, as it would be using *SI* or *MI* frameworks. Instead, it is trained on values that are randomly drawn from the imputations distribution. This process operates as a noise regularization that takes into account the between-imputation uncertainty, resulting in a less biased and more generalized neural network.

Pseudo-code 1 shows the training phase of the *S-HOT* framework. X is the original incomplete dataset, m the amount of imputations to perform, $impute(\cdot)$ the chosen imputation method, $normal(\mu, \sigma)$ is the method that draws each value x_{ij} on the normal distribution parameterized with μ_{ij} an dσ_{ij}, α is the hyper-parameter used to scale the standard deviation parameter, $n\_batches$ is the amount of batches required to span over the whole dataset.

Algorithm 1: Training phase of the *S-HOT* framework

---

input : X, m, $impute(\cdot)$, $normal(\cdot,\cdot)$, α, $n\_batches$
output: A trained neural network
for $i = 1$ *to* m **do**
  | $\tilde{X}^i = impute(X)$;
end
Compute μ and σ from $\tilde{X}$ using equations 1 and 2;
Randomly initialize the neural network;
while *Convergence is not reached* **do**
  | **for** $b = 1$ *to* $n\_batches$ **do**
  | | $batch = normal(\mu[batch\_slice], \alpha \cdot \sigma[batch\_slice])$;
  | | Fit the neural network to extracted $batch$;
  | **end**
end

3.2 Multiple-Hotpatching

Multiple-Hotpatching (*M-HOT*) extends *S-HOT* by using the ensemble paradigm. This framework trains as many learners as imputations are performed, those learners are trained while taking into account between-imputation uncertainty, leading to less biased individual learners. We empirically show that *M-HOT* leads to consistently better results than *MI* while not being computationally more expensive. It is beneficial to use *M-HOT* in situations in which one can afford to train several neural networks in an ensemble manner.

The framework is similar to *S-HOT* with the main difference that we use the ensemble paradigm such as in *MI*. We use the previously defined mathematical notations, $X \in \mathbb{R}^{n \times d}$ is the original incomplete dataset, where each value X_{ij} is either observed or missing. We perform m imputations which leads to m different completed datasets $\tilde{X}^{1...m}$, with $\tilde{X}^k \in \mathbb{R}^{n \times d}$ the k-th completed dataset. We define one learner per completed dataset, leading to m models. We compute the standard deviations σ in the same way as in Eq. 2, we do not need to compute the means. Then, we train the models. To extract a batch $B^k \in \mathbb{R}^{b \times d}$ that will be fed to model k, we draw each value of the batch from a normal distribution parameterized with mean $\tilde{X}_{ij}^k$ and standard deviation $\alpha \cdot \sigma_{ij}$, such as $B_{ij}^k \sim \mathcal{N}(\tilde{X}_{ij}^k, \alpha \cdot \sigma_{ij})$. Thus, all models are trained in an ensemble manner and are seeing imputed values drawn from a normal distribution centered on the corresponding computed imputation with added diversity during their training. This leads to an ensemble of models that are less biased and capable of more generalization than in *MI*, since they take into account the between-imputation uncertainty. *M-HOT* can be used as a substitute to *MI* in any situation in which *MI* is viable. Figure 2 shows, in a more practical way, how Multiple-Hotpatching takes place during the training phase of the neural networks.

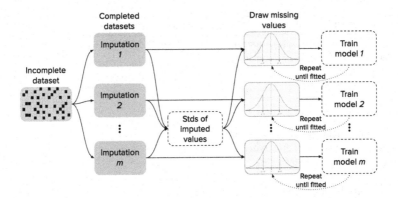

Fig. 2. Multiple-Hotpatching training phase. We perform m imputations and compute the standard deviations of imputed values. We train m models, when a batch is extracted from the k-th completed dataset, missing values are drawn from a normal distribution parameterized using previously computed stds and values from the k-th completed dataset as means. The process is repeated until all models are trained. Those can then be used to obtain predictions in the same way as with *S-HOT* but in an ensemble manner.

4   Experiments

4.1   Experimental Protocol

With our experiments, we show that *S-HOT* obtains significantly better results than *SI* and that *M-HOT* obtains consistently better results than *MI*. In our experiments we are not interested in comparing the performances of the used imputation methods, instead, we are interested in the results we can observe for the same imputation method when using each of the compared imputation frameworks.

We ran part of our experiments on five tabular benchmark classification datasets: the well-known IRIS dataset[2], the STATLOG dataset[3], the WINE Dataset (see footnote 3), the PIMA dataset[4] and the ABALONE dataset (see footnote 3).

We also compared the frameworks on three real-world medical datasets containing a certain amount of missing values. In the NHANES dataset, US National Health and Nutrition Examination Surveys, we used data from studies spanning from years 2000 to 2008, with 95 features and about 33% missing values. The COVID-19 dataset was publicly released with the paper [13], it contains medical information collected in early 2020 on pregnant and breastfeeding women. We based our data preprocessing on the one realized in the original paper, it is composed of 361 patients with 76 features, with about 20% missing data. And the

---

[2] https://scikit-learn.org/stable/modules/generated/sklearets.load_iris.html.

[3] https://archive.ics.uci.edu.

[4] https://rioultf.users.greyc.fr/uci/files/pima-indians-diabetes.

MYOCARDIAL infarction dataset (see footnote 3), composed of 1700 patients with 107 features, with about 5% missing values.

To run our experiments on the benchmark datasets that do not contain any missing values we artificially add various rates of missing values. We then use various imputation methods to impute the missing values following the tested frameworks and perform neural network training to compare the obtained results. We use three different patterns, MCAR, MAR, and MNAR [10], to introduce missing values to the datasets. We use the term "masked" to refer to the added missing values, the missing rate determines the number of values that will be masked. With the MCAR pattern, we introduce missing values completely at random by randomly masking a certain rate of values. For the MAR pattern, a random subset of features is chosen to remain unmasked, this subset is used as an input to a logistic model, and the output of the model is used to mask values out of the remaining features. We used the public implementation of [8]. With the MNAR pattern, the implementation is close to the one of MAR, but the input of the logistic model is masked using an MCAR pattern. Thus, the output of the model depends on values that are indifferently known or masked.

We compare the frameworks using several commonly used imputation methods: MISSFOREST based on iterative random forests [12], SOFTIMPUTE that uses single value decomposition) to iteratively impute missing values [7], GAIN that imputes using two adversarial models [14], MIDA that uses Denoising Autoencoders to impute missing values [3], and SINKHORN that is based on optimal transport for data imputation [8]. Note that we do not aim at comparing performances of those imputation methods, we apply each imputation framework using each imputation method and are only interested in comparing the imputation framework results.

In our experiments we treat MICE, a method that iteratively improves its last imputation using logistic models [1], as an imputation framework to which we compare our frameworks results. Indeed, as described earlier, MICE perform Multiple-Imputation within its own algorithm, it can be considered both an imputation method and an imputation framework.

Our experiments aim to evaluate neural network performances in a supervised-learning setting to check the bias and generalization capacity of the model. We use a simple neural network under the scikit-learn library[5], parameterized using well performing and identical hyper-parameters for each dataset to ensure a fair and unbiased comparison. We used the Adam optimizer with a default learning rate of 0.001. The used neural network is composed of two fully-connected layers in all our experiments, for datasets IRIS, STATLOG, PIMA and ABALONE both layers contains 32 units, for WINE, MYOCARDIAL and NHANES layers contain 64 and 32 units respectively, for the COVID dataset layers are composed of 128 and 32 units. We evaluate the performances of the models using the AUC metric, the Area Under the Receiver Operating Characteristic (ROC) Curve, the balanced accuracy metric, and the F1-score. The

---

[5] https://scikit-learn.org/stable/modules/generated/sklearn.neural_network.
MLPClassifier.html.

AUC corresponds to the area under the ROC curve that is obtained by plotting the true positive rate (recall) against the false positive rate (1-specificity), it reflects the capacity of the model to yield pertinent predictions. The balanced accuracy is defined as the average of recall obtained for each class, which is simply the average of the true positive rate between all the classes. The F1-score combines the precision and recall of a classifier into a single metric by taking their harmonic mean.

To better assess the obtained results we base our comparisons on the Friedman and Nemenyi statistical tests such as described in [2]. The Friedman test is first used to check if the null hypothesis that all compared frameworks are statistically equivalent for a given p-value is rejected or not. It ranks the compared frameworks for each dataset from best to worst, ties are assigned an average rank. The average rank for each compared framework is computed over all datasets and the Friedman test checks whether the measured average ranks are significantly different from the mean rank by computing the Friedman statistic: $\chi_F^2 = \frac{12N}{k(k+1)}(\sum_{j=1}^{k} R_j^2 - \frac{k(k+1)^2}{4})$ (3) , derived in $F_F = \frac{(N-1)\chi_F^2}{N(k-1)-\chi_F^2}$ (4) . The Friedman statistic is distributed according to the F-distribution with $k-1$ and $(k-1)(N-1)$ degrees of freedom, with k the number of frameworks compared and N the number of datasets. If the Friedman statistic is above the critical value of $F(k-1, (k-1)(N-1))$ given the p-value, then the test shows a significant difference between the compared frameworks. If that is the case, a post-hoc test is performed. We use the post-hoc Nemenyi test to compare the frameworks to each other in a pair-wise manner, two frameworks are considered significantly different if their average ranks differ by more than the computed critical distance: $CD = q_\alpha \sqrt{\frac{k(k+1)}{6N}}$, where α is the p-value, q_α comes from [2]. The Nemenyi test can be easily visualized through a simple diagram, which makes its results easy to analyze.

We first perform a comparative study between the four frameworks *SI*, *MI*, *S-HOT*, and *M-HOT* using different imputation methods. We show that no matter the used imputation method and missingness setting, our hotpatching frameworks lead to consistent and good results. We ran our experiments on 9 different missingness settings: MCAR, MAR, and MNAR, and for each pattern missing rates $10\%, 15\%, 25\%$. In each setting, we ran 20 imputations using each of the tested imputation methods and then ran the four frameworks 200 times, each run using a different random seed and train-test set to ensure the results are not biased by the stochastic nature of neural network training. We then compare our results to those of MICE, an imputation method that takes into account imputation uncertainty within its own algorithm, which we consider as a framework. We then experiment on real-world medical datasets containing missing values to evaluate our frameworks results in a real situation. Finally, we compare the required running time for each framework. Since we performed very extensive experiments the complete results can be found in our supplementary material in our GitHub repository.

4.2 Results

Comparative Study Between the Imputation Frameworks. The first step of this experiment is to compute $m = 20$ imputations with each tested imputation method on each dataset and each previously described missingness setting.

Once all imputations have been computed, we run the four compared frameworks *SI*, *MI*, *S-HOT*, and *M-HOT*, following the previously described experimental protocol. Table 1 shows the results of the four compared frameworks when using the MISSFOREST imputation method, the remainder of our results can be found in our supplementary material. We do not observe any influence from the missingness pattern or missing rate on the obtained results, which seems to show that our frameworks are not sensitive to those parameters, and so, can be used in any circumstances. We note that the *M-HOT* framework obtains the best results in the vast majority of cases, while *S-HOT* consistently achieve better results than *SI*. We used Friedman and Nemenyi statistical tests to better assess the obtained results.

Using Eqs. 3 and 4, we compute the Friedman statistic on the average ranks for the results obtained using the MISSFOREST imputation method, we obtain $F_F \approx 344.07$. The amount of compared frameworks is $k = 4$, and $N = 5 \cdot 9 = 45$ for 5 datasets with 9 missingness settings for each. Under a significance level of 0.05 the critical value of $F(3, 132)$ is 2.673, the null-hypothesis is rejected since in our case $F_F > F(3, 132)$. We reiterate the same process to compute the Friedman statistic for each of the other imputation methods used and find that the null hypothesis is rejected in all cases. Thus, we continue with the post-hoc Nemenyi test as described in our protocol. We want to check the significant difference under a p-value of 0.05, given those values we have $q_{0.05} = 2.569$ from table 5 in [2]. We find $CD \approx 0.6992$, two frameworks can be considered as significantly different if their average ranks differ by more than 0.6992.

Figure 3 shows the obtained Nemenyi results that compare the *SI*, *MI*, *S-HOT* and *M-HOT* frameworks using each of the five tested imputation methods. The critical distance is visualized on the top left corner of each diagram, two frameworks can be considered significantly different if they are not linked by the same black bar. We see that in all cases the ordering of the frameworks from worst to best is the same, *SI* performs the worst, then *S-HOT*, followed by *MI*, finally *M-HOT* performs the best. *S-HOT* obtains significantly better results than *SI*, showing that it is a good alternative to *SI* when one want to train a unique large model. We note that *M-HOT* obtains consistently better results than *MI*, which seems to show that it is always a good and viable alternative to *MI*.

Comparative Study Between MICE and Imputation Frameworks. We compare *S-HOT* and *M-HOT* to MICE, we observe that the *M-HOT* framework obtains the best results in most cases. By reiterating the same calculations as before, we find that the Friedman test rejects the null hypothesis. As before we use the post-hoc Nemenyi test, we compute $CD \approx 0.9093$, Fig. 4 shows the

Table 1. Comparison of the results from the four *SI*, *MI*, *S-HOT* and *M-HOT* frameworks using the MISSFOREST imputation method.

Dataset	Pattern		*SI*		*MI*		*S-HOT*		*M-HOT*	
IRIS	MCAR	10%	0.9972681	(4)	0.9976699	(2)	0.9973865	(3)	**0.9976913**	(1)
		15%	0.9942281	(4)	0.9946880	(2)	0.9944422	(3)	**0.9947552**	(1)
		25%	0.9835323	(4)	0.9841479	(2)	0.9839692	(3)	**0.9843596**	(1)
	MAR	10%	0.9975236	(3)	0.9978336	(2)	0.9974886	(4)	**0.9978845**	(1)
		15%	0.9970188	(3)	**0.9973150**	(1)	0.9970006	(4)	0.9973097	(2)
		25%	0.9941960	(4)	0.9943643	(2)	0.9943554	(3)	**0.9943890**	(1)
	MNAR	10%	0.9972209	(4)	**0.9973548**	(1)	0.9972343	(3)	0.9973011	(2)
		15%	0.9937917	(4)	0.9941570	(2)	0.9938451	(3)	**0.9941622**	(1)
		25%	0.9891964	(4)	0.9905233	(2)	0.9902221	(3)	**0.9907251**	(1)
STAT	MCAR	10%	0.9105582	(4)	**0.9132492**	(1)	0.9106386	(3)	0.9132475	(2)
		15%	0.9060337	(4)	0.9091019	(2)	0.9066862	(3)	**0.9091584**	(1)
		25%	0.9047888	(4)	0.9077214	(2)	0.9059597	(3)	**0.9078041**	(1)
	MAR	10%	0.9127799	(3)	**0.9144097**	(1)	0.9127328	(4)	0.9142985	(2)
		15%	0.9079620	(3)	**0.9093156**	(1)	0.9078667	(4)	0.9092031	(2)
		25%	0.8959471	(4)	0.8990894	(2)	0.8968849	(3)	**0.8992881**	(1)
	MNAR	10%	0.9070192	(4)	0.9095510	(2)	0.9071004	(3)	**0.9095576**	(1)
		15%	0.9040681	(4)	**0.9061872**	(1)	0.9042151	(3)	0.9060699	(2)
		25%	0.8951658	(4)	0.8992252	(2)	0.8967363	(3)	**0.8992736**	(1)
WINE	MCAR	10%	0.9987736	(4)	0.9989672	(2)	0.9988008	(3)	**0.9989811**	(1)
		15%	0.9955454	(4)	0.9960062	(2)	0.9956407	(3)	**0.9960307**	(1)
		25%	0.9910498	(4)	0.9919927	(2)	0.9914148	(3)	**0.9923485**	(1)
	MAR	10%	0.9961058	(4)	0.9965056	(2)	0.9961142	(3)	**0.9965060**	(1)
		15%	0.9977720	(4)	**0.9982010**	(1)	0.9978677	(3)	0.9981809	(2)
		25%	0.9952116	(4)	0.9965157	(2)	0.9959576	(3)	**0.9968702**	(1)
	MNAR	10%	0.9987205	(3)	**0.9988244**	(1)	0.9987058	(4)	0.9988131	(2)
		15%	0.9974746	(4)	0.9976465	(2)	0.9974850	(3)	**0.9976683**	(1)
		25%	0.9808498	(4)	0.9841291	(2)	0.9822719	(3)	**0.9844587**	(1)
PIMA	MCAR	10%	0.8193054	(4)	**0.8211340**	(1)	0.8196586	(3)	0.8211193	(2)
		15%	0.8073739	(4)	0.8095505	(2)	0.8078378	(3)	**0.8095824**	(1)
		25%	0.8029002	(4)	0.8060367	(2)	0.8043589	(3)	**0.8065611**	(1)
	MAR	10%	0.8238900	(4)	**0.8257089**	(1)	0.8242472	(3)	0.8256414	(2)
		15%	0.8045918	(4)	0.8080830	(2)	0.8061503	(3)	**0.8083194**	(1)
		25%	0.8017568	(4)	**0.8041203**	(1)	0.8025144	(3)	0.8040062	(2)
	MNAR	10%	0.8280685	(4)	0.8302729	(2)	0.8284115	(3)	**0.8303076**	(1)
		15%	0.8279577	(4)	0.8298929	(2)	0.8283792	(3)	**0.8300464**	(1)
		25%	0.8005008	(4)	0.8043448	(2)	0.8021749	(3)	**0.8047250**	(1)
ABAL	MCAR	10%	0.8737739	(4)	0.8748059	(2)	0.8740180	(3)	**0.8749393**	(1)
		15%	0.8714861	(4)	0.8725539	(2)	0.8717831	(3)	**0.8726250**	(1)
		25%	0.8663833	(4)	0.8674332	(2)	0.8666186	(3)	**0.8675645**	(1)
	MAR	10%	0.8742551	(4)	0.8751972	(2)	0.8743399	(3)	**0.8753671**	(1)
		15%	0.8720722	(4)	0.8731502	(2)	0.8721856	(3)	**0.8731739**	(1)
		25%	0.8697060	(4)	0.8708046	(2)	0.8699619	(3)	**0.8709102**	(1)
	MNAR	10%	0.8760444	(4)	0.8768033	(2)	0.8760447	(3)	**0.8769350**	(1)
		15%	0.8753217	(4)	0.8763271	(2)	0.8754605	(3)	**0.8764456**	(1)
		25%	0.8683507	(4)	0.8696780	(2)	0.8690281	(3)	**0.8697714**	(1)
Average rank			3.8889		1.7556		3.1111		1.2444	

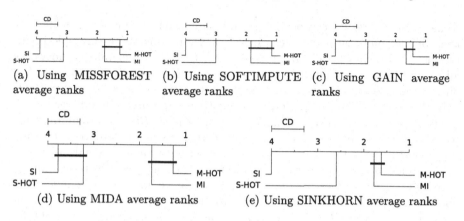

(a) Using MISSFOREST average ranks

(b) Using SOFTIMPUTE average ranks

(c) Using GAIN average ranks

(d) Using MIDA average ranks

(e) Using SINKHORN average ranks

Fig. 3. Nemenyi tests comparing *SI*, *MI*, *S-HOT* and *M-HOT* frameworks using each tested imputation method, $CD \approx 0.6992$.

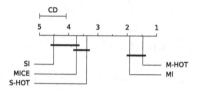

Fig. 4. Nemenyi test comparing the MICE imputation method with the best imputation method for each dataset, $CD \approx 0.9093$.

obtained Nemenyi results that compare the frameworks and MICE. We note that MICE performs largely better than *SI*, despite not being significantly better. *S-HOT* obtains significantly better results than *SI* and slightly better results than MICE. Both *MI* and *M-HOT* frameworks are significantly better than the remaining tested frameworks and methods, once again *M-HOT* performs better than *MI*.

Comparison on Real-World Medical Datasets. We compare our *S-HOT* framework to *SI* and our *M-HOT* framework to *MI* on three real-world medical datasets containing missing values, complete results are in our supplementary material. The results show that our frameworks perform well on real conditions, obtaining better results than their counterpart on the vast majority of cases. *S-HOT* obtains consistently better results than *SI*, which seems to show that *S-HOT* is a good alternative to *SI* in most real-life scenarios. *M-HOT* also performs better than *MI* in those real-world scenarios. Overall, this experiment demonstrates the capacity of our imputation frameworks in real-world situations.

Computational Running Times Comparison. Finally, we compare the computational running time required to execute each framework in each tested scenario, precise results can be found in our supplementary material. In all cases most of the computational time comes from computing the m multiple imputations. *SI* benefits largely from that point, the three remaining frameworks all require the same amount of time to compute the multiple imputations. We observe no difference in the required training time between *SI* and *S-HOT*, only the time required to perform the multiple imputations is different as *SI* needs only one imputation. In the case of *MI* and *M-HOT*, the required time to compute the imputations is the same, there is a small difference in training times. *M-HOT* is a bit slower, of less than a second to a few seconds for larger models and datasets compared to *MI*. The overall difference in total running time between *MI* and *M-HOT* is negligible. Since we have shown that *M-HOT* obtains consistently better results than *MI* most scenarios would benefit from using *M-HOT* over *MI*. When using a fast imputation method, such as SOFT-IMPUTE, the imputation computational time is almost negligible in all cases, using *S-HOT* over *SI* in this context leads to better results for the same overall computational cost.

5   Discussion and Conclusion

Taking account of imputation uncertainty while training a neural network is not typically researched. That is because strong learners, such as neural networks, are naturally able of enough generalization to neglect the consequences of the bias induced by imputation uncertainty. In this paper we researched and proposed two imputation frameworks that can be used to train models while taking this uncertainty level into account, leading to models able of more generalization and better inference results on unseen data.

Our two proposed frameworks, *S-HOT* and *M-HOT* aim to be used as substitutes for Single-Imputation (*SI*) and Multiple-Imputation (*MI*) respectively. We perform extensive experiments to compare imputation frameworks on both benchmark and real-world conditions and show that our frameworks compete with and even outperform other imputation frameworks in many situations. We statistically assess the results using Friedman and Nemenyi tests and show that our frameworks lead to less biased neural networks, improving inference results. We also attentively compare required running times for each framework and conclude that the total running-time difference between *MI* and *M-HOT* is negligible, while *SI* is faster than *S-HOT* but obtains significantly worse results than our *S-HOT* framework. We have shown that *S-HOT* obtains significantly better results than *SI* in all tested scenarios, we conclude that it is beneficial to use *S-HOT* when one needs to train a large and unique neural network. Our experiments show that *M-HOT* obtains consistently better results than *MI* in all tested scenarios for a comparable execution time. When one can afford to train multiple models in an ensemble manner, best results can be obtained using our *M-HOT* framework.

In our frameworks we assume a Gaussian distribution of imputed values. It leads to good empirical results but a Gaussian distribution might not always be pertinent depending on the missing feature nature or used imputation method, future works will focus on this matter. In this paper we have shown that our new frameworks improve generalization capacity of neural networks on imputed tabular data, future works could focus on experimenting with similar frameworks on image or sequential data. This work is a first step towards finding better ways to deal with missing values imputation in machine learning. We hope that it spikes the interest of other researchers throughout the world on this important and often overlooked matter in the machine learning literature.

Acknowledgments. This research is supported by the European Union's Horizon 2020 research and innovation program under grant agreement No 875171, project QUALITOP (Monitoring multidimensional aspects of QUAlity of Life after cancer ImmunoTherapy - an Open smart digital Platform for personalized prevention and patient management).

References

1. van Buuren, S., Groothuis-Oudshoorn, K.: Mice: multivariate imputation by chained equations in R. J. Stat. Softw. **45**(3), 1–67 (2011)
2. Demsar, J.: Statistical comparisons of classifiers over multiple data sets. J. Mach. Learn. Res. **7**, 1–30 (2006)
3. Gondara, L., Wang, K.: MIDA: multiple imputation using denoising autoencoders. In: Phung, D., Tseng, V.S., Webb, G.I., Ho, B., Ganji, M., Rashidi, L. (eds.) PAKDD 2018. LNCS (LNAI), vol. 10939, pp. 260–272. Springer, Cham (2018). https://doi.org/10.1007/978-3-319-93040-4_21
4. Hameed, W.M., Ali, N.A.: Enhancing imputation techniques performance utilizing uncertainty aware predictors and adversarial learning. Period. Eng. Nat. Sci. (PEN) **10**(3), 350–367 (2022)
5. Josse, J., Prost, N., Scornet, E., Varoquaux, G.: On the consistency of supervised learning with missing values. arXiv (2019)
6. Le Morvan, M., Josse, J., Scornet, E., Varoquaux, G.: What's a good imputation to predict with missing values? In: Advances in Neural Information Processing Systems, vol. 34, pp. 11530–11540. Curran Associates Inc. (2021)
7. Mazumder, R., Hastie, T., Tibshirani, R.: Spectral regularization algorithms for learning large incomplete matrices. J. Mach. Learn. Res. JMLR **11**, 2287–2322 (2010)
8. Muzellec, B., Josse, J., Boyer, C., Cuturi, M.: Missing data imputation using optimal transport. In: Proceedings of the 37th International Conference on Machine Learning, pp. 7130–7140. PMLR (2020). ISSN 2640-3498
9. Rubin, D.B., Schenker, N.: Multiple imputation in health-care databases: an overview and some applications. Stat. Med. **10**(4), 585–598 (1991)
10. Rubin, D.B.: Inference and missing data. Biometrika **63**(3), 581–592 (1976)
11. Rubin, D.B.: Multiple Imputation for Nonresponse in Surveys. Wiley, Hoboken (2004)
12. Stekhoven, D.J., Bühlmann, P.: MissForest-non-parametric missing value imputation for mixed-type data. Bioinformatics **28**(1), 112–118 (2012)
13. Yan, L., Zhang, H.-T., et al.: An interpretable mortality prediction model for COVID-19 patients. Nat. Mach. Intell. **2**(5), 283–288 (2020)

14. Yoon, J., Jordon, J., Schaar, M.: GAIN: missing data imputation using generative adversarial nets. In: Proceedings of the 35th International Conference on Machine Learning, p. 5689. PMLR (2018). ISSN 2640-3498
15. Yuan, Y.: Multiple Imputation for Missing Data: Concepts and New Development. SAS Institute Inc. (2005)

Supervised Hybrid Model for Rumor Classification: A Comparative Study of Machine and Deep Learning Approaches

Mehzabin Sadat Aothoi⬤, Samin Ahsan⬤, Najeefa Nikhat Choudhury(✉)⬤, and Annajiat Alim Rasel⬤

BRAC University, 66 Mohakhali, Dhaka, Bangladesh
{mehzabin.sadat.aothoi,samin.ahsan}@g.bracu.ac.bd,
{najeefa.chy,annajiat}@bracu.ac.bd

Abstract. The rapid dissemination of information through the internet has led to the spread of misinformation and rumors. Classifying rumors as fake or real on various web platforms is crucial. While extensive research has been done on rumor classification in English, research for low-resource languages like Bangla is scarce. This paper aims to achieve two objectives. First, to conduct a comparative study of state-of-the-art Machine Learning (ML) and Deep Learning (DL) models for rumor classification in both Bangla and English datasets. Second, to propose ensemble models that can outperform the baseline ML and DL models.

Keywords: Rumor Classification · NLP · Machine Learning · Deep Learning · Hybrid Approach

1 Introduction

Social media is now often preferred for information sharing due to faster propagation and wider reach. Manual verification of information on social media is impractical. Therefore, robust and automated rumor classification is essential. Traditional Machine Learning (ML) approaches often require large labeled datasets and manual feature engineering, which can be time-consuming and resource-intensive. While Deep Learning (DL) models have shown promise in automated feature extraction to achieve high accuracy in rumor classification, they face challenges due to limited data availability and computational resources. Furthermore, research on rumor classification is extensive for English but relatively limited for low-resource languages like Bangla. This research has two objectives: first, to compare the performance of ML and DL models for rumor classification on English and Bangla datasets and second, to propose an ensemble ML model and a hybrid ML-DL model that outperform the baseline ML and DL models.

This paper is organized as follows: Sect. 2 summarizes relevant research on rumor classification. Section 3 describes the datasets and data preprocessing. Section 4 describes the implementation. Section 5 analyzes the experiment results. Finally, Sect. 6 concludes this paper's findings.

© The Author(s), under exclusive license to Springer Nature Switzerland AG 2023
R. Wrembel et al. (Eds.): DaWaK 2023, LNCS 14148, pp. 281–286, 2023.
https://doi.org/10.1007/978-3-031-39831-5_25

2   Related Work

Several attempts have been made to classify fake and real news. The authors of [10] observed that Support Vector Machine (SVM) struggles with noisy data, K-Nearest Neighbor (KNN) is considered a lazy learner with modest accuracy [79.2%], and decision trees (DT) perform well [82.7%] but lack stability. To improve performance, they then employed Multinomial Naive Bayes (MNB) [90.4%], Gradient Boosting (GB) [88.3%] and Random Forest (RF) [86.5%]. In contrast, another research [8] on a Bangla dataset found SVM to outperform MNB in fake news detection. In [13], the authors compared multiple ML and DL models and found SVM with TF-IDF features achieved the highest F1 score [94.39%]. A combination of CNN+BiLSTM achieved an F1 score of 92.01%. The SVM models with Word2vec obtained slightly better results than CNN+BiLSTM.

The BanFakeNews dataset by [7] is highly impactful for Bangla Rumor Classification and contains nearly 50k annotated data. The authors employed several models and SVM outperformed all others. In [2] the researchers proposed a hybrid model to detect Bangla fake news using the BanFakeNews dataset. They employed CNN for feature extraction and traditional ML models for classification. It achieved a high F1 score of 99% for real news and 82% for fake news. However, the dataset's significant class imbalance [97% real news], might have resulted in overfitting. The paper [14] discusses a rumor detection system for detecting rumors using several baseline models and a hybrid ensemble model using LSTM and BERT. In [5] the authors found Recurrent Neural Network(RNN) architecture outperforms pure CNN, RNN-CNN, and CNN-RNN architectures for rumor classification tasks.

3   Datasets and Preprocessing

We utilized three English and one Bangla news classification datasets - Fake or Real News [9], COVID-19 Fake News [11], ISOT Fake News [3,4] and BanFakeNews [7] respectively. All the datasets have binary labels- "Real" or "Fake". We extracted the tweet body for COVID-19 Fake News Dataset, while for other datasets we extracted the title and text of the news articles. Table 1 represents

Table 1. Summary of the datasets used

Name	Language	Source	Data	Labels	
				Real	Fake
Fake or Real News	English	Kaggle	6,060	50%	50%
COVID-19 Fake News	English	Kaggle	6,420	52%	48%
ISOT Fake News	English	University of Victoria	44,898	48%	52%
BanFakeNews	Bangla	Kaggle	49,977	97%	3%

the summary of the datasets used. All datasets except for BanFakeNews exhibit a balanced distribution of real and fake labels. To reduce disparity in BanFakeNews, we randomly discarded "real" data, thus slightly adjusting the ratio [87.5% real, 12.5% fake].

For preprocessing, in the English datasets, null rows and unnecessary columns were dropped. LabelEncoder was used to assign 0 for Fake, and 1 for Real labels. Separate fake and real news CSV files were merged and shuffled. NLTK was used on the English datasets to divide the content into X (features) and Y (labels), convert it to lowercase, remove stop words, and apply stemming. The same tasks for the BanFakeNews dataset were done using the bnlp_toolkit's BasicTokenizer and corpus. TfidfVectorizer was used to convert textual data into feature vectors for all datasets. Finally, dense matrix representations of the vectorized data of each dataset were created for the ML models. Keras preprocessors and tokenizers were utilized for the DL models.

4 Implementation

The model implementations involved two steps: fitting the models with preprocessed data, and 5-fold cross-validation for accurate results.

4.1 Traditional ML Approaches

The selected traditional ML models were Support Vector Machine (SVM), Decision Tree (DT), Naive Bayes (NB), and Random Forest (RF). Scikit-learn [12] library was used for implementation. The DecisionTreeClassifier class- conducts binary and multi-class classifications, RandomForestClassifier- creates an ensemble of decision trees, GaussianNB- assumes a Gaussian distribution of features, SVC with the default RBF kernel- proficient in binary classification was utilized for the DT, RF, NB and SVM models respectively. For SVM, we increased the *cache_size* parameter to 2000MB and set *gamma* to 100 to enhance the decision boundary and reduce runtime.

4.2 DL Approaches

The chosen DL models were Recurrent Neural Network (RNN), Convolutional Neural Network (CNN), and Bi-directional Encoder Representations from Transformers (BERT). The DL models were constructed using the Keras API [6] and the TensorFlow library [1]. BERT was implemented using the BERT preprocessor and encoder from Tensorflow-hub. The data was divided into a train-test ratio of 80:20. The BERT model consisted of input, *bert_preprocess*, *bert_encoder*, dropout with a 0.2 rate, and 2 dense layers: consisting of 10 neurons and 1 neuron, respectively. The first employed a ReLU activation function and the second used a sigmoid activation function. CNN involved text tokenization with Keras preprocessor and padded sequence conversion. The model consisted of 32-dimensional embedding, convolutional, and max-pooling layers. We then added dense layers and flattened those matrices into vectors. Lastly, RNN employed padding, 32-dimensional embedding, bidirectional wrappers, dense layers, and dropout. All models were trained using the Adam optimizer, binary cross-entropy loss, and accuracy metrics over five epochs.

4.3   The Ensemble Stack ML Model

The first proposed ensemble approach (Fig. 1a) involved stacking the four traditional ML models: predictions from DT, SVM, and NB were passed to RF as the final estimator. StackingClassifier from *sklearn.ensemble* was used, performing 5-fold cross-validation for evaluation.

4.4   The Hybrid ML-DL Model

RNN (best performing DL model) and RF (best performing ML model) were chosen for the second ensemble hybrid ML-DL model (Fig. 1b). Loss function-"binary cross-entropy", optimizer-"Adam", and error measure-"F1-score" was used. Custom functions were developed for F1-score, recall and precision calculation as they are not available in Keras. The RNN model was saved and loaded, and the predictions by RNN- the weak learner with comparatively lower accuracy, were passed as input to RF- the meta-learner or final predictor with the highest accuracy, for final classification.

5   Results and Analysis

As shown in Table 2, RF outperformed all other traditional ML models, except in the case of the "UVIC ISOT" dataset. RF achieved the highest average accuracy [93.56%], and SVM had the worst average accuracy [61.15%] but scored slightly higher for the imbalanced Bangla dataset. RF's accuracy was higher than DT's as it aggregates outputs from multiple decision trees. NB assumes continuous and independent features, and so was less accurate for the selected datasets. SVM, utilizing an RBF kernel performed poorly and took the longest to train and test. Setting a gamma value of 100 and a higher cache size did not

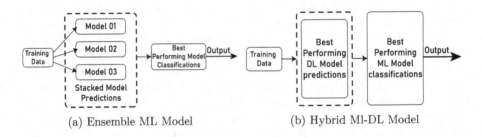

(a) Ensemble ML Model   (b) Hybrid Ml-DL Model

Fig. 1. Ensemble Model Structures

Table 2. Accuracy and F1-Score of the models[1]

Models		Datasets								Average	
		FakeOrReal		COVID-19		ISOT		BanFakeNews			
		Acc	F1	Acc	F1	Acc	F1	Acc	F1	Acc	F1
Traditional ML	DT	81.05	81.05	88.02	88.59	98.72	98.72	92.23	92.23	90	90.15
	RF	89.85	89.77	92.72	93.06	98.36	98.35	93.31	93.31	93.56	93.62
	NB	79.36	80.11	86.57	87.25	82.23	81.75	85.88	85.88	83.51	83.75
	SVM	51.47	67.46	53.82	69.39	52.67	68.77	88.04	88.04	61.5	73.42
DL	BERT	72.37	73	82.39	81.25	84.3	84.5	95.67	85.5	83.68	81.06
	RNN	86.74	86.09	91.67	91.5	96.9	97	95.14	97.18	92.61	92.94
	CNN	87.21	86.84	91.58	91.5	95.4	96.5	96.24	96.73	92.61	92.89
Ensemble	Ensemble ML	78.73	78.26	88.32	88.76	**98.93**	**98.97**	90.97	94.98	89.24	90.24
	Hybrid ML-DL	**92.34**	**82.79**	**94.47**	**94.6**	98.1	97.5	**96.25**	**97.85**	**95.29**	**95.69**

[1] FakeOrReal = Fake or Real News, COVID-19 = COVID-19 Fake News, ISOT = UVIC ISOT;
Acc = Accuracy (%), F1 = F1-Score (%)

improve the performance. A linear kernel could improve SVM results but default implementations from Scikit-learn were used.

In case of the DL models, BERT had the lowest average accuracy [83.68%], except for the "BanFakeNews" dataset where a pre-trained BanglaBERT model was used. BERT's fixed "meaning" representation may have limited its impact on rumor classification. RNN [92.61%] and CNN [92.60%] performed similarly on average. CNN is usually better fitted for spatial data like images, while RNN is better for text classification and showed higher accuracy in each dataset.

The proposed hybrid ML-DL model [95.29%] consistently outperformed the baseline models, while utilizing fewer resources. The F1 scores of the models from Table 2 further establish the hybrid ML-DL models' [95.68%] superiority over the other models. However, for the ISOT dataset, the Ensemble ML model was the best [98.97%]. Despite that, the overall higher score of the hybrid ML-DL model proves it has better precision and recall than the other baseline models. In our proposed model, RF accurately classified rumors using predictions by RNN. This combination improved classification accuracy by leveraging the strengths of both ML and DL approaches.

6 Conclusion and Future Work

Our research had two objectives: to compare the performance of traditional ML and DL models for rumor classification using English and Bangla datasets and to propose ensemble ML and hybrid ML-DL approaches for improved performance. In the comparative study, the ensemble ML model only outperformed NB and SVM, scoring similarly to DT. The hybrid ML-DL model achieved superior performance with 95.29% accuracy and 95.68% F1 score. This study provides a comprehensive summary of rumor classification and lays a foundation for further exploration, such as constructing Bangla datasets to compensate for data scarcity, utilizing techniques to handle imbalanced datasets, involving more models in the comparative study, experimenting with additional hybrid models, and enhancing under-performing models.

References

1. Abadi, M., Agarwal, A., Barham, P., Brevdo, E., Chen, Z., Citro, C., et al.: TensorFlow: large-scale machine learning on heterogeneous systems (2015). https://www.tensorflow.org/
2. Adib, Q.A.R., Mehedi, M.H.K., Sakib, M.S., Patwary, K.K., Hossain, M.S., Rasel, A.A.: A deep hybrid learning approach to detect bangla fake news (2021). https://doi.org/10.1109/ISMSIT52890.2021.9604712
3. Ahmed, H., Traore, I., Saad, S.: Detection of online fake news using N-gram analysis and machine learning techniques. In: Traore, I., Woungang, I., Awad, A. (eds.) ISDDC 2017. LNCS, vol. 10618, pp. 127–138. Springer, Cham (2017). https://doi.org/10.1007/978-3-319-69155-8_9
4. Ahmed, H., Traore, I., Saad, S.: Detecting opinion spams and fake news using text classification. Secur. Privacy 1(1), e9 (2018). https://doi.org/10.1002/spy2.9
5. Al-Sarem, M., Boulila, W., Al-Harby, M., Qadir, J., Alsaeedi, A.: Deep learning-based rumor detection on microblogging platforms: a systematic review. IEEE Access 7, 152788–152812 (2019). https://doi.org/10.1109/ACCESS.2019.2947855
6. Chollet, F., et al.: Keras (2015). https://keras.io
7. Hossain, M.Z., Rahman, M.A., Islam, M.S., Kar, S.: Banfakenews: a dataset for detecting fake news in bangla (2020)
8. Hussain, M.G., Hasan, M.R., Rahman, M., Protim, J., Hasan, S.A.: Detection of bangla fake news using MNB and SVM classifier (2020)
9. Jillani, M.G.: Fake or real news (2022). https://www.kaggle.com/datasets/jillanisofttech/fake-or-real-news
10. Kumar, A., Sangwan, S.R.: Rumor detection using machine learning techniques on social media. In: Bhattacharyya, S., Hassanien, A.E., Gupta, D., Khanna, A., Pan, I. (eds.) International Conference on Innovative Computing and Communications. LNNS, vol. 56, pp. 213–221. Springer, Singapore (2019). https://doi.org/10.1007/978-981-13-2354-6_23
11. Patwa, P., Sharma, S., Pykl, S., Guptha, V., Kumari, G., Akhtar, M.S., et al.: Fighting an infodemic: Covid-19 fake news dataset. arXiv:2011.03327, vol. 1402, pp. 21–29 (2021). https://doi.org/10.1007/978-3-030-73696-5_3
12. Pedregosa, F., Varoquaux, G., Gramfort, A., Michel, V., Thirion, B., Grisel, O., et al.: Scikit-learn: machine learning in python. J. Mach. Learn. Res. 12, 2825–2830 (2011)
13. Sharif, O., Hossain, E., Hoque, M.M.: Combating hostility: Covid-19 fake news and hostile post detection in social media (2021)
14. Tafannum, F., Sharear Shopnil, M.N., Salsabil, A., Ahmed, N., Rabiul Alam, M.G., Tanzim Reza, M.: Demystifying black-box learning models of rumor detection from social media posts. In: 2021 IEEE 12th Annual Ubiquitous Computing, Electronics & Mobile Communication Conference (UEMCON), pp. 0358–0364 (2021). https://doi.org/10.1109/UEMCON53757.2021.9666567

Attention-Based Counterfactual Explanation for Multivariate Time Series

Peiyu Li$^{(\boxtimes)}$, Omar Bahri , Soukaïna Filali Boubrahimi ,
and Shah Muhammad Hamdi

Department of Computer Science, Utah State University, Logan, UT 84322, USA
{peiyu.li,omar.bahri,soukaina.boubrahimi,s.hamdi}@usu.edu

Abstract. In this paper, we propose Attention-based Counterfactual
Explanation (AB-CF), a novel model that generates post-hoc counterfactual explanations for multivariate time series classification that narrow the attention to a few important segments. We validated our model
using seven real-world time-series datasets from the UEA repository. Our
experimental results show the superiority of AB-CF in terms of validity, proximity, sparsity, contiguity, and efficiency compared with other
competing state-of-the-art baselines.

Keywords: EXplainable Artificial Intelligence (XAI) · Counterfactual
Explanation · Multivariate Time Series · Attention based

1 Introduction

Over the past decade, artificial intelligence (AI) and machine learning (ML)
systems have achieved impressive success in a wide range of applications. The
challenge of many state-of-art ML models is a lack of transparency and interoperability. To deal with this challenge, the EXplainable Artificial Intelligence
(XAI) field has emerged. A lot of efforts have been made to provide post-hoc
XAI for image, vector-represented data, and univariate time series data while
significantly less attention has been paid to multivariate time series data [2].
The high dimensional nature of multivariate time series makes the explanation
models one of the most challenging tasks [4]. In this work, we propose a model-
agnostic counterfactual explanation method for multivariate time series data.
According to the recent literature on counterfactual explanations for various
data modalities [4], an ideal counterfactual explanation should satisfy the following properties: validity, proximity, sparsity, contiguity, and efficiency. Our
method is designed to balance these five optimal properties.

2 Related Work

In the post-hoc interpretability paradigm, the counterfactual explanation
method proposed by Wachter et al. [10] and the native guide counterfactual

© The Author(s), under exclusive license to Springer Nature Switzerland AG 2023
R. Wrembel et al. (Eds.): DaWaK 2023, LNCS 14148, pp. 287–293, 2023.
https://doi.org/10.1007/978-3-031-39831-5_26

(NG-CF) XAI method proposed by Delaney et al. [4] are the two most popular methods. The first one aims at minimizing a loss function to encourage the counterfactual to change the decision outcome and keep the minimum Manhattan distance from the original input instance. Based on this method, several optimization-based algorithms that add new terms to the loss function to improve the quality of the counterfactuals were proposed as an extension [5,6]. NG-CF uses Dynamic Barycenter (DBA) averaging of the query time series and the nearest unlike neighbor from another class to generate the counterfactual instance. Recently, several shapelet-based and temporal rule-based counterfactual explanation methods have been proposed to provide interpretability with the guide of mined shapelets or temporal rules [2,3,7]. However, these aforementioned counterfactual explanation methods suffer from generating counterfactuals that are low sparsity, low validity, or high cost of processing time to mine the shapelets or temporal rules. To deal with these challenges, in our work, we propose to focus on minimum discriminative contiguous segment replacement to generate more sparse and higher-validity counterfactuals efficiently.

3   Methodology

3.1   Notation

We define a multivariate time series dataset $D = \{X_0, X_1, ..., X_n\}$ as a collection of n multivariate time series where each multivariate time series has mapped to a mutually exclusive set of classes $C = \{c_1, c_2, ..., c_n\}$. A multivariate time series classification model f is pretrained using the dataset D. Dataset D comes with a pre-defined split ratio. We train each model only on the training data, then explain their predictions for all test instances. We define the instances from the test set as query instances. For each query instance that is associated with a class $f(X_i) = c_i$, a counterfactual explanation model $\mathcal{M}$ generates a perturbed sample with the minimal perturbation that leads to $f(X'_i) = c'_i$ such that $c_i \neq c'_i$. We define the perturbed sample X'_i as the counterfactual explanation instance, and c'_i as the target class we want to achieve for X'_i.

3.2   Proposed Method

In this section, we introduce our proposed method in detail. Figure 1 shows the process of the proposed method.

For each query instance, to generate its counterfactual explanation, we try to discover the most important top k segments at the very first step. Specifically, we use a sliding window to obtain the subsequences of the query instance, each subsequence will be considered a candidate segment. For our case of multivariate time series data, after applying a sliding window to an input multivariate time series with d dimensions $X = [x^1, x^2, ..., x^d]$, we can obtain a set of all candidates segments as follows:

$$subsequences = \{\{x_i^1, x_{i+1}^1, ..., x_{i+L-1}^1\}, ..., \{x_i^d, x_{i+1}^d, ..., x_{i+L-1}^d\}\}, \quad (1)$$

where $1 \leq i \leq m - L + 1$ is the starting position of the sliding window in the time series and $L = 0.1 \times m$ is the sliding window width, stride $= L$.

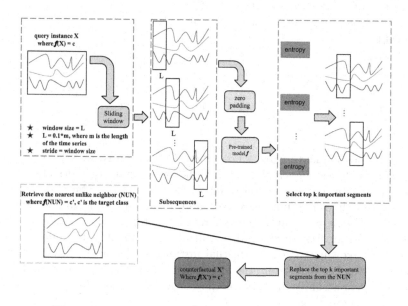

Fig. 1. Proposed Counterfactual Explanation Method

After we obtain the set of candidate segments, we fit the *subsequences* to a pretrained model f. It is noteworthy that f is a pre-trained model on the training set data of full length, therefore we used the padding technique to concatenate the segments with zeroes until the length is equal to the full length of the original time series data.

Next, each candidate is predicted by the pre-trained model, and the outcome is a probability vector $[p_1, ..., p_n]$ for the classification of n classes. We use Shannon entropy, introduced by Shannon [9], as listed in Eq. 2 to measure the information embedded in each segment given the probability distribution.

$$E = -\sum_{c=1}^{|n|} p_c log_2(p_c) \geq 0 \tag{2}$$

where p_c is the class probability of class c the pre-trained model f predicted and $|n|$ is the total number of classes.

The information entropy E determines whether the candidate is discriminative. According to the information theory [9], if the information entropy is maximized, indicates that the prediction probabilities tend to follow a uniform distribution, which further indicates that this candidate fails to provide discriminative information for the model to make a prediction. On the opposite, the closer E to 0, the more important information that the candidate subsequence

Table 1. UEA datasets metadata

ID	Dataset Name	TS length	DS train size	DS test size	Dimensions	classes
0	ArticularyWordRecognition	144	275	300	9	25
1	BasicMotions	100	40	40	6	4
2	Cricket	1197	108	72	6	12
3	Epilepsy	206	137	138	3	4
4	ERing	65	30	270	4	6
5	NATOPS	51	180	180	24	6
6	RacketSports	30	151	152	6	4

carries for the model to make a prediction. Consequently, we selected the top k distinguishable segment candidates with the lowest values of information entropy. k is a parameter representing the number of focused parts to be considered in the proposed method. k is initialized to 1 and increased during the counterfactual explanation generation until the generated counterfactual explanation satisfies the validity property. In the end, k will be determined as the minimum value that can make sure the counterfactual instance is classified to the target class. The final step is to substitute the top k segments from the nearest unlike neighbor. This nearest unlike neighbor is the 1-nearest neighbor from the target class in the training dataset.

4   Experiments

4.1   Datasets

We evaluated our proposed method on the publicly-available multivariate time series data sets from the University of East Anglia (UEA) MTS archive [1]. In particular, we selected seven datasets from the UEA archive that demonstrate good classification accuracy on state-of-the-art classifiers as reported in [8]. This ensures the quality of our generated counterfactual instances. Table 1 shows the metadata of the seven datasets.

4.2   Baseline Methods

We evaluated our proposed method with the following two baselines:

- **Native guide counterfactual (NG-CF):** NG-CF uses Dynamic Barycenter (DBA) averaging of the query time series **X** and the nearest unlike neighbor from another class to generate the counterfactual example [4].
- **Alibi Counterfactual (Alibi):** The Alibi follows the work of Wachter et al. [10], which constructs counterfactual explanations by optimizing an objective function,

$$L = L_{pred} + \lambda L_{dist}, \tag{3}$$

4.3   Experimental Result

In this section, we utilize the following four evaluation metrics to compare our proposed method with the other two baselines concerning the desirable properties of counterfactual instances according to the literature review.

The first one is *target probability*, which is used to evaluate the validity property. We define the **validity** metric by comparing the target class probability for the prediction of the counterfactual explanation result. The closer the target class probability is to 1, the better. The second one is the *L1 distance*, which is used to evaluate the **proximity** property. We measure the L1 distance between the counterfactual instance and the query instance, a smaller L1 distance is desired. Then we calculate the percent of data points that keep unchanged after perturbation to show the **sparsity** level. A high sparsity level that is approaching 100% is desirable, which means the time series perturbations made in $\mathbf{X}$ to achieve $\mathbf{X}'$ is minimal. Finally, we compare the *running time* of the counterfactual instances generation to verify the **efficiency**, the faster a valid counterfactual instance can be generated the better.

Figure 2 shows our experimental results, the error bar shows the mean value and the standard deviation over each dataset set. From Fig. 2b, we note that our proposed method achieves a competitive L1 distance compared with NG-CF. For the ALIBI method, even though it achieves the lowest L1 distance, the counterfactuals' target probability achieved by ALIBI is lower than 50% (see Fig. 2a), which means the counterfactuals generated by ALIBI are not even valid. In addition, from Fig. 2d, we can notice that the running time of using ALIBI is much higher than our proposed method and NG-CF. In contrast, our proposed method can achieve the highest target probability within one second. Our approach also demonstrates clear advantages in terms of sparsity compared to the two baselines (see Fig. 2c). This is because we only replace a few segments during perturbation, while almost half of the data points remain unchanged. In summary, our proposed method generates counterfactual explanations that balance the five desirable properties while NG-CF and ALIBI tend to maximize one property with the cost of compromising the others. The source code of our model and more visualization results figures are available on our AB-CF project website[1].

---

[1] https://sites.google.com/view/attention-based-cf.

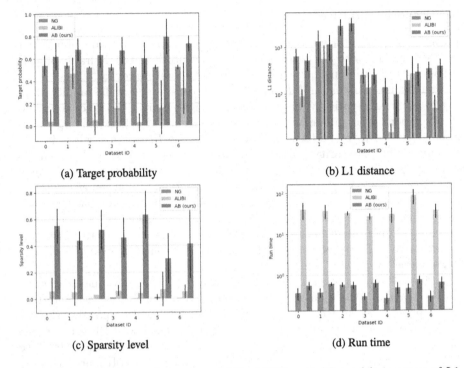

Fig. 2. Comparing the performances of NG, ALIBI, and AB models in terms of L1 distance, sparsity, target probability, and run time

5   Conclusion

In this paper, we propose to extract the most distinguishable segments from high-dimensional data and only focus on those distinguishable segments during perturbation to avoid changing the whole time series data to obtain valid counterfactuals. Our experiments demonstrate that our proposed method stands out in generating counterfactuals that balance the validity, sparsity, proximity, and efficiency compared with the other baselines. Our method also shows a high contiguity level since only several single contiguous segments need to be replaced.

Acknowledgments. This project has been supported in part by funding from GEO Directorate under NSF awards #2204363, #2240022, and #2301397 and the CISE Directorate under NSF award #2305781.

References

1. Bagnall, A., et al.: The UEA multivariate time series classification archive. arXiv preprint arXiv:1811.00075 (2018)
2. Bahri, O., Boubrahimi, S.F., Hamdi, S.M.: Shapelet-based counterfactual explanations for multivariate time series. arXiv preprint arXiv:2208.10462 (2022)

3. Bahri, O., Li, P., Boubrahimi, S.F., Hamdi, S.M.: Temporal rule-based counterfactual explanations for multivariate time series. In: 2022 21st IEEE International Conference on Machine Learning and Applications (ICMLA), pp. 1244–1249. IEEE (2022)
4. Delaney, E., Greene, D., Keane, M.T.: Instance-based counterfactual explanations for time series classification. In: Sánchez-Ruiz, A.A., Floyd, M.W. (eds.) ICCBR 2021. LNCS (LNAI), vol. 12877, pp. 32–47. Springer, Cham (2021). https://doi.org/10.1007/978-3-030-86957-1_3
5. Filali Boubrahimi, S., Hamdi, S.M.: On the mining of time series data counterfactual explanations using barycenters. In: Proceedings of the 31st ACM International Conference on Information & Knowledge Management, pp. 3943–3947 (2022)
6. Li, P., Bahri, O., Boubrahimi, S.F., Hamdi, S.M.: SG-CF: Shapelet-guided counterfactual explanation for time series classification. In: 2022 IEEE International Conference on Big Data (Big Data), pp. 1564–1569. IEEE (2022)
7. Li, P., Boubrahimi, S.F., Hamd, S.M.: Motif-guided time series counterfactual explanations. arXiv preprint arXiv:2211.04411 (2022)
8. Ruiz, A.P., Flynn, M., Large, J., Middlehurst, M., Bagnall, A.: The great multivariate time series classification bake off: a review and experimental evaluation of recent algorithmic advances. Data Min. Knowl. Disc. 35(2), 401–449 (2021)
9. Shannon, C.E.: A mathematical theory of communication. ACM SIGMOBILE Mob. Comput. Commun. Rev. 5(1), 3–55 (2001)
10. Wachter, S., Mittelstadt, B., Russell, C.: Counterfactual explanations without opening the black box: automated decisions and the GDPR. Harv. JL Tech. 31, 841 (2017)

DRUM: A Real Time Detector for Regime Shifts in Data Streams via an Unsupervised, Multivariate Framework

Adnan Bashir$^{(\boxtimes)}$ (iD) and Trilce Estrada (iD)

Computer Science Department, University of New Mexico, Albuquerque, NM, USA
{abashir,estrada}@cs.unm.edu

Abstract. In this work we present DRUM, an unsupervised approach that is based on statistical properties of multivariate data streams to identify regime shifts in real time. DRUM processes streams in small chunks, learns their statistical properties, and makes generalizations as time goes by. We show how this straightforward approach requires minimal computation and reaches state of the art accuracy, making it ideal for embedded and cyber physical systems.

Keywords: Multivariate data streams · Change point detection · Statistical analysis · Unsupervised · Real time · Online detection

1 Introduction

A regime shift refers to a sudden or significant change over time in the behavior of a system. It can occur in various natural and social systems, ranging from ecosystems and climate patterns to economic and social systems. Regime shifts in cyber physical systems can indicate potential problems or pattern changes that need to be dealt with in real time [3]. Regime Shift Detection (RSD), or Change Point Detection (CPD) refers to the identification of such changes in the underlying distribution of data streams [5]. Detection of regime shifts is of vital importance in many real-world problems such as weather monitoring, early detection of cyber security threats, medical monitoring, speech analysis, market analysis, human activity monitoring and many more [23]. For dynamic cyber physical systems, real time detection of regime shifts is essential for prevention and mitigation of system failures. Understanding when a regime shift is underway enables informed decision making, improved risk management, and accurate policy development, leading to increased resilience of the overall system.

In this paper we present DRUM (A real time **D**etector for **R**egime shifts in data streams via an **U**nsupervised and **M**ultivariate framework). DRUM has the following properties that make it suitable for solving regime shift detection in data streams: **Unsupervised** - Our method doesn't require labels or human intervention. **Real time responsiveness** - Due to very low computation, it offers minimum lag in detecting change points. **Online** - Our method processes data as a stream and does not require retraining or batch processing to adapt to changes over time. **General** - Our method has been tested on various univariate and multivariate datasets, and it was able to perform as well or with higher accuracy as other frameworks.

2 Related Work

Offline Methods - Offline CPD refers to those methods that require a complete data stream for detection [5]. Numerous offline CPD algorithms have been proposed in the past few decades. CUSUM method [21] was one of the earliest change point detectors that identified points in time series with their cumulative sum exceeding a threshold value. DRE-CUSUM [2] is built upon CUSUM which splits the data stream from an arbitrary point and compares the distribution on both sides to detect any change. Binary segmentation was introduced as an approximate change point detection method that greedily splits the data stream into disjointed segments based on a cost function. To reduce the higher time complexity of binary segmentation, pruning was added [14]. Wild Binary Segmentation extended the original binary segmentation method by applying CUSUM on subsets of data stream [10].

Online Methods - Online CPD refers to those methods that process the data as it arrives. They can be viewed as ϵ-real time algorithms [5]. Where ϵ is the delay in timestamps required by the algorithm to accurately detect a change point. To incorporate multivariate data streams and online requirement of CPD, Binary Online Changepoint Detection (BOCD) was proposed [1]. BOCD introduced a term called run-length, which denotes data stream since the most recent change point. BOCD approximates the probability distribution over run-length and compares it with previously approximated distribution for change point detection. Lately, several methods were proposed that were based on BOCD namely BOCPDMS [16] and RBOCPDMS [15]. We have compared all variants of BOCP due to its online detection ability.

Supervised Methods - Supervised learning algorithms have proven to be very effective when it comes to change point detection but they require labeled data and a training phase. Traditional machine learning methods like SVM [8], random forests [6], decision trees [22], hidden Markov models [20], nearest neighbors [19], and other binary classifiers [4] have been used to deal with this problem as a supervised classification problem. Among them, RFPOP [9] detects change points in an offline setting by engineering the cost function via dynamic programming algorithm. Another such approach is PELT [14] which minimizes the detection delay in univariate time series. Facebook has proposed PROPHET [25] which works as a regression model which requires domain expertise for tuning. This is not feasible for unsupervised and online applications.

Deep Learning and Neural Network-based Models - Neural Networks have been readily used for time series analysis. But they also suffer some limitations when it comes to online CPD. As stated by Zhang et al. [27], the network structure, training method and sample data may affect the performance and accuracy of Neural Network based methods. Also training Neural Networks is a time-consuming process and usually requires a big chunk of data in training. For example, GAN (Generative Adversarial Networks [11]) based models require entire variable set concurrently to tune the hidden layer parameters (such as MAD-GAN [18]). Re-tuning these models is again a time-consuming process that compromises the real time response of these detectors. Hence a majority of Neural Network based CPD algorithms outperform when trained in an offline setting.

3 DRUM

In this section we describe our approach (DRUM) which consists of two modules: a scoring module, which computes statistics of the individual data streams; and a detection module, which quantifies the rate of change and determines if a regime shift has occurred.

Scoring Module: this module takes a multivariate data stream and calculates our Lobo Change Score (LCS), which is the weighted sum of change in mean (Δm), change in standard deviation (Δs), and change in fluctuations across the running mean (Δfrm) between two data windows of the data stream. Formally, the data stream is an infinite sequence of values generated by an underlying system and is denoted as $S = \{x_1, x_2, ..., x_{t-1}, x_t, x_{t+1}, ...\}$, where for a multivariate time-series with n variables, x_t is the n-dimensional data vector at time t [26]. LCS is computed for small chunks of data that we call windows. That is, given variable i of the stream, for a window of size $d + 1$ starting in position j, our algorithm processes the sequence $w_{i,[j,j+d]} = \{x_{i,j}, x_{i,j+1}, x_{i,j+2}, ..., x_{i,j+d}\}$. LCS is composed of three components:

$$\Delta m_{i,j,k} = \left| m(w_{i,[j,j+d]}) - m(w_{i,[k,k+d]}) \right| \tag{1}$$

$$\Delta s_{i,j,k} = \left| s(w_{i,[j,j+d]}) - s(w_{i,[k,k+d]}) \right| \tag{2}$$

$$\Delta frm_{i,j,k} = \left| frm(w_{i,[j,j+d]}) - frm(w_{i,[k,k+d]}) \right| \tag{3}$$

where for two windows: $w_{i,[j,j+d]}$ starting at position j, and $w_{i,[k,k+d]}$ starting at position k; both of them corresponding to variable i, $\Delta m_{i,j,k}$ is the absolute difference between their mean; $\Delta s_{i,j,k}$ is the absolute difference in their standard deviation; and Δfrm is the absolute difference of fluctuation across the running mean. frm is computed by counting the number of intersections between the running mean and the signal itself within the current window. Then, LCS for windows $[j, j + d]$ and $[k, k + d]$ is the weighted sum of the three components described above, as shown by Eq. 4:

$$LCS_{j,k} = \alpha \sum_{i=1}^{n} \Delta m_{i,j,k} + \beta \sum_{i=1}^{n} \Delta s_{i,j,k} + \gamma \sum_{i=1}^{n} \Delta frm_{i,j,k} \tag{4}$$

where α, β and γ are weight coefficients, used to differentiate the contribution of the three components. They can have values between 0 and 1, these values can be updated adaptively depending on the data characteristics, or can be kept constant to simplify hyperparameter selection. As shown in Algorithm 1, we use two types of windows for calculating LCS. LCS_D is calculated using disjointed windows, that is, windows of size $d + 1$ that do not overlap, they track coarse changes over a period of time. While LCS_S uses consecutive sliding windows to detect with finer precision when a change occurs. Figure 1 visually explains the two types of windows.

Algorithm 1: DRUM

Input: Multivariate data stream S; window length $d + 1$; α, β, γ
Output: LCS_D, LCS_S
$t \leftarrow 0; l \leftarrow t + 1; j \leftarrow d + 1; k \leftarrow 1;$
$LCS_D = \{\}$,$LCS_S = \{\}$
while $S \neq \{\}$ **do**
   while $t + + \leq d$ **do**
      | $\forall_i w_{i,[k,k+d]} \leftarrow x_{i,t}$ /* gather data for window $[k, k + d]$ */
   end
   $t + +$ /* use t as the timestamp counter   */
   /* compute LCS_D for disjoint windows j and k */
   if $j \leq t \leq j + d$ **then**
      | $\forall_i w_{i,[j,j+d]} \leftarrow x_{i,t}$ /* gather data for window $[j, j + d]$ */
   end
   if $t = j + d$ **then**
      | $LCS_D \leftarrow \forall_i \alpha \Delta m_{i,j,k} + \beta \Delta s_{i,j,k} + \gamma \Delta frm_{i,j,k}$
      | $k \leftarrow j$
      | $j \leftarrow j + d + 1$
   end
   /* compute LCS_S for consecutive sliding windows ending at t */
   update $\forall_i w_{i,[t-d,t-1]}$
   update $\forall_i w_{i,[t-d+1,t]}$
   $LCS_s \leftarrow \forall_i \alpha \Delta m_{i,t-d,t-d+1} + \beta \Delta s_{i,t-d,t-d+1} + \gamma \Delta frm_{i,t-d,t-d+1}$
   LCS_S.append(LCS_s)
   if $detectChange(LCS_D)$ **then**
      | **return** maxChange(LCS_S)
   end
end

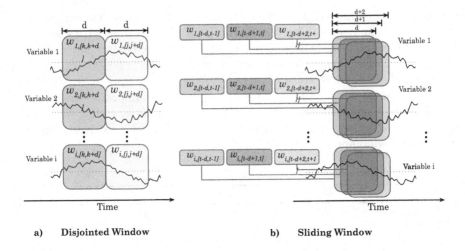

a)   **Disjointed Window**   b)   **Sliding Window**

Fig. 1. Visualization of a) Disjointed Windows and b) Sliding Windows

Detection Module: Once LCS is calculated for disjointed (LCS_D) and sliding windows (LCS_S), it is evaluated by the detection module. The first step in detecting regime shift is to identify the window in which the change has occurred. If LCS_D has increased or decreased by 5% or more, the window is marked as a change point window. Once the change window is identified, we use LCS_S to further identify the specific timestamp with the maximum change score. This timestamp is returned as a change point.

4   Evaluation

To evaluate the accuracy and performance of our method, we compared its behavior on datasets labeled for CPD (see Table 1) with respect to other state of the art methods (see Table 2). Additional details regarding these methods were presented in Sect. 2. We used three metrics for comparison: NAB score, F1 score, and execution time. Below we present the results of such comparisons.

Table 1. Datasets and Benchmarks used for Evaluation

DATASET	Description	# of Variables	Length	# of Series
TCPD [7]	Real-world dataset for bench-marking CPD algorithms	1–5	15–991	42
TSSB [24]	Subset of UCR TS datasets	1	572–15970	66
SKAB [13]	Sensors on a testbed	8	666–1327	34
TEP [12]	Industrial chemical process	52	500–960	22
MDS	Mesa Del Sol microgrid	18	518960	1

Table 2. Properties of CPD frameworks used for evaluation

METHOD	Year	Online	No Training	Unsupervised	Multivariate
DRUM (ours)	2023	✔	✔	✔	✔
RFPOP [9]	2019	✘	✔	✘	✘
RBOCPDMS [16]	2018	✔	✔	✔	✔
PROPHET [25]	2018	✘	✘	✘	✘
BOCPDMS [15]	2018	✔	✔	✔	✔
WBS [10]	2012	✘	✔	✔	✘
PELT [14]	2012	✘	✔	✔	✘
BOCPD [1]	2007	✔	✔	✔	✔

NAB Score. NAB [17] was proposed in 2015 by Numenta[1]. It was the first anomaly and change point detection benchmark designed to evaluate unsupervised and real time algorithms. It rewards early detection and uses a scaled sigmoidal function to score false positives and false negatives. Figure 2 shows the NAB standard profile score. Our approach outperforms other methods in four out of five datasets (TCPBD, TEP, MDS, SKAB), and performs competitively in TSSB. It is important to note that NAB score allows the framework to perform a change point detection within a window (by default it is 5% of the length of time series).

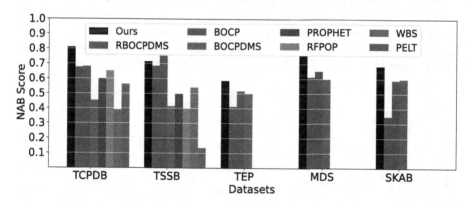

Fig. 2. NAB Standard Profile Score, Empty bars show that the framework is not multivariate capable

F1 Score. When a framework has a binary output (in this case change or no change), F1-Score has been readily used by the research community to evaluate such frameworks. F1 scores for the different methods are presented in Table 3 with the highest score per dataset in bold. Our approach outperforms other methods in four out of five datasets.

---

[1] https://numenta.com/.

Table 3. F1 Score. Empty fields mean that the framework is not designed for multivariate time series data

METHOD	TCPDB	TSSB	SKAB	TEP	MDS
DRUM (ours)	0.804	**0.715**	**0.634**	**0.664**	**0.714**
BOCPD [1]	**0.818**	0.345	0.568	0.608	0.640
BOCPDMS [15]	0.620	0.589	0.618	0.652	0.687
RBOCPDMS [16]	0.447	0.548	0.623	0.325	0.686
PELT [14]	0.787	0.563	-	-	-
PROPHET [25]	0.534	0.446	-	-	-
WBS [10]	0.533	0.214	-	-	-
RFPOP [9]	0.531	0.197	-	-	-

Execution Time. We also compared the runtime of CPD frameworks both on univariate and multivariate time series. All Experiments were run on an Intel Core i7-8750H @ 2.20 GHz with 6 cores and 32 GB of RAM. Implementation was done using Python 3.7 on Windows 11. Fig. 3a shows the execution time for univariate time series from 10,000 to 100,000 instances, sampled at intervals of 10,000. Our approach has the lowest execution time with approximately linear behavior, while others show an exponential response. Figure 3b shows the execution time for multivariate time series with a fixed length of 10,000 but with the number of features ranging from 2 to 20. Computation time for BOCPD and its variants depends on change point location: if a change point is not detected more samples are stored to approximate the distribution of the data stream. This results in longer execution over time. While our approach always used a constant amount of data and exhibits linear scalability.

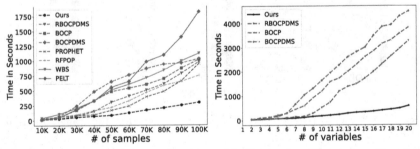

(a) Univariate time series of increasing length

(b) Multivariate time series of length 10K with an increasing number of features

Fig. 3. Comparison of execution time

5 Conclusion

In this work, we present an unsupervised change point detector that works on streams and scales linearly with the number of variables being tracked. It requires minimal computations and can be applied toward detecting different events happening in a system in real time. Such events include system failures, concept drift, and state transitions. Since it is lightweight, it can be run on IoTs with low compute power.

Acknowledgments. This work was supported by National Science Foundation (NSF) EPSCoR grant number OIA-1757207.

References

1. Adams, R.P., MacKay, D.J.C.: Bayesian Online Changepoint Detection (2007). http://arxiv.org/abs/0710.3742
2. Adiga, S., Tandon, R.: Unsupervised change detection using dre-cusum. arXiv preprint arXiv:2201.11678 (2022)
3. Ahamed, R., Lavin, A., Purdy, S., Agha, Z.: Unsupervised real-time anomaly detection for streaming data. Neurocomputing **262**, 134–147 (2017). https://doi.org/10.1016/j.neucom.2017.04.070
4. Aminikhanghahi, S., Cook, D.J.: A survey of methods for time series change point detection. Knowl. Inf. Syst. **51**(2), 339–367 (2016). https://doi.org/10.1007/s10115-016-0987-z
5. Aminikhanghahi, S., Wang, T., Cook, D.J.: Real-time change point detection with application to smart home time series data. IEEE Trans. Knowl. Data Eng. **31**(5), 1010–1023 (2019). https://doi.org/10.1109/TKDE.2018.2850347
6. Athey, S., Tibshirani, J., Wager, S.: Generalized random forests (2019)
7. van den Burg, G.J., Williams, C.K.: An evaluation of change point detection algorithms. arXiv, pp. 1–33 (2020)
8. Camci, F.: Change point detection in time series data using support vectors. Int. J. Pattern Recognit. Artif. Intell. **24**(01), 73–95 (2010)
9. Fearnhead, P., Rigaill, G.: Changepoint detection in the presence of outliers. J. Am. Stat. Assoc. **114**(525), 169–183 (2019)
10. Fryzlewicz, P.: Wild binary segmentation for multiple change-point detection. Ann. Stat. **42**(6), 2243–2281 (2014)
11. Goodfellow, I., et al.: Generative adversarial nets. In: Advances in Neural Information Processing Systems, pp. 2672–2680 (2014)
12. Katser, I., Kozitsin, V., Lobachev, V., Maksimov, I.: Unsupervised offline change-point detection ensembles. Appl. Sci. **11**(9), 1–19 (2021). https://doi.org/10.3390/app11094280
13. Katser, I.D., Kozitsin, V.O.: Skoltech anomaly benchmark (SKAB) (2020). https://www.kaggle.com/dsv/1693952.   https://doi.org/10.34740/KAGGLE/DSV/1693952
14. Killick, R., Fearnhead, P., Eckley, I.A.: Optimal detection of changepoints with a linear computational cost. J. Am. Stat. Assoc. **107**(500), 1590–1598 (2012)
15. Knoblauch, J., Damoulas, T.: Spatio-temporal Bayesian on-line changepoint detection with model selection. In: International Conference on Machine Learning, pp. 2718–2727. PMLR (2018)

16. Knoblauch, J., Jewson, J.E., Damoulas, T.: Doubly robust Bayesian inference for non-stationary streaming data with divergences. In: Advances in Neural Information Processing Systems, vol. 31 (2018)

17. Lavin, A., Subutai, A.: Numenta anomaly benchmark. In: International Conference on Machine Learning and Applications, vol. 14 (2015)

18. Li, D., Chen, D., Shi, L., Jin, B., Goh, J., Ng, S.K.: MAD-GAN: multivariate anomaly detection for time series data with generative adversarial networks. arXiv, vol. 1, pp. 703–716 (2019)

19. Liu, Y.W., Chen, H.: A fast and efficient change-point detection framework based on approximate k-nearest neighbor graphs. arXiv preprint arXiv:2006.13450 (2020)

20. Miller, D.J., Ghalyan, N.F., Mondal, S., Ray, A.: Hmm conditional-likelihood based change detection with strict delay tolerance. Mech. Syst. Signal Process. **147**, 107109 (2021)

21. Page, E.S.: Continuous inspection schemes. Biometrika **41**(1/2), 100–115 (1954)

22. Reddy, S., Mun, M., Burke, J., Estrin, D., Hansen, M., Srivastava, M.: Using mobile phones to determine transportation modes. ACM Trans. Sen. Netw. **6**(2) (2010). https://doi.org/10.1145/1689239.1689243

23. Reeves, J., Chen, J., Wang, X.L., Lund, R., Lu, Q.Q.: A review and comparison of changepoint detection techniques for climate data. J. Appl. Meteorol. Climatol. **46**(6), 900–915 (2007)

24. Schäfer, P., Ermshaus, A., Leser, U.: ClaSP - time series segmentation. In: CIKM (2021)

25. Taylor, S.J., Letham, B.: Business time series forecasting at scale. PeerJ Preprints 5:e3190v2 **35**(8), 48–90 (2017)

26. Tran, D.H.: Automated change detection and reactive clustering in multivariate streaming data. In: 2019 IEEE-RIVF International Conference on Computing and Communication Technologies (RIVF), pp. 1–6. IEEE (2019)

27. Zhang, G., Patuwo, B.E., Hu, M.Y.: Forecasting with artificial neural networks: the state of the art. Int. J. Forecast. **14**(1), 35–62 (1998). https://doi.org/10.1016/S0169-2070(97)00044-7

Hierarchical Graph Neural Network with Cross-Attention for Cross-Device User Matching

Ali Taghibakhshi[1,2]([envelope]), Mingyuan Ma[1], Ashwath Aithal[1], Onur Yilmaz[1], Haggai Maron[1], and Matthew West[2]

[1] NVIDIA, Santa Clara, USA
[2] Department of Mechanical Science and Engineering,
University of Illinois at Urbana-Champaign, Urbana, IL, USA
alit2@illinois.edu

Abstract. Cross-device user matching is a critical problem in numerous domains, including advertising, recommender systems, and cybersecurity. It involves identifying and linking different devices belonging to the same person, utilizing sequence logs. Previous data mining techniques have struggled to address the long-range dependencies and higher-order connections between the logs. Recently, researchers have modeled this problem as a graph problem and proposed a two-tier graph contextual embedding (TGCE) neural network architecture, which outperforms previous methods. In this paper, we propose a novel hierarchical graph neural network architecture (HGNN), which has a more computationally efficient second level design than TGCE. Furthermore, we introduce a cross-attention (Cross-Att) mechanism in our model, which improves performance by 5% compared to the state-of-the-art TGCE method.

Keywords: Graph neural network · User matching · Cross-attention

1 Introduction

Ensuring system security and effective data management are critical challenges in the modern day [3,4]. In this regard, data integration plays a vital role in facilitating data management, as it enables the integration of data from diverse sources to generate a unified view of the underlying domain. One of the primary challenges in data integration is the problem of entity resolution, which involves identifying and linking multiple data records that correspond to the same real-world entity. The problem of entity resolution arises in a wide range of domains, including healthcare, finance, social media, and e-commerce. Entity resolution is a challenging problem due to various factors, including the presence of noisy and ambiguous data, the lack of unique identifiers for entities, and the complexity of the relationships between different entities.

A. Taghibakhshi—This work was done while Ali Taghibakhshi was an intern at NVIDIA.

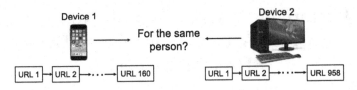

Fig. 1. Cross-device user matching problem: only based the URL visit logs of two different devices, determine whether or not they belong to the same real-world person.

Among entity resolution tasks, cross-device user matching is of significant importance. This task involves determining whether two separate devices belong to the same real-world person based on their sequential logs. The device sequential logs are time-stamped actions taken by the user over a relatively long period of time, say a few months. These actions are often in the form of browsing a Uniform Resource Locator (URL), and almost always, user identifications are not available due to privacy reasons. Refer to Fig. 1 for an illustration of the cross-device user matching task.

It is a common occurrence for users to engage in online activities across multiple devices. However, businesses and brands often struggle with having insufficient user identities to work with since users are perceived as different individuals across different devices due to their unique activities. The ability to automatically identify the same user across multiple devices is essential for gaining insights into human behavior patterns, which can aid in applications such as user profiling, online advertising, improving system security. In recent years, the has been a flourishing amount of studies focusing on cross-device user matching [9].

With the advent of machine learning-based methods for entity resolution, several studies have focused on learning distributed embeddings for the devices based on their URL logs [6,11,12]. The earlier studies focused on utilizing unsupervised feature learning techniques [7], developing handcrafted features for the device logs, or relied on co-occurrence of key attributes of URL logs in pairwise classification [12].

Methods that utilize deep learning have a greater ability to convey dense connections among the sequential device logs. For instance, researchers have utilized a 2D convolutional neural network (CNN) framework to encode sequential log representations to understand the relationship between two devices [16]. However, this model primarily captures local interactions within user sequence logs, limiting its ability to learn the entire sequence or a higher-level pattern. Recently, there has been further emphasis on the effectiveness of sequential models like recurrent neural networks (RNNs) and attention-based techniques in modeling sequence patterns and achieving promising results in numerous sequence modeling tasks [5,13,15]. Although these methods work well for sequence modeling, they are not specifically designed for user-matching tasks and may not be optimal for learning sequential log embeddings.

Recently, researchers proposed a two-tier graph contextual embedding (TGCE) network for the cross-device user matching [6] task. While previous methods for the task often failed at long-range information passing along the sequence logs, TGCE leverages a two-level structure that can facilitate information passing beyond the immediate neighborhood of a device log. This was specifically achieved by considering a random walk starting from every node in a device log, connecting all of the visited nodes to the original node, and performing a round of message passing using the newly generated shortcut edges.

Although the two-tier structure seems to enable long-range information sharing, we note two major limitations with the existing method. First, in the device graph, the random walk on the URL nodes may randomly connect two URLs that have been visited at two far-away time-stamps. Intuitively, two different URLs browsed by a device with weeks of gap in between share less information than two URLs visited in a shorter time frame. Second, at the end of the TGCE architecture, for the pairwise classification task, the generated graph embeddings for two devices are entry-wise multiplied and sent through a fully connected network to determine if they belong to the same person. However, there could be significant key features in the learned embeddings that may be shared between the devices, which can alternatively get lost if the architecture does not compare them across one another.

To address the above two issues, we propose a new hierarchical graph neural network (HGNN) inspired by the star graph architecture [10]. In the terminology of HGNN, we refer to the URL nodes as *fine* nodes, and in an unraveled sequence of URL logs, HGNN assigns a *coarse* node to every K consecutive fine nodes. The message passing between the coarse and fine nodes enables effective long-range message passing without the need to excessively add edges, as in the random walk method. Moreover, for the pairwise classification task, we utilize a cross-attention mechanism inspired by Li *et al.* [8], which enables entry-wise cross-encoding of the learned embeddings. The main contributions of this paper are summarized as follows:

- We model a given device log as a hierarchical heterogeneous graph, which is 6x faster than the previous state-of-the-art while keeping a competitive level of accuracy and performance.
- We employ a cross-attention mechanism for pairwise matching of the graphs associated with a device log, which improves the accuracy of the overall method by about 5%.

2   Related Work

The cross-device user matching task was first introduced in the CIKM Cup 2016[1] on the Data Centric Alliance (DCA) dataset, and the first proposed methods for the task mainly considered hand-crafted features. For instance, the runner-up solution [9] produces sub-categories based on the most significant URLs to

---

[1] http://cikm2016.cs.iupui.edu/cikm-cup/.

generate detailed features. Furthermore, the competition winner solution proposed by Tay *et al.* [14] utilizes "term frequency inverse document frequency" (TD-IDF) features of URLs and other related URL visit time features. However, their manually designed features did not fully investigate more intricate semantic details, such as the order of behavior sequences, which restricted their effectiveness. Aside from the hand-crafted features, the features that are developed from the structural information of the device URL visit data are also crucial for accomplishing the task of user matching. To further process sequential log information, studies have applied LSTM, 2D-CNN, and Doc2vec to generate semantic features for a sequence visited by a device [11,12,16].

Sequence-based machine learning models have also been employed for different entity resolution tasks; for instance, recurrent neural networks (RNN) have been utilized to encode behavior item sequential information [5]. Nevertheless, long-range dependencies and more advanced sequence features are not well obtained using sequence models [6]. With the advent of graph neural networks (GNN), studies have leveraged their power for many entity resolution tasks. By utilizing neighborhood-based aggregation, GNNs effectively capture and propagate structural information, which enables them to perform excellently in numerous tasks such as node and graph classification. In order to employ GNNs, researchers have modeled device logs as individual graphs where nodes and edges represent visited URLs and transitions between URLs. Each node and/or edge has an initial feature vector obtained from the underlying problem, and the layers of GNNs are then employed to update these features based on information passing in the local neighborhood of every node, such as the SR-GNN paper [17]. Another example is the LESSR [1] method for recommendation systems where the method is capable of long-range information capturing using an edge-order preserving architecture. However, these methods are specifically designed for the recommendation task and do not necessarily achieve desirable results on the cross-device user matching task.

Recently, researchers have proposed TGCE [6], a two-tier GNN for the cross-device user matching task. In the first tier, for every device log, each URL is considered as a node, and directional edges denote transitions between URLs. In the second tier, shortcut edges are formed by starting a random walk from every node and connecting all of the visited nodes to it. After a round of message passing in the first tier, the second tier is supposed to facilitate long-range information sharing in the device log. After the second tier, a position-aware graph attention layer is applied, followed by an attention pooling, which outputs the learned embedding for the whole graph. For the final pairwise classification, these learned embeddings for each of the devices are multiplied in an entry-wise manner and are sent to a fully connected deep neural network to determine whether they belong to the same user.

3   Hierarchical Graph Neural Network

In this section, we discuss how we employ a two-level heterogeneous graph neural network for the cross-device user matching problem.

3.1   Problem Definition

The aim of the cross-device user matching problem is to determine whether two devices belong to the same user, given only the URL visits of each device. Denote a sequence of visited URLs by a device v by $\mathcal{S}_v = (s_1, s_2, ..., s_n)$, where s_i denotes the i'th URL visit by the device (note that s_i and s_j are not necessarily different, for $i \neq j$). We build a hierarchical heterogeneous graph, G_v, based on the sequence $\mathcal{S}_v$ as follows: for a visited URL, s_i, consider a *fine* node in G_v and denote it by f_i. Note that if multiple URL visits (s_i) correspond to the same URL, we only consider one node for it in G_v. Then, we connect nodes corresponding to consecutively visited URLs by directed edges in the graph (preferred over undirected edges to emphasize on the temporal order of URL visits); we connect f_i and f_{i+1} by a directional edge (if f_i and f_{i+1} correspond to the same URL, the edge becomes a self-loop). Up to this point, we have defined the fine-level graph, and we are ready to construct the second level, which we call the *coarse* level.

To construct the second level, we partition the sequence $\mathcal{S}_v$ into non-overlapping subgroups of K URLs (where K is a hyperparameter), where each subgroup consists of consecutively visited URLs (the last subgroup may have less than K URLs). For every subgroup j, we consider a coarse node, c_j, and connect it to all of the fine nodes corresponding to the URLs in subgroup j via undirected edges.

3.2   Fine Level

In the fine level of the graph G_v, for every node f_i, we order the nodes corresponding to the URLs that have an incoming edge to f_i according to their position in $\mathcal{S}_v$. We denote this ordered sequence of nodes by $N_i = (f_{j_1}, f_{j_2}, ..., f_{j_\kappa})$. Also, we denote the feature vector of the fine node f_i by x_i. The l-th round of message passing in the fine-level graph updates the node features according to the following update methods:

$$M_i^{(l)} = \Phi^{(l)}([x_{j_1}, x_{j_2}, ..., x_{j_\kappa}, x_i]), \tag{1}$$

$$x_i^{(l+1)} = \Psi^{(l)}(x_i^{(l)}, M_i^{(l)}), \tag{2}$$

where $\Phi^{(l)}$ is a sequence aggregation function (such as sum, max, GRU, LSTM), for which we use GRU [2] (to leverage temporal order of URLs in encoding), and $\Psi^{(l)}$ is a function for updating the feature vector (e.g., a neural network), for which we use a simple mean.

3.3   Coarse Level

In every round of heterogeneous message passing between fine and coarse level nodes, we update both the fine and coarse node features. Consider the coarse node c_j, and denote its feature by $\tilde{x}_j$. Also, denote the fine neighbor nodes of c_j

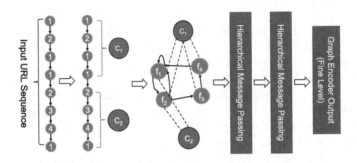

Fig. 2. From left to right: heterogeneous (fine and coarse) graph modeling from a given URL sequence. The hierarchical message passing blocks consist of message passing on the fine nodes with a GRU aggregation function. Next, the coarse node features are updated using a mean aggregation function. Finally, the fine node features are updated using their previous feature vector as well as an aggregated message from their associated coarse nodes obtained via an attention mechanism between coarse and fine level nodes.

by $\tilde{\mathcal{N}}(c_j)$. In the l-th layer of heterogeneous message passing, the coarse node feature update is as follows:

$$\tilde{x}_j^{(l+1)} = \underset{i \in \tilde{\mathcal{N}}(c_j)}{\square} (W_1^{(i)} x_i), \tag{3}$$

where $W_1^{(l)}$ is a learnable matrix and $\square$ is an aggregation function (such as mean, max, sum), for which we use mean (which was the most effective in our case). Denote by $\mathcal{N}(f_i)$ the set of coarse nodes connected to the fine node f_i. We first learn attention weights for the heterogeneous edges, and then we update fine nodes accordingly. In the l-th round of heterogeneous message passing, the fine node features are updated as follows:

$$e_{i,j}^{(l)} = \phi(W_2^{(l)} x_i^{(l)}, W_3^{(l)} \tilde{x}_j^{(l)}), \tag{4}$$

$$\alpha_{i,j}^{(l)} = \frac{\exp(e_{i,j}^{(l)})}{\sum_{j \in \mathcal{N}(f_i)} \exp(e_{i,j}^{(l)})}, \tag{5}$$

$$x_i^{(l+1)} = \xi(x_i^{(l)}, \sum_{j \in \mathcal{N}(f_i)} \alpha_{i,j}^{(l)} \tilde{x}_j^{(l)}), \tag{6}$$

where $W_2^{(l)}$ and $W_3^{(l)}$ are learnable matrices, and ξ and ϕ are update functions (such as a fully connected network). Figure 2 shows the overall architecture of fine and coarse level message passing.

3.4   Cross Attention

After the message passing rounds in the fine level and long-range information sharing between fine and coarse nodes, we extract the learned fine node embeddings and proceed to cross encoding and feature filtering, inspired by the GraphER architecture [8]. We consider two different device logs v and w, and their learned fine node embeddings as a sequence, ignoring the underlying graph structure. We denote the learned fine node embeddings for device logs v and w as $X_v \in \mathbb{R}^{m_v \times d}$ and $X_w \in \mathbb{R}^{m_w \times d}$, where m_v and m_w are the number of nodes in the fine level of G_v and G_w, respectively. We learn two matrices for cross-encoding X_v into X_w and vice-versa. Consider the i-th and j-th rows of X_v and X_w, respectively, and denote them by $x_{v,i}$ and $x_{w,j}$. The entries $\hat{\alpha}_{i,j}$ of the matrix $A_{v,w}$ for cross-encoding X_v into X_w are obtained using an attention mechanism (and similarly for $A_{w,v}$):

$$\hat{e}_{i,j} = \zeta(W_3 x_{v,i}, W_3 x_{w,j}), \tag{7}$$

$$\hat{\alpha}_{i,j} = \frac{\exp(\hat{e}_{i,j})}{\sum_{k=1}^{m_w} \exp(\hat{e}_{i,k})}, \tag{8}$$

where ζ is an update function (such as a neural network), for which we use a simple mean. After obtaining the cross-encoding weights, we apply feature filtering, a self-attention mechanism that filters important features. The filtering vector is obtained as $\beta_v = \text{sigmoid}(W_4 \tanh(W_5 X_v^T))$, where W_4 and W_5 are learnable weights (β_w is obtained similarly). We apply the feature-filtering vector to the cross-encoding matrix as follows:

$$L_{v,w} = [\text{diag}(\beta_v)(A_{v,w} X_w - X_v)] \odot [\text{diag}(\beta_v)(A_{v,w} X_w - X_v)], \tag{9}$$

where $\odot$ denotes the Hadamard product ($L_{w,v}$ is also obtained similarly). The $L_{v,w} \in \mathbb{R}^{m_v \times d}$ and $L_{w,v} \in \mathbb{R}^{m_w \times d}$ matrices come from the Euclidean distance between the cross-encoding of X_v into X_w and X_w into X_v, and therefore are a measure of the closeness of the original sequence logs of v and w.

To obtain a size-independent comparison metric, we apply a multi-layer perceptron (MLP) along the feature dimension of L matrices (the second dimension, d), followed by a max-pooling operation along the first dimension. Finally, we apply a dropout and a ReLU nonlinearity. This yields vectors $r_{v,w}$ and $r_{w,v}$ that have a fixed size for any pair of v and w. For the final pairwise classification task, we concatenate $r_{v,w}$ and $r_{w,v}$ and pass it through an MLP followed by a sigmoid activation to determine if the two devices belong to the same user or not:

$$\hat{y} = \text{sigmoid}(\text{MLP}(r_{v,w} || r_{w,v})). \tag{10}$$

The overall cross attention architecture employed in this paper is illustrated in Fig. 3.

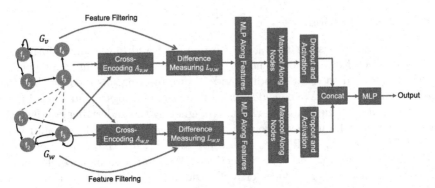

Fig. 3. Pairwise device graph matching: After the message passing, the two device graphs are cross-encoded via an attention mechanism followed by an attention-based feature filtering. The resulting matrix for each graph is then passed through an MLP layer, acting along the feature, followed by a maxpool operator along the nodes. Next, the obtained vectors pass through a dropout layer followed by an activation function. Finally, the resulting vectors of the two graphs are concatenated and passed through an MLP to obtain the final output.

4   Experiment

In this section, we will describe the dataset, training details, and discuss how our method outperforms all other baselines, including TGCE [6], the previous state-of-the-art.

4.1   Training Details

We studied the cross-device user matching dataset made publicly available by the Data Centric Alliance[2] for the CIKM Cup 2016 competition. The dataset consists of 14,148,535 anonymized URL logs of different devices with an average of 197 logs per device. The dataset is split into 50,146 and 48,122 training and test device logs, respectively. To obtain the initial embeddings of each URL, we applied the same data preprocessing methods as in [6,11]. We used a coarse-to-fine node ratio of $K = 6$, a batch size of 800 pairs of device logs, a learning rate of 10^{-3}, and trained the model for 20 epochs (the hyperparameters are fixed for optimal performance of the model using grid search). We used the binary cross-entropy (BCE) loss function for training our model. The training, evaluation, and test were all executed on an A100 NVIDIA GPU. The BCE loss during training as well as the validation F1 score are shown in Figs. 4 and 5, respectively.

---

[2] https://competitions.codalab.org/competitions/11171.

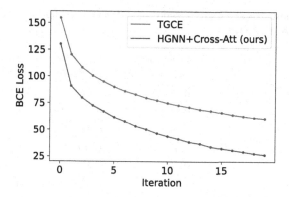

Fig. 4. Binary cross-entropy loss of our proposed method against that of TGCE. During training, our method obtains strictly better loss values.

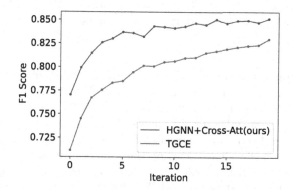

Fig. 5. Validation F1 score during training. Throughout the training, our method achieves strictly better F1 scores for the validation set compared to that of TGCE.

4.2 Results

In this section, we evaluate the precision, recall, and F1 score of our method on the test set and compare it to available baselines. All of the baselines have been obtained similarly as described in [6]. We present two variants of our method; the first one, which we label "HGNN", only differs from TGCE in the design of the second tier, i.e., we use the hierarchical structure presented in Subsects. 3.2 and 3.3, followed by the rest of the TGCE architecture. The second variant, which we label "HGNN+Cross-Att", uses the hierarchical structure in Subsects. 3.2 and 3.3, and also utilizes the cross-attention mechanism presented in Subsect. 3.4 after the hierarchical structure. As shown in Table 1, the "HGNN+Cross-Att" variant outperforms all of the baselines on the F1 score metric, including the second-best method (TGCE) by 5% on the test data.

Table 1. Precision, Recall, and F1 score of different methods for cross-device user matching on DCA dataset.

	Precision at Best F1 Score	Recall at Best F1 Score	Best F1 Score
TF-IDF	0.33	0.27	0.26
Doc2vec	0.29	0.21	0.24
SCEmNet	0.38	0.44	0.41
GRU	0.37	**0.49**	0.42
Transformer	0.39	0.47	0.43
SR-GNN	0.35	0.34	0.34
LESSER	0.41	**0.48**	0.44
TGCE	0.49	0.44	0.46
HGNN (ours)	0.48	0.43	0.45
HGNN+Cross-Att (ours)	**0.57**	**0.48**	**0.51**

We also compare the training time of the two variants of our method with that of TGCE. As shown in Table 2, our hierarchical structure is significantly more efficient than that of TGCE while keeping a competitive F1 score. Table 2 essentially indicates that by simply replacing the second-tier design of TGCE with our hierarchical structure (presented in Subsects. 3.2 and 3.3), the method becomes 6x faster while almost keeping the same performance. This is due to the large number of artificial edges generated in the random walk passes in the creation of the second tier of TGCE. Moreover, although including cross-attention slows down the model, we can still obtain the same training time as TGCE and achieve 5% better overall F1 score.

Table 2. Best F1 score and end-to-end training time of HGNN (without Cross-Att), HGNN+Cross-Att, and TGCE. The HGNN model is 6x faster than TGCE with a slight trade-off (about 1%) on the accuracy side. The HGNN+Cross-Att model has the same training time as TGCE while achieving 5% better F1 score.

	Best F1 Score	End-to-end Training Time	Number of Epochs
TGCE	0.46	60 h	20
HGNN (ours)	0.45	**10 h**	20
HGNN+Cross-Att (ours)	**0.51**	60 h	**6**

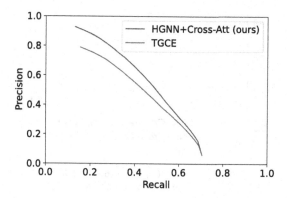

Fig. 6. Precision-Recall curve of the proposed method and that of TGCE on the test data.

Figure 6 shows the precision-recall curve of our method (the HGNN+Cross-Att variant, trained for 6 epochs) with that of TGCE (trained for 20 epochs). As shown in the figure, the precision-recall curve of our method is strictly better than that of TGCE. In other words, for every recall score, our method has a better precision. Additionally, we further trained the HGNN+Cross-Att variant for 20 epochs (the same number of epochs TGCE was trained for) to study if any further improvement is achieved on the test set. We also plot the F1 score with different thresholds (from 0 to 1 incremented by 0.01) for our model trained for 6 and 20 epochs and compare it to that of TGCE. As shown in Fig. 7, our model trained for 20 epochs strictly outperforms TGCE (also trained for 20 epochs) for every threshold for obtaining the F1 score. We note that for the full 20 epoch training, our model would require about three times more training time, and achieves F1 score of just about two percent higher than TGCE. Nevertheless, after training our model for six epochs, it achieves the highest overall F1 score, surpassing TGCE by a remarkable 5%. This achievement is particularly noteworthy considering that, as indicated in Table 2, our model requires the same amount of time to train as TGCE. Hence, we conclude that our model only needs 6 training epochs to achieve the best perfomance and training it for 20 epochs results in overfitting as discussed.

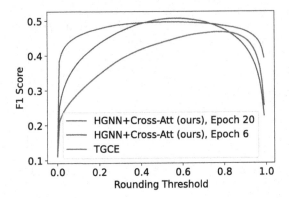

Fig. 7. F1 score against rounding threshold for our method (both for networks trained for 6 and 20 epochs) compared to that of TGCE (trained for 20 epochs).

5   Conclusions

In this paper, we present a novel graph neural network (GNN) architecture for a demanding entity resolution task: cross-device user matching, which determines if two devices belong to the same user based only on their anonymized internet logs. Our method comprises of designing an effective hierarchical structure for achieving long-range message passing in the graph obtained from device URL logs. After passing device logs through such a hierarchical GNN, we employ a cross-attention mechanism to effectively compare device logs against each other to determine if they belong to the same user. We demonstrate that our method outperforms available baselines by at least 5%, while having the same training time as the previous state-of-the-art method, establishing the effectiveness of our proposed method.

Acknowledgements. This research was supported by NVIDIA Corporation.

References

1. Chen, T., Wong, R.C.W.: Handling information loss of graph neural networks for session-based recommendation. In: Proceedings of the 26th ACM SIGKDD International Conference on Knowledge Discovery & Data Mining, pp. 1172–1180 (2020)
2. Chung, J., Gulcehre, C., Cho, K., Bengio, Y.: Empirical evaluation of gated recurrent neural networks on sequence modeling. arXiv preprint arXiv:1412.3555 (2014)
3. Gharaibeh, A., et al.: Smart cities: a survey on data management, security, and enabling technologies. IEEE Commun. Surv. Tutor. **19**(4), 2456–2501 (2017)
4. Gholizadeh, N.: Iec 61850 standard and its capabilities in protection systems (2016)
5. Hidasi, B., Karatzoglou, A., Baltrunas, L., Tikk, D.: Session-based recommendations with recurrent neural networks. arXiv preprint arXiv:1511.06939 (2015)

6. Huang, H., et al.: Two-tier graph contextual embedding for cross-device user matching. In: Proceedings of the 30th ACM International Conference on Information & Knowledge Management, pp. 730–739 (2021)

7. Le, Q., Mikolov, T.: Distributed representations of sentences and documents. In: International Conference on Machine Learning, pp. 1188–1196. PMLR (2014)

8. Li, B., Wang, W., Sun, Y., Zhang, L., Ali, M.A., Wang, Y.: Grapher: token-centric entity resolution with graph convolutional neural networks. In: Proceedings of the AAAI Conference on Artificial Intelligence, vol. 34, pp. 8172–8179 (2020)

9. Lian, J., Xie, X.: Cross-device user matching based on massive browse logs: the runner-up solution for the 2016 cikm cup. arXiv preprint arXiv:1610.03928 (2016)

10. Pan, Z., Cai, F., Chen, W., Chen, H., De Rijke, M.: Star graph neural networks for session-based recommendation. In: Proceedings of the 29th ACM International Conference on Information & Knowledge Management, pp. 1195–1204 (2020)

11. Phan, M.C., Sun, A., Tay, Y.: Cross-device user linking: url, session, visiting time, and device-log embedding. In: Proceedings of the 40th International ACM SIGIR Conference on Research and Development in Information Retrieval, pp. 933–936 (2017)

12. Phan, M.C., Tay, Y., Pham, T.A.N.: Cross device matching for online advertising with neural feature ensembles: first place solution at cikm cup 2016. arXiv preprint arXiv:1610.07119 (2016)

13. Qiu, R., Yin, H., Huang, Z., Chen, T.: Gag: global attributed graph neural network for streaming session-based recommendation. In: Proceedings of the 43rd International ACM SIGIR Conference on Research and Development in Information Retrieval, pp. 669–678 (2020)

14. Ramos, J., et al.: Using tf-idf to determine word relevance in document queries. In: Proceedings of the First Instructional Conference on Machine Learning, vol. 242, pp. 29–48. Citeseer (2003)

15. Sun, F., et al.: Bert4rec: sequential recommendation with bidirectional encoder representations from transformer. In: Proceedings of the 28th ACM International Conference on Information and Knowledge Management, pp. 1441–1450 (2019)

16. Tanielian, U., Tousch, A.M., Vasile, F.: Siamese cookie embedding networks for cross-device user matching. In: Companion Proceedings of the the Web Conference 2018, pp. 85–86 (2018)

17. Wu, S., Tang, Y., Zhu, Y., Wang, L., Xie, X., Tan, T.: Session-based recommendation with graph neural networks. In: Proceedings of the AAAI Conference on Artificial Intelligence, vol. 33, pp. 346–353 (2019)

Data Management

Unified Views for Querying Heterogeneous Multi-model Polystores

Lea El Ahdab[✉], Olivier Teste, Imen Megdiche, and Andre Peninou

Université de Toulouse, IRIT, Toulouse, France
{lea.el-ahdab,olivier.teste,imen.megdiche,andre.peninou}@irit.fr

Abstract. Data storage, in various SQL and NoSQL systems brings complexity to data querying when entities are fragmented because data is not always stored in the same system, plus heterogeneous structures can appear for entities. A unique query language is not sufficient to address data distribution and heterogeneity. Considering vertically distributed data, this work implements a framework capable of rewriting a user query addressed over a unified view to access all data and provide results with transparency. Our framework works with a conceptual model producing unified views to guarantee polystore querying without having to know data distribution nor data heterogeneity. It complements the initial query with intermediate operations. It is applied on an e-commerce scenario (UniBench benchmark) distributed vertically between relational and document-oriented databases. Performance results and the low impact of query rewriting process are illustrated in this work.

Keywords: Polystore · Heterogeneity · Data distribution

1 Introduction

Various storage systems have emerged and constitute polystore systems that federate SQL and NoSQL data stores. Querying a polystore without a unique model is complicated due to databases diversity and data distribution. Solutions have appeared focusing on vertical data distribution [1–3]. The distribution of one entity class over several databases is not considered in such works. In this article, we introduce a framework for querying a multi-model polystore system with vertically distributed entities. The framework provides unified logical views of the polystore in relational or document model. The user queries over one of these logical views which serves as a pivot representation for translating user queries into the different paradigms of the multi-model polystore, guaranteeing transparency of data distribution and to data heterogeneity. In Sect. 2 of this paper, we explain our scope with a motivating example based on an e-commerce scenario. Section 3 discusses existing solutions and their limits. Section 4 defines query construction process and Sect. 5 shows results of our experiments on real data. In the last section, we conclude on this work and we give some perspectives about the future ones.

R. Wrembel et al. (Eds.): DaWaK 2023, LNCS 14148, pp. 319–324, 2023.
https://doi.org/10.1007/978-3-031-39831-5_29

2   Motivating Example

Based on an E-Commerce scenario from the multi-model benchmark UniBench
[11], we consider four entity classes distributed in two family systems: two relational databases (DB1 and DB2) and one document-oriented database (DB3) .
DB1 contains Customers entity, DB3 contains Reviews Entity. Entities *Product*
and *Orders* are vertically fragmented into DB2 and DB3 and their ids product_id
and order_id are the fragmenting key. The relationship Order_Line is included
in Orders of DB3 (Fig. 1). The consideration of document-oriented databases
brings possible data heterogeneity. In this example, entity classes Orders and
Reviews contains different structures: different structures for nested values for
order_line, optional values for feedback.

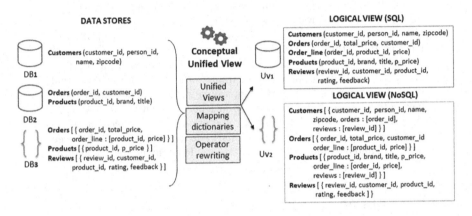

Fig. 1. Our multi-model framework based on logical views and one unified internal
conceptual view ensuring data location and data model equivalences

Let us consider the query: analysis of order prices and brand per customer
within the best rating products ($>=5$). Such query requires the user to query
SQL tables and documents collections to retrieve and join both Customers,
Reviews and build additional joins to retrieve Products and Orders/order_lines.
Our approach is based on unified logical views that present all the data either
in relational (UV_1 in Fig. 1) or document-oriented (UV_2 in Fig. 1) form. These
logical views are used to hide data distribution in the polystore and their various
modeling paradigms. The user builds a query against one of these logical views.
Our system works to generate executable sub-queries on the different databases
which are connected using joins over a specific property of the fragmented entities (Fig. 2). This process potentially induces data transfers. It works on the
algebraic tree of the query and transforms it to insert necessary joins to resolve
data distribution and "rebuild" fragmented entities when necessary. A final step
transforms the algebraic tree to insert data transfers and transformations. The
final result is presented in the form of the unified logical view used for querying.

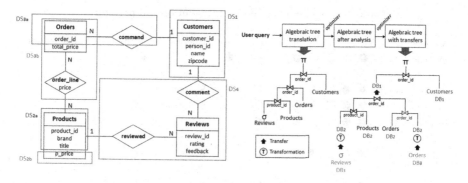

Fig. 2. Application of our framework on the presented use case's query using the associated conceptual model

3   Related Work

Combining SQL and NoSQL systems in one infrastructure, called polystore, brings the notions of multi-store, heterogeneity and data distribution. With vertical distribution where one entity class is found in one datastore of the polystore, unary operators are executed on one system and the binary operator *join* is executed outside DBMS with an external function [1,5,12]. HydRa [10], a framework, mentions entity fragmentation but do not explain how to considerate it for querying purposes. Inferring schemas is proposed to unify querying the polystore's data. It can be a graph representation [6,7] or a u-schema model [5] illustrating structural variations. It brings the issues of query language(s) to access data and the modification of data storage each time data is manipulated. Changing data representation impacts users and modifies the initial paradigm presented to them. A unifying model does not work on data heterogeneity. Some works focus on semantic heterogeneity [3,8] or syntactic [6] issues provided by the multi-storage environment. Structural heterogeneity is set aside but they consider matching techniques to find equivalences between attributes. Table 1 illustrates the differences found between our works and others working on vertical data distribution inside polystores. ● is when the characteristic is fully presented, ◑ is when some cases are missing and ○ is when the characteristic is not addressed in the paper. We compare the considered systems inside the multi-model polystores (relational R, document-oriented D, column-oriented C and graph G), data heterogeneity (structural, semantic and syntactic), the query language(s) of the polystore and if it is question of entity class distribution in one or several system (fragmentation). In our paper, we work on relational and document-oriented system where the user is able to query a polystore in a SQL or a NoSQL language (MongoDB) in a context of vertical distribution where one entity class can be distributed in multiple databases from both systems. Our rewriting system take into account data transfer and transformation and favors the use of DBMS operators as well as its performance.

Table 1. A comparison of existing solutions on polystores

Authors	R	D	C	G	Struct.	Sem.	Syn.	Query	Entity class fragmentation
El Ahdab et al	●	●	○	○	●	◐	◐	*SQL* *MongoDB*	●
Barret et al [6]	●	●	○	●	○	○	●	SparkQL	○
Candel et al [5]	●	●	○	●	○	○	○	SQL	○
Ben Hamadou et al [7]	●	●	●	○	●	●	○	SQL MongoDB	○
Hai et al [8]	●	●	○	●	○	●	○	SQL JSONiq	○
Duggan et al [3]	●	●	●	○	○	●	○	Declarative	○
Forresi et al [12]	●	●	●	○	○	○	○	Spark	○

4   The Proposed Framework

Our framework is based on querying against unified views of a polystore. Unified views are deducted from the entity relationship model of data which highlights entity classes, attributes of entities, entity keys that can serve as distribution key, relationship roles, and relationship attributes. A logical view U_V is the factorization of all distributed entities inside the polystore, according to a fragmentation key. There is one logical view per polystore system (relational or document). We follow converting rules between the conceptual model and the logical models seen by the user: one entity corresponds to one dataset (relation or collection) and the relationships are implemented according to their cardinality. For (N, M) cardinality, in SQL a new relation is created that contains the relations keys of the N and M side (along with relationship attributes); in collections nested values are added inside the two linked collections. For (1, N) and (0, N) cardinality, in SQL a foreign key is created in the 0/1 side relation; in collections, a foreign key is created in the 0/1 side collection and nested values with foreign key are added in N-side collection. Unified views also hide data heterogeneity. To manipulate these variations of attributes, we use an existing mapping technique [7] using a dictionary for each dataset grouping for each attribute (entity, relationship) and for each key of the conceptual model, their equivalences in the unified views and in the real data implementation inside the polystore. The user can build a query against one logical view (Fig. 1). We consider a user query Q_{user} on the unified view U_V referring to one DB of the polystore PL. It is composed of operators from a non-closed set of $\{\sigma, \pi, \bowtie\}$ that manipulate datasets (relations or collections). The main objective is to query the polystore by analyzing Q_{user} as an algebraic graph to generate sub queries on all systems of PL. The steps of our rewriting engine can be described by algebraic tree transformations (Fig. 2): i) build an algebraic tree of the query against the unified view, ii) locate each dataset of the query in the polystore to know whose databases contains it, iii) reconstruct fragmented entities when need by adding necessary joins, iv) add transfer and/or transform operation when needed in the tree. Finally, to deal with structural heterogeneity, the engine uses this dictionary and processes to rewrite each query operator of the algebraic tree. In case of multiple correspondences in the dictionary, our solution privileges the equivalent attribute from the same database of the system interrogated. When the final rewritten query is executed, results are presented to the user in the data format of the system selected depending on the queried logical view (relational or document).

5   Experiments

We use UniBench dataset presented in a context of multi model DBMS (http:// udbms.cs.helsinki.fi/?projects/ubench). We have adapted data distribution as explained in Sect. 2 and in Fig. 2 between two SQL databases (MySQL) and one document-oriented database (MongoDB). Queries were classified according to their operators composition and to the number of dataset needed to rewrite the operation ("Monotable", "Multitable"). Our evaluation focuses on the comparison of rewriting time on each logical view (relational and document-oriented) and the impact of data distribution for one entity class per database and for one entity class in multiple databases. The join operator by itself presents the lowest rewriting time (0.0003 s). It is due to the presence of the entity key in every fragment inside polystores. For the selection and projection, they work more with attributes than keys. The average rewriting time of a query with a combination of all operators is higher (0.0041 s) than the average one for mono operator operations (0.0003 s). For every attribute found in the sub-queries, the dictionary is went through in order to find the exact position in the polystore and then to create the intermediate joins. Data distribution inside polystore impacts rewriting time: with two relation databases and one document-oriented one, it is easier to find the attribute in a simple structure than in nested values as we can find inside the NoSQL system. The logical view considered is the only effect to the query rewriting time in this case. Considering the execution of each query, the average execution time is close to 10 s, including data transformation and data transfers. Adding the rewriting time does not impact the global query time since the rewriting time does not extend 0.0041 s.

6   Conclusion

In this paper we focus on polystore systems with relational and document-oriented systems, where entity classes are vertically distributed between datastores and may be vertically fragmented. We define unified logical views in one data model (relational or document) that cover all the real datasets in the polystore. We define a query rewriting mechanism able to access data in all databases of the polystore according to a dictionary. Considering SPJ operators, the user can transparently query both relational and document-oriented databases with heterogeneous datasets. We have conducted experiments on a Unibench dataset, showing the effectiveness of the rewriting solution. Considering our future work on polystore systems, we will focus on experimenting data transfers and data transformation optimisation.

Acknowledgments. This work was supported by the French Gov. in the framework of the Territoire d'Innovation program, an action of the *Grand Plan d'Investissement* backed by France 2030, Toulouse Métropole and the GIS neOCampus.

References

1. Kolev, B., Valduriez, P., Bondiombouy, C., et al.: CloudMdsQL: querying heterogeneous cloud data stores with a common language. Distrib. Parallel Datab. **34**, 463–503 (2016)
2. Bogyeong, K., Kyoseung, K., Undraa, E., Sohyun, K., Juhun, K., Bongki, M.: M2Bench: a database benchmark for multi-model analytic workloads. PVLDB **16**(4), 747–759 (2022)
3. Duggan, J., Elmore, A.J., Stonebraker, M., et al.: The bigdawg polystore system. ACM Sigmod Rec. **44**(2), 11–16 (2015)
4. Karnitis, G., Arnicans, G.: Migration of relational database to document-oriented database: Structure denormalization and data transformation. In: 7th International Conference on Computational Intelligence, Communication Systems and Networks, pp. 113–118. IEEE (2015)
5. Candel, C.J.F., Ruiz, D.S., García-molina, J.J.: A unified metamodel for NoSQL and relational databases. Inf. Syst. **104**, 101898 (2022)
6. Barret, N., Manolescu, I., Upadhyay, P.: Abstra: toward generic abstractions for data of any model. In: 31st ACM International Conference on Information & Knowledge Management, pp. 4803–4807 (2022)
7. Ben Hamadou, H., Gallinucci, E., Golfarelli, M.: Answering GPSJ queries in a polystore: a dataspace-based approach. In: Laender, A.H.F., Pernici, B., Lim, E.-P., de Oliveira, J.P.M. (eds.) ER 2019. LNCS, vol. 11788, pp. 189–203. Springer, Cham (2019). https://doi.org/10.1007/978-3-030-33223-5_16
8. Hai, R., Quix, C., Zhou, C.: Query rewriting for heterogeneous data lakes. In: Benczúr, A., Thalheim, B., Horváth, T. (eds.) ADBIS 2018. LNCS, vol. 11019, pp. 35–49. Springer, Cham (2018). https://doi.org/10.1007/978-3-319-98398-1_3
9. Papakonstantinou, Y.: Polystore query rewriting: the challenges of variety. In: EDBT/ICDT Workshops (2016)
10. Gobert, M., Meurice, L., Cleve, A.: HyDRa a framework for modeling, manipulating and evolving hybrid polystores. In: IEEE International Conference on Software Analysis, Evolution and Reengineering (SANER), pp. 652–656. IEEE (2022)
11. Zhang, C., Lu, J., Xu, P., Chen, Y.: UniBench: a benchmark for multi-model database management systems. In: Proceedings of the Technology Conference on Performance Evaluation and Benchmarking (TPCTC 2018), Rio de Janeiro, Brazil, pp. 7–23 (2018)
12. Forresi, C., Gallinucci, E., Golfarelli, M., Hamadou, H.B.: A dataspace-based framework for OLAP analyses in a high-variety multistore. VLDB J. **30**(6), 1017–1040 (2021). https://doi.org/10.1007/s00778-021-00682-5

RODD: Robust Outlier Detection in Data Cubes

Lara Kuhlmann[2,3](✉) ⓘ, Daniel Wilmes[1](✉) ⓘ, Emmanuel Müller[1,4] ⓘ,
Markus Pauly[2,4] ⓘ, and Daniel Horn[2,4] ⓘ

[1] Department of Computer Science, TU Dortmund University, Dortmund, Germany
{daniel.wilmes,emmanuel.mueller}@cs.tu-dortmund.de
[2] Department of Statistics, TU Dortmund University, Dortmund, Germany
lara.kuhlmann@tu-dortmund.de, {pauly,dhorn}@statistik.tu-dortmund.de
[3] Graduate School of Logistics, Department of Mechanical Engineering, TU
Dortmund University, Dortmund, Germany
[4] Research Center Trustworthy Data Science and Security, TU Dortmund University,
Dortmund, Germany

Abstract. Data cubes are multidimensional databases, often built from
several separate databases, that serve as flexible basis for data analysis.
Surprisingly, outlier detection on data cubes has not yet been treated
extensively. In this work, we provide the first framework to evaluate
robust outlier detection methods in data cubes (RODD). We introduce a
novel random forest-based outlier detection approach (RODD-RF) and
compare it with more traditional methods based on robust location estimators. We propose a general type of test data and examine all methods
in a simulation study. Moreover, we apply ROOD-RF to real-world data.
The results show that RODD-RF leads to improved outlier detection.

Keywords: Outlier Detection · Data Cubes · Categorical Data ·
Random Forest

1 Introduction

The amount of data created, captured, copied, and consumed worldwide annually
has increased rapidly over the past years: from 2 zettabytes in 2010 to around 79
zettabytes in 2021 [16,25]. Both in public data and in internal data of companies,
the detection of rare events in the form of outliers can provide valuable insights
into customer opinions, market developments and processes [15]. Thereby, the
data is often not available in a single, but in several databases and the data
is merged in a separate decision support database called *data warehouse*. In
contrast to a traditional database management system, which serves to record
transactions, a data warehouse allows the extraction of knowledge from the data
through analysis. A data warehouse is commonly modeled via a so-called *data
cube* [12].

L. Kuhlmann and D. Wilmes—These authors contributed equally to this work.

R. Wrembel et al. (Eds.): DaWaK 2023, LNCS 14148, pp. 325–339, 2023.
https://doi.org/10.1007/978-3-031-39831-5_30

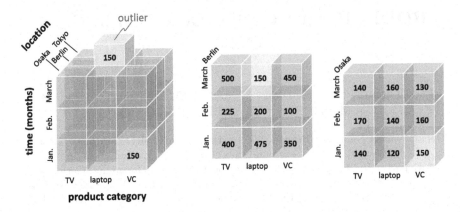

Fig. 1. Left part: Example of a data cube; middle part: a slice of the data cube displaying the sales in Berlin; right part: a slice of the data cube displaying the sales in Osaka.

A data cube allows us to view the data in multiple dimensions. An example of a data cube is given in Fig. 1. Consider the sales data of an electronics company viewed in the categorical dimensions of product category, month and city. We construct the following toy example: the type of product can either be a laptop, a vacuum cleaner or a TV; a month can be (for sake of clarity) January, February or March of 2022; the considered cities are Osaka, Tokyo and Berlin. The values of the categorical dimensions define a cell of a data cube in which a numerical value y, in our case the concrete sales number, is stored: e.g. the cell marked in green, names the sales number of all vacuum cleaners in March 2022 in Osaka, which is in our example 150. In this example, another cell (laptop, March, Berlin; marked in yellow) has the same sales number 150. How can we determine whether one of these data points is an outlier or not? What should methods for robust outlier detection in data cubes (RODD) look like?

We do not have a ground truth of outliers and thus, only methods of unsupervised learning are applicable. One data-driven way is to compare each point with 'related' points. In the middle part of Fig. 1, a slice of the data cube shows the sales in Berlin for the different months and products and in particular contains the marked yellow cell. Although the sales number 150 is not the lowest value in the slice, it is considerably lower than the numerical values in the column and row of the considered cell: The sales of TVs and vacuum cleaners in March and also, the sales of laptops in February and January are higher. Thus, the sales of laptops in Berlin for March would be expected to be a lot higher than 150 and it is reasonable to view this cell as an outlier. Note that for the green cell (right-hand side of Fig. 1), the view of the data cube regarding Osaka does not show any of the peculiarities mentioned above: the cell's value does not deviate much from the other values of the view. Thus, the value $y = 150$ is anomalous only for the cell marked in yellow, not for the one marked in green. Because of this, statistical methods like simple extreme value analysis [36] fail in this setting: 150 is no extreme value. Clustering the values in the slice regarding the city Berlin (middle part of Fig. 1) into the clusters $\{100, 150, 200, 225\}$ and $\{350, 400, 450, 475, 500\}$, would rather yield that $y = 150$ is contained in a cluster. Hence, also techniques such as distance-based outlier detection [19] and density-based outlier detection like local outlier factor (LOF) [4] fail. This poses a challenge to the detection of outliers in data cubes.

Another challenge is that many algorithms for outlier detection are only applicable in the presence of numerical values [37]. In our data cubes scenario, however, there are categorical values that determine the position of a cell in the data cube and there is only one numerical value, in our case a sales value.

Given these observations, an intuitive way for RODD is the following: use any sensible prediction method to estimate each cell's sales number. If the estimated value $\hat{y}$ for a cell differs too much from the actual value y, mark it as outlier. Based on this intuition, [30] resembles an ANOVA approach [34] for RODD. Using trimmed mean estimates $\hat{y}$, the deviation between the actual and the estimated sales value, $r := |y - \hat{y}|$, is normalized by the spread σ of the data. This yields the so-called *SelfExp* value which defines the outlierness of an event: $\text{SelfExp}(y) := \dfrac{r}{\sigma}$. All SelfExp values that exceed a certain threshold τ are considered as an outlier. Whereas [30] only considers the usage of a specific trimmed mean for the computation of $\hat{y}$, the approach actually works for other sensible estimators as well. In particular, we propose to use a random forest (RF) regressor [3] for the computation of $\hat{y}$ as it was proven to be a robust method for many regression tasks, see [7,9,17]. We call this method RODD-RF.

The paper is structured as follows: Sect. 2 presents related work. Section 3 introduces a framework for RODD. Thereafter, we discuss different approaches for estimating the value for a data cube cell (Sect. 4). A simulation study evaluates the performance of the different estimators (Sect. 5). The results of the study and its limitations are discussed (Sect. 6) and the RODD-RF method is validated on a real-world data set (Sect. 7). Finally, a conclusion and an outlook are provided (Sect. 8).

2   Related Work

Outlier detection methods can be differentiated into supervised, semi-supervised, and unsupervised methods. Supervised outlier detection methods require the labeling of anomalous training data. Based on the labeled data, classifiers are developed, which can identify outliers. Examples of RF classifiers are given by [21,35]. When examples of outliers are available but not an entire labeled data set, semi-supervised methods can be applied. Based on the examples, they deduce rules for classification [23,29]. Semi-supervised outlier detection is commonly used for fraud detection [11]. In the case of data cubes applied in an industrial context, neither supervised nor semi-supervised outlier detection methods can be used. Companies usually have no information about outliers in their data prior to applying an outlier detection method. However, unsupervised methods do not require the labeling of anomalies [5].

[37] distinguishes between distance-based, density-based, clustering-based, graph-based, learning-based, statistical-based, and ensemble-based methods. Distance-based methods, density-based methods, such as Local Outlier Factor (LOF) [4,18] and Kernel density estimation methods [20,26], clustering-based methods, and graph-based methods are designed for data sets that only contain numerical values and are thus not applicable to data cubes. Ensemble-based simply means that two or more methods are combined. Learning-based methods such as autoencoders have achieved promising results in outlier detection [1,6,13,40]. However, they are very difficult to explain [32] and subsequently, important information on what differentiates an outlier from the rest of the data set would be lost, if these methods were applied for outlier detection in data cubes. Statistical-based methods assume a certain distribution of the data and identify outliers based on a comparison of the actual data and the assumed distribution. An example of a model-based method is a Gaussian mixture model [39], whose parameters are learned via Expectation maximization [8]. A second example is regression-based methods. Here the prediction of the regressor is compared with the actual value of the data point [24,37]. Due to the above-mentioned characteristics of data cubes, we focus on regression-based methods in this paper. Our goal is to compare the performance of different regressors, such as the RF regressor.

3   A Framework for Outlier Detection in Data Cubes

A data cube is an n-dimensional database DB. Let $D := \{d_1, \ldots, d_n\}$ be the set of all dimensions where each dimension is categorical. Each data point $p \in DB$ gets assigned a numerical value $y := f(p)$ which is stored in the data cube. For example, $f(p)$ can be a sales number. In the following, we abstract a framework for outlier detection in data cubes from the technique presented in [30]: In a high-level view, one computes for each numerical value y a corresponding estimated value $\hat{y}$ and also the spread of the data. This is incorporated into an outlierness score $\text{score}(y) = |y - \hat{y}|/\sigma$. If $\text{score}(y)$ exceeds a threshold τ, the value y and its corresponding data point p is considered an outlier. Below we present a general way for the determination of $\hat{y}$. To this end, let us introduce some notation: Given a data point $p \in DB$, the projection of p to the dimensions $D' \subsetneq D$ is noted as $p' := \text{proj}(p, D')$. We now resemble a general analysis of variance (ANOVA) approach [34] as follows: To describe the effect of the subset of dimensions D' on the sales number $y(p)$, we assign a model coefficient $\gamma(p')$ to each projection. Here, a typical choice for the model coefficients $\gamma(p')$ is the (trimmed) mean over all values in the projection p', see the illustrative example below. As sales number are often of multiplicative nature [30], we consider its logarithmic value: We thus estimate the logarithmic sales number for p by summing the effects $\gamma(p')$ for every projection, i.e. via

$$\log \hat{y}(p) = \sum_{p' \in \text{proj\_set}(p)} \gamma(p'), \tag{1}$$

where the sum runs through all possible projections p′ of p. Each model coefficient takes into account a subset $D' \subsetneq D$ of dimensions and a corresponding selection of data cells. A model coefficient can be viewed as an estimated value over the data cells which are picked by it. Let us show this via an example: Assume in Fig. 2 we have the same setting as in our toy example of Sect. 1 and want to compute the estimated numerical value (sales value) $\hat{y}$ of the data point p=(laptop, March, Berlin), marked yellow, which we abbreviate p $= (i, j, k)$. Each model coefficient now computes an estimated value over a *view*, which means a subset of dimensions and data points of the data cube.

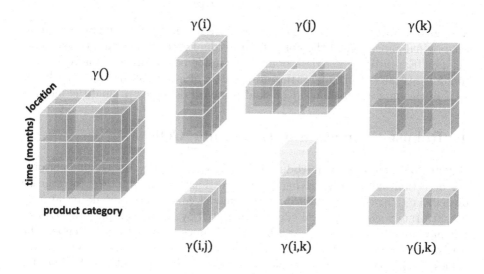

Fig. 2. Example: Different views on a data cube. Each of the seven views on the data cube displays the relevant cells which are used to compute a corresponding model coefficient γ. All seven model coefficients are then needed to determine the estimated value $\hat{y}$ of the yellow cell.

For our tuple p $= (i, j, k)$ the set proj_set of all projections of p is given by proj_set(p) $= \{(i, j), (i, k), (j, k), (i), (j), (k), ()\}$. Note that the empty projection () is allowed. We then collect all model coefficients in the set $\{\gamma(p') \mid p' \in$ proj_set(p)$\}$ $= \{\gamma(i, j), \gamma(i, k), \gamma(j, k), \gamma(i), \gamma(j), \gamma(k), \gamma()\}$. Now that we know which model coefficients are to be computed, we have to determine for every model coefficient the corresponding set of data points on which it is computed. Given a subset of dimensions D', the view of a model coefficient $\gamma(p')$ is then defined as view($\gamma(p')$) $:= \{p \in DB \mid p'$ is projection of p$\}$ (cf. Figure 2 with data point p $= (i, j, k)$). We now compute $\log \hat{y}(p)$, and by this the deviation $|y(p) - \hat{y}(p)|$ by finding all sufficient model coefficients and then computing every model coefficient on the corresponding set of data points. E.g., for $\gamma()$ we consider the full toy data cube as the empty tuple (). It is a projection of every data point and thus resembles a global effect. Finally, we compute:

$$\log \hat{y}(\mathrm{p}) = \gamma(i,j) + \gamma(i,k) + \gamma(j,k) + \gamma(i) + \gamma(j) + \gamma(k) + \gamma(). \qquad (2)$$

To obtain a standardized outlier score, we divide the resulting absolute residual $|y(\mathrm{p}) - \hat{y}(\mathrm{p})|$ by a suitable measure of dispersion $\sigma = \sigma(\mathrm{p})$, see Sect. 4 for a specific choice. The comparison with a treshhold τ leads to the outlierness score

$$\mathrm{SelfExp}(y(\mathrm{p})) = \max\left(\frac{|y(\mathrm{p}) - \hat{y}(\mathrm{p})|}{\sigma(\mathrm{p})} - \tau, 0\right). \qquad (3)$$

Though our framework is more general we have retained the notion SelfExp from the original paper [30]. If the normalized residual exceeds a certain threshold τ, i.e. if $\mathrm{SelfExp}(y(\mathrm{p})) > \tau$, a data point p is considered as outlier. Usually, τ is set to 2.5. This refers to the fact, that in the case of a normal distribution almost 99 % of the data are within the 2.5 fold standard deviation around the mean.

4   Robust Estimation of Data Cube Cells

Building on the equations introduced in Sect. 3, we can now define different outlier detection methods by giving concrete estimators for $\hat{y}$ and σ. In the case of SelfExp [30], the computation of the model coefficients is done by averaging. Starting with the general effect represented by $\gamma()$ we compute the 12.5 %-trimmed mean (tm) of the logarithm of all values $y(\mathrm{p})$ in the data cube. The tm is used for sake of robustness and neglects the 12.5 % largest and smallest values. Going back to our example with $\mathrm{p} = (i,j,k)$ the model coefficients for the projections on one dimension can be computed as

$$\gamma(i) = \operatorname*{tm}_{\mathrm{p}'=(j,k)} \log y(\mathrm{p}) - \gamma(),$$

and for a 2-dimensional coefficient $\gamma(i,j)$ as

$$\gamma(i,j) = \operatorname*{tm}_{\mathrm{p}'=(k)} \log y(\mathrm{p}) - \gamma(i) - \gamma(j) - \gamma().$$

All other coefficients are computed in a similar fashion, e.g. coefficient $\gamma(i,k)$ is obtained by averaging over the logarithm of all aggregated sales data of the dimension time and then subtracting the global coefficient $\gamma()$ and all coefficients that depend on i, respectively k. As already mentioned, the way of computing $\hat{y}(\mathrm{p})$ resembles an ANOVA, more concretely for this example a three-fold ANOVA with two-fold interaction effects [31]. The same construction pattern is applied to higher dimensional coefficients.

Following [30], we define $\sigma(\mathrm{p})$ via

$$\sigma(\mathrm{p})^2 = (\hat{y}(\mathrm{p}))^\rho. \qquad (4)$$

Here, ρ is chosen using the maximum likelihood principle under the assumption that the data are normally distributed with a mean value $\hat{y}(\mathrm{p})$. From this we can derive that ρ has to satisfy the following equation [30]:

$$\sum_{\mathrm{p}} \left(\frac{(y(\mathrm{p}) - \hat{y}(\mathrm{p}))^2}{(\hat{y}(\mathrm{p}))^\rho} \cdot \log(\hat{y}(\mathrm{p})) - \log(\hat{y}(\mathrm{p})) \right) = 0, \tag{5}$$

To our knowledge, the SelfExp estimator was the only instantiation of robust outlier detection in data cubes so far. But it has never been systematically evaluated nor challenged. The trimmed mean as an estimator is intuitive but there might be more effective ways to calculate a robust estimator. We discuss further approaches for RODD in the sequel. To distinguish the SelfExp approach from the others, we introduce the notation $\hat{y}_{S75}$ for the SelfExp (sales value) estimator, as it neglects 25% observations when computing the trimmed means.

Since the SelfExp publication in 1998 more advanced machine learning regression methods were developed. Machine learning models were reported to often outperform linear regression methods (that are underlying an ANOVA) and gained popularity among both researchers and practitioners [2]. One important novelty was the introduction of the RF [10]. The RF by Breiman [3] is an ensemble regressor that applies a bagging approach. It consists of multiple decision trees, which are trained on samples from the data set. Each tree predicts a value and in the last step, an average across all these values is calculated and serves as the final prediction. The RF regressor was already applied in several use cases with promising results, e.g. [7,9,17,22]. For our purpose we train the RF on the whole data cube. We propose the RF-based estimator as an instantiation of $\hat{y}_{RF}$ and based on this we implemented our RODD-RF detection algorithm.

A downside of $\hat{y}_{RF}$ is that it is computationally more intensive than mean-based estimators. We therefore also consider other less computationally intensive estimators as our main competitors. As mentioned before, the trimmed mean is very intuitive. Moreover, it is easy to compute and popular in robust statistics [38]. The SelfExp method works with a trimming percentage of 12.5% and thus uses 75% of the data for the mean calculation. The choice for this trimming percentage was not justified by [30] and other choices might be at least equally plausible. For example, the trimmed mean that uses 90% of the data was shown to achieve good results in several studies [14,28,33]. We thus propose $\hat{y}_{S90}$, a trimmed mean estimator with 5% trimming percentage. We additionally evaluate a trimmed mean estimator $\hat{y}_{S60}$ with 20% trimming percentage. Finally, we also consider the median as robust location measure $\hat{y}_{Median}$.

5 Simulation Study – Experimental Setup

To evaluate the performance of all the presented RODD instantiations, we conducted a simulation study. We used synthesized data since using a real-word data cube is not possible because of the missing ground truth of outliers. The study focused on comparing the quality of the estimators $\hat{y}_{S75}$, $\hat{y}_{RF}$, $\hat{y}_{S90}$, $\hat{y}_{S60}$

and $\hat{y}_{\text{Median}}$ for predicting the value of a data cube cell. We calculated the classification metrics sensitivity, specificity and accuracy. However, the disadvantage of these three evaluation metrics is that they are dependent on a threshold value, which in this case was set to 2.5. To avoid dependency on this threshold value, we also measured the area under the ROC curve (AUC). The ROC curve plots the trade-off between sensitivity and (1 - specificity) varying the threshold parameter τ. If the AUC score is 1, the classification is perfect, if the score is 0.5, the classification is not better than random guessing.

Moreover, we used an ANOVA for statistical comparison of the outlier detection methods [31]. Their effect and the effect of the further experimental parameters on the AUC score was tested. An ANOVA is a statistical method to determine if the means of groups differ significantly from each other [31]. It is similar to the t-test, only that it is capable of comparing more than two groups.

In the following, the synthesization of data cubes is explained and for better understanding, an example is provided. Eight data cubes, each with three categorical dimensions, were synthesized. The number of dimensions was chosen according to a common application example of outlier detection in sales data (see Sect. 3). The amounts of different values per dimension were chosen based on a real-world data set from a German household device-selling company. Table 1 displays step-by-step how the data cube cells were constructed. First, for every value of each dimension, an expected value was selected from a random range of numbers (see Table 1a). The range varied for each dimension and for each data cube. In the example, the expected value of monthly sales of the product vacuum cleaner is 98, of the city Osaka 110 and for the month of January 91. Due to interaction effects, which can be found in real examples of sales figures, expected values for combinations of values of two different dimensions were synthesized (see Table 1b). The arithmetic mean of the expected values and the interaction effects were calculated and rounded (see Table 1c). For some combinations, there is an interaction effect, but for others, there is not, as the noise was rounded and often resulted in zero. For the calculation of the expected value of a vacuum cleaner in Osaka, the expected values of the dimensions, 98 and 110 were added. Then, the interaction effect of 4 was added and this sum was divided by 2. The rounded result was 106. The expected value for the sales of vacuum cleaners in January is simply the rounded mean of 98 and 91 because there is no interaction effect between this specific product and the month. To calculate the expected value of sales in Osaka in January 110, 91 and 4 were added and then divided by 2, resulting in 103. Finally, the value for one cell was calculated and rounded considering the expected values of each category and the interaction effects (see Table 1c). As this value perfectly matched the expected value for the cell, it can be described as noiseless. Due to the calculation process, the expected values approximately follow a normal distribution.

For the creation of outliers, a small sample was taken from the data cubes. Based on the remaining cells, the arithmetic mean and the interquartile range were calculated separately for each value of each categorical variable. 1.5 times the interquartile range was subtracted and added to the respective arithmetic

Table 1. Example for the construction of an artificial, noiseless data cube.

Dimension	Category	Value
Product	VC	98
Location	Osaka	110
Location	Berlin	87
Time	January	91
Time	February	93

(a) Random selection of expected values for each dimension.

Combination	Interaction	Value
VC ∩ Osaka	4	106
VC ∩ Berlin	0	93
VC ∩ January	0	95
VC ∩ February	-3	94
Osaka ∩ January	4	103
Osaka ∩ February	0	102
Berlin ∩ January	0	89
Berlin ∩ February	5	93

(b) Random selection of interaction effects.

Product	Location	Time	Calculation	Value
VC	Osaka	January	(98 + 110 + 91 + 106 + 95 + 103) / 6	101
VC	Osaka	February	(98 + 110 + 93 + 106 + 94 + 102) / 6	101
VC	Berlin	January	(98 + 87 + 91 + 93 + 95 + 89) / 6	92
VC	Berlin	February	(98 + 87 + 93 + 93 + 94 + 93) / 6	93

(c) Calculation of the noiseless data cube, cell values are rounded.

mean, resulting in two 'outlier boundaries' per value of a dimension. For each cell, the rounded outlier boundary that was closest to the original noiseless cell value was selected and replaced the original value. Subsequently, the values in the sample would be classified as outliers according to the Interquartile Range technique, when looking at each dimension separately. In order to make the data more realistic, integers were added as noise to both outliers and inliers. The noise was calculated as the product of a random number and a fraction $\in \{0.25\%, 1\% or 5\%\}$ of the standard deviation of all values in the data cube. In order to consider the distribution of the sales values when adding noise, the standard deviation was calculated for each data cube. Then, the standard deviation was divided by 2.5, 5, 7.5, 10 and 12.5, respectively, afterwards it was multiplied with a random integer between -10 and 10. For the split of the data cubes into outliers and inliers 30 different random seeds were used. This resulted in 3 600 data cubes, on which the RODD methods were tested. There are more data generation methods, but we decided to use the above described because, in our opinion, it resembles the structure of a real-world data cube.

For our RODD-RF method, we used the implementation of the RF regressor from the Python library sklearn [27]. For computational reasons, the hyper-parameter tuning of the RF was performed on a small subset of 120 datacubes using 50 Random Search on six parameters (see below). The subset was chosen randomly to achieve representativeness. The resulting best parameters were used for all other RFs. This is realistic since performing a new parameter tuning before each application of RODD-RF would be too time-consuming. The tuned RF consisted of 1 500 trees. The number of features was limited to the square root of the total number of features in the data set. The maximal depth of the RF was set to 60, the minimal sample split to 5. The minimum number of samples required to be at a leaf node was 1. For the other parameters, the default values were used. When building the trees, bootstrap samples were used. This led to a large forest consisting of many deep trees. Each (deep) tree tends to overfit the sample it was trained on a lot. In this way, many patterns within the data were captured by a tree. Only if the sample contained outliers, a distortion in the estimation of the cell value could be caused. However, this overfitting is later corrected by averaging over the large amount of trees, leading to a good detection of the outliers.

6   Simulation Study – Results

The experiments showed that there was no single superior method, but for every metric, another RODD method was best for outlier detection (see Table 2). The highest sensitivity was achieved using $\hat{y}_{S60}$. A nearly perfect specificity was the result of using $\hat{y}_{Median}$. The accuracy was equally high for both the $\hat{y}_{S75}$ and $\hat{y}_{S90}$ approaches. The AUC score was best when applying RODD-RF. Differences between $\hat{y}_{S60}$, $\hat{y}_{S75}$ and $\hat{y}_{S90}$ were generally small.

Table 2. Evaluation metrics for the five different RODD methods.

Estimator	Sensitivity	Specificity	Accuracy	AUC Score
$\hat{y}_{S75}$	0.2079	0.9900	**0.9817**	0.6985
$\hat{y}_{S60}$	**0.2116**	0.9984	0.9813	0.6998
$\hat{y}_{S90}$	0.1899	0.9993	**0.9817**	0.6947
$\hat{y}_{Median}$	0.0032	**0.9998**	0.9790	0.5848
$\hat{y}_{RF}$	0.2012	0.9951	0.9778	**0.7222**

An factorial ANOVA was run to evaluate the influence of the experimental parameters. Due to the high number of simulations, all p-values were numerically almost zero. The F values showed that the chosen estimator $\hat{y}$, the basis data set, the noise and the percentage of outliers had an impact on the outlier detection performance. Following up on the result of the ANOVA, t-tests were performed for each parameter. The results of the t-tests on the effect of the estimator on the AUC score and the level of noise are displayed in Table 3 (we omit the test results for the basis data set here). The effects are respectively measured in comparison to a benchmark scenario, thus they can be positive or negative. Both $\hat{y}_{S60}$ and $\hat{y}_{RF}$ achieved a higher AUC score than $\hat{y}_{S75}$, which served as a reference model. However, $\hat{y}_{RF}$ is the estimator with the largest positive effect. Moreover, its results even differ significantly from those of $\hat{y}_{S75}$ at an error level of 0.1%.

The p-value of the $\hat{y}_{S60}$ is 0.2376 and thus, it is not significantly different from $\hat{y}_{S75}$.

For the t-test on the impact of noise on the AUC score, the subset with very much noise served as the reference. Naturally, the AUC score increases with decreasing amount of noise. For the amount of outliers in the data set, the subset with 0.25% serves as reference. As expected, the more outlier are present the more complicated it is to find them, i.e., the AUC score decreases.

Table 3. t-tests on the effect of the chosen RODD method on the AUC (left) the level of noise (middle) and the percentage of outliers (right). *** mark a significant effect at the level of 0.1% and ** at the level of 1%.

Estimator	Effect
$\hat{y}_{S60}$	0.0014
$\hat{y}_{S90}$	−0.0038**
$\hat{y}_{Median}$	−0.1136***
$\hat{y}_{RF}$	0.0256***

Noise	Effect
much noise	0.0598***
moderate noise	0.1476***
little noise	0.2379***
very little noise	0.3033***

Outlier Percentage	Effect
1%	-0.0011
5%	−0.0206***

In order to visualize the different AUC scores, they were adapted. As shown by the ANOVA and the t-tests, the experimental parameters have an impact on the AUC score and thus, might deter the results. For every parameter combination, the mean AUC over all methods and random seeds was calculated. Then, for every individual AUC score, the corresponding mean was subtracted. The resulting comparison is illustrated in Fig. 3. The boxplots visualize the similarity of the estimators $\hat{y}_{S75}$, $\hat{y}_{S60}$ and $\hat{y}_{S90}$ and also show the superiority of RODD-RF ($\hat{y}_{RF}$). The estimator $\hat{y}_{Median}$ achieved a notably lower AUC score.

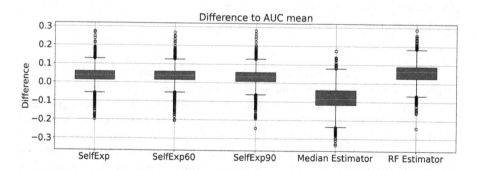

Fig. 3. Boxplot comparing the AUC scores of the five different RODD methods.

Besides the performance of RODD methods, their run time was also evaluated. The $\hat{y}_{S75}$, $\hat{y}_{S60}$, $\hat{y}_{S90}$ and $\hat{y}_{Median}$ estimators had very similar run times. Their minimal run time was 0.13 s and on average they took between 0.39 and

0.43 s. The maximal run time was between 1.00 and 1.49 s. The calculation of $\hat{y}_{RF}$ required considerably more time. The run time was at minimum 4.08 s, on average 37.92 s and at maximum 1 592.05 seconds. RFs always require a non-neglectable amount of run time, but the hyper-parameter of the RF models used in the simulation study contribute immensely to increasing the computation time. For example, the default value for the amount of trees in a forest is 100, but based on the hyper-parameter tuning, we set it to 1 500. However, without the parameter tuning the performance of the RF was considerably lower. Real-world data cubes might be substantially larger than the test data cubes. In this case, it is advisable to evaluate if the increase in the AUC score is worth the increase in computational effort. A run time, which can rise up to 1 592 seconds (26 minutes), might make the method unusable, especially if the application occurs frequently.

7   Application to Real Data

We also validated the RODD method on a real-world data set provided by a German household-device selling company. The first five columns of Table 4 give an overview of the provided information on a subset of the data.

Table 4. Subset of the household-device sales data cube together with expected sales and SelfExp values by the RODD-RF method.

Year	Month	Product	Distribution Channel	Actual Sales	Expected Sales	SelfExp	Deviation
2017	January	product3	channel1	5182	3917	2.8	+32.3%
2017	February	product3	channel1	4288	2547	3.9	+68.4%
2017	March	product3	channel1	5056	3675	3.1	+37.6%
2017	April	product2	channel1	4099	2822	2.8	+45.3%
2019	November	product1	channel1	3287	5069	3.9	-35.2%
2019	November	product1	channel2	4883	2379	5.5	+105.3%

The data contained information about the sales numbers of three different products, sold via three different distribution channels (e.g. local store and online shop). The sales from the years 2017, 2018 and 2019 were aggregated on a monthly basis, resulting in 324 data points. As the company itself had no information available about outliers in their data, no ground truth could be assumed and we could only do a plausibility check on the identified outliers. The RODD-RF method found 17 outliers, 12 times the sales were higher than expected and 5 times lower. When investigating the outliers, experts from the company found plausible explanations for each of them. It happened twice that a marketing campaign shifted the sales from one channel to another, leading to exceptionally high sales of a product in one channel, while decreasing the sales of the same product in another channel. In the first case, the sales increased by 105% in channel 1 compared to the expected value while it was reduced by

35% in channel 2 (see rows 5 and 6 of Table 4). Moreover, a direct discount on a product and the reduction of the price of a set proved to be very effective marketing measures. The direct discount increased the sales by 45% (see row 4 of Table 4) and the set offer by between 32% and 68% (see rows 1–3 of Table 4).

8 Conclusion

We propose RODD, a general framework for robust outlier detection in data cubes. RODD subsumes the current gold-standard, the so-called *SelfExp* method [30], as special case. As first publication we present a systematic comparison with four new robust estimators for outlier detection. Each differs in the statistical estimation of the cell value. While [30] uses a trimmed mean with a 12.5% trimming percentage ($\hat{y}_{S75}$) we analyze approaches based on 5% ($\hat{y}_{S90}$) and 20% trimming ($\hat{y}_{S60}$), the median ($\hat{y}_{Median}$) as well as a Random-Forest ($\hat{y}_{RF}$, RF-RODD). We compared all five methods in a simulation study on 3 600 data cubes. Using AUC score as a quality measure we observed the best results with our new RODD-RF. Compared to traditional *SelfExp* [30], this effect was highly significant (p value $< 0.1\%$). For the other methods, the positive effect on the AUC was not so pronounced ($\hat{y}_{S60}$) or even negative ($\hat{y}_{90}$, $\hat{y}_{Median}$).

Taking these results into account, we apply ROOD-RF to a real world data set. No labels for the anomalies are available for this data set. However, it turns out that for all anomalies found by the algorithm, an explanation could also be found by human expert and domain knowledge.

For future work, we plan to incorporate higher dimensional data cubes with more than three categorical dimensions. Moreover, another measure for the standardization of the difference between the actual and the estimated value could be considered, e.g. a quantile based approach. Furthermore, we see high potential in hierarchical learners that are less explored in machine learning literature.

Acknowledgements. This work was supported by the Research Center Trustworthy Data Science and Security, an institution of the University Alliance Ruhr.

References

1. Andrews, J.T., Morton, E.J., Griffin, L.D.: Detecting anomalous data using auto-encoders. Int. J. Mach. Learn. Comput. **6**(1), 21 (2016)
2. Ardabili, S., Mosavi, A., Várkonyi-Kóczy, A.R.: Advances in machine learning modeling reviewing hybrid and ensemble methods. In: Várkonyi-Kóczy, A.R. (ed.) INTER-ACADEMIA 2019. LNNS, vol. 101, pp. 215–227. Springer, Cham (2020). https://doi.org/10.1007/978-3-030-36841-8_21
3. Breiman, L.: Random forests. Mach. Learn. **45**(1), 5–32 (2001)
4. Breunig, M.M., Kriegel, H.P., Ng, R.T., Sander, J.: Lof: identifying density-based local outliers. In: Proceedings of the 2000 ACM SIGMOD International Conference on Management Of Data, pp. 93–104 (2000)
5. Campos, G., et al.: On the evaluation of unsupervised outlier detection: measures, datasets, and an empirical study. Data Min. Knowl. Disc. **30**(4), 891–927 (2016). https://doi.org/10.1007/s10618-015-0444-8

6. Chen, J., Sathe, S., Aggarwal, C., Turaga, D.: Outlier detection with autoencoder ensembles. In: Proceedings of the 2017 SIAM International Conference on Data Mining, pp. 90–98. SIAM (2017)
7. Cootes, T.F., Ionita, M.C., Lindner, C., Sauer, P.: Robust and accurate shape model fitting using random forest regression voting. In: Fitzgibbon, A., Lazebnik, S., Perona, P., Sato, Y., Schmid, C. (eds.) ECCV 2012. LNCS, vol. 7578, pp. 278–291. Springer, Heidelberg (2012). https://doi.org/10.1007/978-3-642-33786-4_21
8. Dempster, A.P., Laird, N.M., Rubin, D.B.: Maximum likelihood from incomplete data via the em algorithm. J. Roy. Stat. Soc.: Ser. B (Methodol.) **39**(1), 1–22 (1977)
9. El Mrabet, Z., Sugunaraj, N., Ranganathan, P., Abhyankar, S.: Random forest regressor-based approach for detecting fault location and duration in power systems. Sensors **22**(2), 458 (2022)
10. Friedman, J.H.: Recent advances in predictive (machine) learning. J. Classif. **23**(2), 175–197 (2006)
11. Gao, J., Cheng, H., Tan, P.N.: Semi-supervised outlier detection. In: Proceedings of the 2006 ACM Symposium on Applied Computing, pp. 635–636 (2006)
12. Gray, J., et al.: Data cube: a relational aggregation operator generalizing group-by, cross-tab, and sub-totals. Data Min. Knowl. Disc. **1**(1), 29–53 (1997)
13. Hawkins, S., He, H., Williams, G., Baxter, R.: Outlier detection using replicator neural networks. In: Kambayashi, Y., Winiwarter, W., Arikawa, M. (eds.) DaWaK 2002. LNCS, vol. 2454, pp. 170–180. Springer, Heidelberg (2002). https://doi.org/10.1007/3-540-46145-0_17
14. Hill, M., Dixon, W.: Robustness in real life: A study of clinical laboratory data. Biometrics, pp. 377–396 (1982)
15. Hochkamp, F., Rabe, M.: Outlier detection in data mining: Exclusion of errors or loss of information? In: Hamburg International Conference of Logistics (HICL) 2022. In: Proceedings of the Hamburg International Conference of Logistics (HICL) (2022)
16. Holst, A.: Volume of data/information created, captured, copied, and consumed worldwide from 2010 to 2025. Statista, June (2021)
17. Huang, H., Pouls, M., Meyer, A., Pauly, M.: Travel time prediction using tree-based ensembles. In: Lalla-Ruiz, E., Mes, M., Voß, S. (eds.) ICCL 2020. LNCS, vol. 12433, pp. 412–427. Springer, Cham (2020). https://doi.org/10.1007/978-3-030-59747-4_27
18. Jin, W., Tung, A.K.H., Han, J., Wang, W.: Ranking outliers using symmetric neighborhood relationship. In: Ng, W.-K., Kitsuregawa, M., Li, J., Chang, K. (eds.) PAKDD 2006. LNCS (LNAI), vol. 3918, pp. 577–593. Springer, Heidelberg (2006). https://doi.org/10.1007/11731139_68
19. Knorr, E.M., Ng, R.T.: A unified notion of outliers: Properties and computation. In: KDD. vol. 97, pp. 219–222 (1997)
20. Latecki, L.J., Lazarevic, A., Pokrajac, D.: Outlier detection with kernel density functions. In: Perner, P. (ed.) MLDM 2007. LNCS (LNAI), vol. 4571, pp. 61–75. Springer, Heidelberg (2007). https://doi.org/10.1007/978-3-540-73499-4_6
21. Mohandoss, D.P., Shi, Y., Suo, K.: Outlier prediction using random forest classifier. In: 2021 IEEE 11th Annual Computing and Communication Workshop and Conference (CCWC), pp. 0027–0033. IEEE (2021)
22. Nakashima, H., Arai, I., Fujikawa, K.: Passenger counter based on random forest regressor using drive recorder and sensors in buses. In: 2019 IEEE International Conference on Pervasive Computing and Communications Workshops (PerCom Workshops), pp. 561–566. IEEE (2019)

23. Oliver, A., Odena, A., Raffel, C.A., Cubuk, E.D., Goodfellow, I.: Realistic evaluation of deep semi-supervised learning algorithms. In: Advances in Neural Information Processing Systems, vol. 31 (2018)

24. Park, C.M., Jeon, J.: Regression-based outlier detection of sensor measurements using independent variable synthesis. In: International Conference on Data Science. pp. 78–86. Springer (2015)

25. Pauleen, D.J., Wang, W.Y.: Does big data mean big knowledge? km perspectives on big data and analytics. J. Knowl. Manage. **21**(1) (2017)

26. Pavlidou, M., Zioutas, G.: Kernel density outlier detector. In: Akritas, M.G., Lahiri, S.N., Politis, D.N. (eds.) Topics in Nonparametric Statistics. SPMS, vol. 74, pp. 241–250. Springer, New York (2014). https://doi.org/10.1007/978-1-4939-0569-0_22

27. Pedregosa, F., et al.: Scikit-learn: machine learning in Python. J. Mach. Learn. Res. **12**, 2825–2830 (2011)

28. Rocke, D.M., Downs, G.W., Rocke, A.J.: Are robust estimators really necessary? Technometrics **24**(2), 95–101 (1982)

29. Ruff, L., et al.: Deep semi-supervised anomaly detection. arXiv preprint arXiv:1906.02694 (2019)

30. Sarawagi, S., Agrawal, R., Megiddo, N.: Discovery-driven exploration of olap data cubes. In: International Conference on Extending Database Technology. pp. 168–182. Springer (1998). https://doi.org/10.1007/bfb0100984

31. Searle, S.R., Gruber, M.H.: Linear models. John Wiley & Sons (2016)

32. Shankaranarayana, S.M., Runje, D.: ALIME: autoencoder based approach for local interpretability. In: Yin, H., Camacho, D., Tino, P., Tallón-Ballesteros, A.J., Menezes, R., Allmendinger, R. (eds.) IDEAL 2019. LNCS, vol. 11871, pp. 454–463. Springer, Cham (2019). https://doi.org/10.1007/978-3-030-33607-3_49

33. Spjøtvoll, E., Aastveit, A.H.: Comparison of robust estimators on data from field experiments. Scandinavian J. Stat. **7**, 1–13 (1980)

34. St, L., Wold, S., et al.: Analysis of variance (anova). Chemom. Intell. Lab. Syst. **6**(4), 259–272 (1989)

35. Vargaftik, S., Keslassy, I., Orda, A., Ben-Itzhak, Y.: Rade: Resource-efficient supervised anomaly detection using decision tree-based ensemble methods. Mach. Learn. **110**(10), 2835–2866 (2021)

36. Walfish, S.: A review of statistical outlier methods. Pharm. Technol. **30**(11), 82 (2006)

37. Wang, H., Bah, M.J., Hammad, M.: Progress in outlier detection techniques: a survey. Ieee Access **7**, 107964–108000 (2019)

38. Welsh, A.: The trimmed mean in the linear model. Ann. Stat. **15**(1), 20–36 (1987)

39. Yang, X., Latecki, L.J., Pokrajac, D.: Outlier detection with globally optimal exemplar-based gmm. In: Proceedings of the 2009 SIAM International Conference on Data Mining, pp. 145–154. SIAM (2009)

40. Zhou, C., Paffenroth, R.C.: Anomaly detection with robust deep autoencoders. In: Proceedings of the 23rd ACM SIGKDD International Conference on Knowledge Discovery and Data Mining, pp. 665–674 (2017)

Data-driven and On-Demand Conceptual Modeling

Damianos Chatziantoniou[1(✉)] and Verena Kantere[2]

[1] Department of Management Science and Technology,
Athens University of Economics and Business, Athens, Greece
damianos@aueb.gr
[2] School of Electrical Engineering and Computer Science, University of Ottawa,
Ottawa, Canada
vkantere@uottawa.ca

Abstract. The data infrastructure of an organization involves data stored in different models and systems, spreadsheets, files, data mining models, streams, even output of stand-alone programs. In many cases, it is necessary to build on demand, agilely a virtual schema on top of this data infrastructure. We propose a novel graph-based conceptual model, called data virtual machine (DVM), which is designed bottom-up. DVM nodes represent attribute domains and edges represent programs that (through their output) relate values between domains. The data engineer defines a collection of such programs – which can be done quickly, on demand – and the schema, a graph, is derived. This is the inverse process of traditional approaches, where the schema is a priori set and the data engineer designs how to "fit" existing data to this schema. In a DVM, nodes are entities and attributes at the same time, which is useful in many settings. In this paper we also formalize the translation of the relational model to DVM and we present an extended use case that showcases the novelty of the DVM model and the new capabilities it comes with.

Keywords: Data virtualization · data-driven and on-demand modeling

1 Introduction

Modern organizations collect, store and analyze a wealth of data from different sources and applications, used in a variety of data analysis tasks, such as BI, data exploration, analytics etc, to provide a competitive advantage to the business. Data integration is necessary to provide the data scientist with a "holistic" view of the enterprise's *data infrastructure* (or part of it.) The term 'data infrastructure' encompasses much more than data persistently stored in DBMSs, SQL, NoSQL, or otherwise. It also involves flat files, spreadsheets, transient data handled by stream engines and programs with an output useful to analysis, e.g. computation of the social influence or the churn category of a customer - these are also attributes of the customer and should be captured in a data model.

R. Wrembel et al. (Eds.): DaWaK 2023, LNCS 14148, pp. 340–355, 2023.
https://doi.org/10.1007/978-3-031-39831-5_31

Traditional data integration focuses on settling on a schema and either define appropriate data processing tasks to populate/refresh this schema, in case of data warehousing, or define wrappers to bind data with a virtual schema, in case of mediators/virtual databases [8,11,13]. In both cases, *existing* data is "fitted" to a *predefined* schema. Similarly, there are numerous works on the creation of Virtual Knowledge Graphs via mappings as the means of integrating and accessing heterogeneous data sources [2]. These approaches are rigorous and time-consuming but offer semantics (i.e. a data model to explore). An alternative approach followed in many production environments when time to build a model or a report is of essence, is the use of a programming language to extract, transform and assemble data within a program, in an ad-hoc fashion. While agility is the clear benefit of this approach, data exploration for the end-user is non-existent: the creation of a new dataset requires a new program. An interesting question is whether we can find a sweet spot between the two alternatives.

We define a novel data-driven conceptual model, build agilely and on demand, the Data Virtual Machine (DVM) [3,6]. The DVM is a graph based on entities and attributes - concepts that simple users understand well. DVM nodes represent attribute domains and edges represent mappings between these domains, as depicted by the (two-dimensional) output of programs. For example, the output of the SQL query "SELECT custID, age FROM Customers" provides a mapping between the domain of attribute custID and that of age, and vice versa. A DVM node is an attribute and an entity, at the same time. The schema is not known a priori: given a program P with output (u, v) where $u \in A$ and $v \in B$, two nodes A and B are created in DVM - if do not already exist - and two edges $A \rightarrow B$ and $B \rightarrow A$ labeled with P are created. A DVM is generated given a collection of such programs, (hereafter *data processing tasks*), which can be quickly defined by data engineers and with the aid of visual tools.

A DVM: (a) can be easily developed and maintained by data engineers, with the definition of data processing tasks, (b) is easily understood by data scientists and simple users, (c) enables the latter to intuitively, visually, express advanced analytical queries that would otherwise require complex SQL or Python programs written by skilled data engineers, (d) treats any node in the graph both as an attribute and an entity, which allows effortless schema-reorientation (single view of any other entity), and (e) can straightforwardly materialized to different logical models (e.g. relational, semi-structured.) We have developed a research prototype, DataMingler [4] that demonstrates all of the above. In this paper we focus on (a), (b) and partly (c). DVM creation is a *data virtualization* [10] technique, a significant enterprise trend. The goal is to allow data retrieval and manipulation without requiring technical details about the data and the provision a single customer view (or single view of any other entity) of the overall data [21]. All major DB vendors offer products in this field, e.g. [7,17].

In the following, Sect. 2 describes the constituents and the users of analytics environments, provides a real-life motivating example and proposes requirements for data virtualization. Section 3 discusses the ER and DVM relationship and defines DVMs. Section 4 provides a formal translation of the relational model to DVM and in Sect. 5 presents a case study on the virtualization of a relational source together with other sources. Section 6 concludes the paper.

2   Motivation and Problem Definition

2.1   Modern Data Landscapes

A modern data landscape involves a plethora of data management systems and processes, useful in data analysis. An analytics environment encompasses several roles (data stakeholders). We identify below four distinct roles.

Data Engineers: Computer science people with knowledge on data management principles and techniques, relational or otherwise. Their main duty is to extract and transform data from multiple sources and then integrate these to a structure deemed appropriate for input to a data analysis task. They would like to have a high-level data layer where they can easily and quickly map data and programs' output (e.g. models) onto it. This layer could also be used to share data with other data stakeholders in a consistent manner.

Data Scientists/Statisticians: Statisticians and/or computer science people with knowledge on statistical modeling, and/or machine learning techniques. They usually build a model for an entity, using the features (attributes) of that entity. We cannot assume that these people are DB-literate, knowing SQL, relational modeling, programming, etc. They would like to have at their disposal a simple conceptual model that makes clear the entities and their attributes, so they can easily select/experiment with those of interest. Possibly they also want to transform these attributes with built-in aggregate functions or plug-in functions written in Python or R. The end-result is a table (usually a dataframe) that will be used - in most cases - as input for a learning algorithm.

Data Subjects/Data Contributors: this is an emerging role, though a not well-defined one yet. Under European GDPR law, data subjects (customers, suppliers, employees, users in apps, whoever an organization keeps data for) are entitled to their data - including the results of models built using their data. These data have to be delivered to data subjects in a machine-readable format (i.e. somehow structured, such as in excel, JSON, relations, etc.) Data subjects may use these data to get a better rate from a competitor, sell them to marketing companies or hand them to credit bureaus for some kind of rating. A new market will be built around data exchange, data integration and model-building and fuel the data-driven economy. Data subjects would like to easily select (parts of) their data and export them to a model of their choice (e.g. relational, semi-structured, etc.) In addition, they should be able to link their datasets from different organizations in a simple manner to create their "data portfolios". Some sort of "self-service" data integration should be easily attainable [12].

Data Protection Officers: business/legal/Information Systems people with some good technical skills. Their primary role is to ensure that their organization processes personal data of individuals according to applicable data protection laws. They want to see data provenance and consent at a fine granular level.

2.2   Motivating Example

In a recent project for a major telecom provider, we had to predict customers' churn in the presence of structured and unstructured data, residing at different systems, relational and non-relational, and file systems. For this project a predictive model had to be designed and implemented taking into account the many possible variables (attributes or features) characterizing the customer (demographic, interactions with call center, emails, social data, etc.) The goal was to equip the data scientist with a simple tool that would enable fast and interactive experimentation with data: (a) agilely define features (e.g. age, emails, last bill's amount, conversations with call center agents) of a customer from multiple data sources, (b) easily define transformations, selections and aggregations on these features, possibly plugging in functions from different programming languages (e.g. R, Python) for these transformations, and (c) efficiently combine these features into a tabular structure (data frame) to feed a learning algorithm.

One approach would be to build a data warehouse for this project, something that would provide semantics (in relational terms) to the end-users with the clear benefit in data exploration (use of a query language.) Schema could change and be reused. The drawbacks in this approach were (a) initial wait to deploy the data warehouse (i.e. no agility), (b) rigidity in schema modifications, (c) end-to-end involvement of data engineers (to form queries), since statisticians do not know SQL in most cases, and, (d) difficult to plug-in functions written in different programming languages (lack of polyglotism) in dataframe queries.

Another approach would be to use a programming language such as Python or R to form dataframes with a clear benefit in agility: dataframes would be rapidly defined and, thus, models could be quickly tested. However, there is no data modeling involved - no semantics - and each new dataframe is a new program (difficult to reuse, comprehend and maintain). In addition, the problems of end-to-end involvement of data engineers and lack of polyglotism remain.

The question was whether there is any approach that combines the semantics that data modeling provides with the agility of programming languages. Can a data scientist get involved in dataframing formation, alleviating data engineers from this task? The idea was to create a "virtual data desktop", on which schema designers (data engineers) could rapidly map customer's attributes, and data scientists could easily (visually) define transformations on attributes and combine them into dataframes. The key-observation was that an entity (e.g. a customer) can be associated with an attribute via a program, the output of which relates the entity's primary key values and the attribute values. This could be an SQL query (e.g. SELECT custID, age FROM Customers), a Python program that reads a csv file that stores customer's emails and outputs (custID, email_text) pairs, or a cypher query over a graph database that outputs (custID, friendID) pairs.

2.3   Related Work

Classical data integration deals with the definition of a mediated, global schema on top of heterogeneous relational data sources using a mapping between the

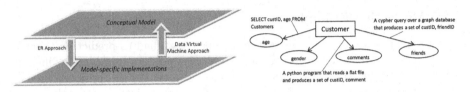

Fig. 1. DVM vs ER-modeling   **Fig. 2.** An entity with attributes

global and the local schemata [8]. Answering queries using views [11] aims to retrieving data and optimizing the query. Data exchange [13] aims to transform data from a source to a target schema based on mappings with a focus on the existence of a solution. In both problems query containment and equivalence are of great importance and investigated with respect to the open and closed world assumption. Our attention is not on the original queries, neither with respect to definition, nor evaluation. Finding alternative queries to extract data, determining the query equivalence, assessing data quality or performing optimization, is also out the scope of building a DVM.

Virtual Knowledge Graphs [16] focus on the definition of mappings [2] with a balanced trade-off of expressiveness and complexity. The W3C proposed OWL 2.0 [18] based on DL-Lite [1]. Many techniques perform semantic reasoning and inference [15] using an ontology, or create ontologies incrementally e.g. [14]. Tools that serve this purpose are Protégé [19], and FaCT++ [9]. Oppositely to OWL 2.0 and semantic reasoners, DVM is not meant for the expression of complex conceptual relations or for reasoning. Thus, we do not focus in defining sophisticated mappings and balancing expressiveness with complexity.

3   Data Virtual Machines

A DVM describes entities and attributes in a graphical way, much like the traditional ER model. However, while ER uses a top-down approach, from a conceptual design to actual data, DVM uses a bottom-up approach, from existing data back to a conceptual model. Figure 1 shows DVM- vs ER-modeling.

3.1   ER and Data Virtual Machines

The key idea in a DVM is to make it easy to add an attribute to an entity from a variety of data sources (relational databases, flat files, excel files, NoSQL, etc.) For instance, for a customer entity, examples of attributes include his/her age, gender and income, but also his/her emails, images, locations and transactions. An attribute of an entity can have one or more values - for example, the age of a customer is a single value, but the emails of a customer can be many - in ER diagrams these are called *multi-valued* attributes. In addition, in ER diagrams, attributes can be *derived*. A derived attribute is an attribute where its value is produced by some computational process, i.e. there exists a process that maps

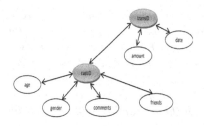

Fig. 3. A simple DVM example

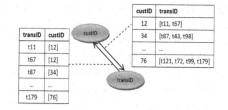

Fig. 4. Attributes and edges in DVMs

one or more values to the entity. In a DVM, since we map *existing* data to entities, we can only have derived attributes. For example, the query ''SELECT custID, age FROM Customers" can be used to bind an age to the customer entity (via the primary key of the entity, custID). The computational process that "defines" the attribute (in this case, the SQL statement) labels the edge connecting the entity and the attribute. In this way, one can represent any data processing task onto a DVM, as long as it has a 2-dimensional output. Examples involve the SQL statement mentioned above, but also a MongoDB query, a Cypher query, programs that read from a flat or an excel file, even programs that assign a churn probability to a customer, as mentioned in Sect. 1. This data processing task can be intra- or inter-organization. Figure 2 shows attributes for the customer entity (for simplicity we draw attributes with a solid line rather than a dashed, double line, as ER theory dictates).

Given that a data process can only associate two attributes, entities can be represented in the model by their primary key(s), i.e. an attribute. In Fig. 3, the 'customer' entity is represented by the custID attribute. The 'transactions' of a customer (represented by transID) is another (multi-valued) attribute of 'customer'. It is also an entity itself, with its own set of attributes. Considering entities as attributes and vice versa and allowing mult-valued attributes, eliminates the need for relationships as a model construct.

Let us consider again the query ''SELECT custID, age FROM Customers''. While it maps an age to a custID, it also maps one or more custIDs to a specific age value. In other words, a data processing task with an output $\{(u, v) : u \in U, v \in V\}$ (multi-set semantics) provides *two* mappings, one from U to V and one from V to U, i.e. edges in a DVM are bidirectional (Fig. 3). In this respect, all nodes are equal, i.e. there is no hierarchy, and all connections are symmetrical, i.e. there are no primary keys. However, attributes that are connected with multiple other attributes can be viewed as entities, and are shown in different color.

3.2 DVM: Theoretical Framework

A DVM represents a collection of *mappings* (edges) between *attribute domains* (nodes), where mappings are manifested as data processing tasks with a 2-dimensional output (the attribute domains) over existing data.

Definition 1. *[Key-list Structure] A key-list structure (KL-structure) K is a set of (key, list) pairs, $K = \{(k, L_k)\}$, where L_k is a list of elements or the empty list and $\forall\ (k_1, L_{k_1}), (k_2, L_{k_2}) \in K,\ k_1 \neq k_2$. Both keys and elements of the lists are strings. The set of keys of KL-structure K is denoted as $keys(K)$; the list of key k of KL-structure K is denoted as $list(k, K)$. If $k \notin keys(K)$, the value of $list(k, K)$ is null. The schema of a KL-structure K, denoted as $K(A, B)$ consists of two labels, A and B. A is the role of the key and B is the role of the list in the key-list pairs.*   □

A key-list structure is a multi-map data structure, where mapped values to a key are organized as a list.

Definition 2. *[Data Virtual Machines] Assume a collection $\mathcal{A}$ of n domains $A_1, A_2, \ldots, A_n$, called attributes. Assume a collection S of m multisets, $S_1, S_2, \ldots, S_m$, where each multiset S has the form: $S = \{(u, v)\ :\ u \in A_i, v \in A_j, i, j \in \{1, 2, \ldots, n\}\}$, called data processing tasks. For each such $S \in \{S_1, S_2, \ldots, S_m\}$ we define two key-list structures, K_{ij}^S and K_{ji}^S as:*

K_{ij}^S*: for each u in the set $\{u : (u, v) \in S\}$ we define the list $L_u = \{v : (u, v) \in S\}$ and (u, L_u) is appended to K_{ij}^S.*
K_{ji}^S *is similarly defined.*

The data virtual machine corresponding to these attributes and data processing tasks is a multi-graph $G = \{\mathcal{A}, \mathcal{S}\}$ constructed as follows:

- *each attribute becomes a node in G*
- *for each data processing task S we draw two edges $A_i \rightarrow A_j$ and $A_j \rightarrow A_i$, labeled with K_{ij}^S and K_{ji}^S respectively.*

The key-list structure that corresponds to an edge $e : A_i \rightarrow A_j$ is denoted as $KL(e)$, with schema (A_i, A_j).   □

The term 'data processing task' refers to its output. The terms 'attributes' and 'nodes' of a DVM are used interchangeably in the remaining of the paper.

Example 1. Assume two attributes, `custID` and `transID` and the output of the SQL query ''SELECT custID, transID FROM Customers that maps transactions to customers and vice versa. The attributes, edges and the respective key-list structures are shown in Fig. 4.   □

3.3   Queries over DVMs

Data scientists/statisticians usually form dataframes in Python, R or Spark. A dataframe is a table built incrementally, column-by-column, around an entity. The first column(s) is usually the key of the entity (e.g. customer ID) and the remaining ones are related attributes. An attribute can be processed (aggregated, filtered or transformed via a user-defined function in Python or R) before

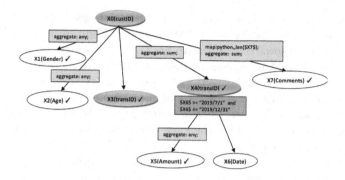

Fig. 5. A dataframe query example

added as a column to the dataframe. The latter often serves as input to learning algorithms or for ad hoc reporting. It is important to facilitate this process in a simple and intuitive, visual manner. One should easily select (possibly along a path) attributes, express conditions and define transformations, using built-in methods or plug-in functions in some programming language of his/her choice (polyglotism) to form a dataframe.

A dataframe query over a DVM is defined as a tree, consisting of nodes and edges of the DVM [6] - *not necessarily* a subtree of the DVM. The same node/edge could appear multiple times in a dataframe query. For example, the query: "for each customer (custID), show his/her gender and age, the list of his/her transactions, the total amount of these transactions in the second semester of 2019 and the total number of characters of his/her comments" is shown in Fig. 5. To evaluate a dataframe query, the transformations on the edges have to be applied and the edges have to be combined, either along a path or across the same level. This evaluation (as well as optimization) take place within an algebraic framework equipped with operators that take as input one or more edges (i.e. key-list structures) and have as output an edge (another key-list structure.) We refer interested readers to a prototype tool implementing DVMs, called DataMingler [4,5].

4 Mapping of a Relational Database to a DVM Schema

The goal of a DVM schema is to represent uniformly part of the information of a set of given relational databases (or other types of databases) that is of interest to their users. The interest of users concerning source information is expressed as a set of queries over the source databases. Therefore if the source is a relational database, the information of interest could be encapsulated in a set of SQL or relational algebra queries. In this section we focus on the assumption that there is a given relational database, for which we need to create a DVM schema. For this case, we proceed by first mapping the database schema to a DVM. Then, the DVM can be augmented with more conceptual information that is conveyed by user queries ontop of the database schema.

There is a unique total mapping function that can translate elements of the relational model to elements of the DVM model. Figure 6 shows a relational schema and Fig. 7 shows the respective DVM. The mapping function takes as input a relational database schema as well as a queries over a relational database instance that adheres to the former, and produces a DVM schema.

We base the definition and creation of the mapping of the relational model (RM) to DVM on previous works that map relational to RDF, with or without an OWL vocabulary [20]. Along the same lines, we define the mapping of RM to DVM as a total function of RM elements to DVM elements.

In the following we describe the creation of the mapping in two steps, first based on the database schema and then based on user queries. Specifically:

Definition 3. *A relational database schema is a pair $\{\mathcal{R}, \Sigma\}$. $\mathcal{R}$ is a finite set of relation names, $|\mathcal{R}| = n$, and $\forall R_i \in \mathcal{R}$, $i = 1, ..., n$ R_i is a relation name that corresponds 1-1 with a set of named attributes $attr(R_i)$, $attr(R_i) = (A_i 1, ..., A_i k)$, for some integer k. Every attribute A is a set of a name and a domain, i.e. $A = \{A_{name}, A_D\}$. Σ isa set of integrity constraints on $\mathcal{R}$.*

Definition 4. *A relational database is a triple of a relational database schema, its integrity constraints Σ and an instance I, $\{\mathcal{R}, \Sigma, I\}$ s.t. $I \models \Sigma$.*

The integrity constraints Σ is the set of all the primary keys and the foreign keys of the database schema, as these are traditionally defined. Due to lack of space we do not redefine them here, but we use the symbols $PK(\mathcal{A}, R)$, $FK(\mathcal{A}, R)$ to represent that a set of attributes $\mathcal{A}$ constitutes a primary key or a foreign key of a relational schema $R \in \mathcal{R}$, respectively.

Definition 5. *A mapping between a relational database and a DVM schema is a total function M that takes as input a relational database $\{\mathcal{R}, \Sigma, I\}$ and a set of queries $\mathcal{Q}$ on it and produces a DVM schema $G = \{\mathcal{A}, \mathcal{S}\}$ as defined earlier, i.e. $M : \{\mathcal{R}, \Sigma, I\} \rightarrow \mathcal{G}$, where $\mathcal{G}$ is all possible $G = \{\mathcal{A}, \mathcal{S}\}$ that adhere to the definition of the DVM. Each query $Q \in \mathcal{Q}$ is an ordered set of two sets of attributes, namely the attributes constituting the key $\mathcal{K}$ and the attributes constituting the list $\mathcal{L}$, i.e. $Q = \{\mathcal{K}, \mathcal{L}\}$.*

In the following we present the specifics of the mapping function M. Without loss of generality, we use SQL for the expression of queries used as input to the mapping. Concerning a query Q, the key $\mathcal{K}$ can be a combination of more than one attributes, as is the case for the list $\mathcal{L}$. The mapping $M((\mathcal{R}, \Sigma, I), \mathcal{Q})$ is a set of logical rules that produce the DVM elements, given the relational database $(\mathcal{R}, \Sigma, I)$ and given a set of queries $\mathcal{Q}$ over the database schema. The rules construct the DVM schema in two steps, first by mapping the database schema to a initial form of a DVM schema, and then by augmenting the latter with DVM elements that are contributed by the queries.

4.1 Step 1: Mapping Construction Based on the Database Schema

The rules that compute the total mapping $M((\mathcal{R}, \Sigma, I), \mathcal{Q})$ create a DVM node for every attribute of every relation in the database schema, except for the

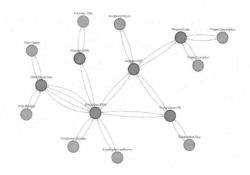

Employee (<u>SSN</u>, Gender, LastName, <u>DeptCode</u>)

Dept(<u>DeptCode</u>, Name, Budget, MgrSSN)

Assigned (<u>SSN</u>, <u>Code</u>, Hours)

Project (<u>Code</u>, Description, Location)

Supervision (<u>SSN</u>, <u>Code</u>, <u>Order</u>, Supervisor)

Founder (<u>SSN</u>, Title)

Fig. 6. A relational database schema to be mapped to a DVM

Fig. 7. The DVM corresponding to the database schema

attributes that are part of primary or foreign keys. If there is a composite primary key in a relation schema, a new corresponding DVM node is also added. Hence, if a primary key is a single attribute, no more nodes are added. Next, the edges are added: every pair of nodes that correspond to the primary key of a relation schema and another attribute of the same relation (therefore this is an attribute that is part of neither the primary nor a foreign key) are connected with a pair of edges that correspond to queries that create the two key-list structures, one from the primary key to the attribute, and one the opposite way. These structures are built eventually as described in the definition of the DVM. Finally, for every pair of a primary key of one relation schema and a respective foreign key of another, a pair of edges is added between the DVM nodes that correspond to the primary keys of these two relations (note that there are no nodes for attributes that are parts of foreign keys, therefore these edges connect nodes corresponding to respective primary keys). The following sketches in a formal manner these rules.

Definition 6. *A mapping between a relational database and a DVM schema is a total function M that takes as input a relational database $\{\mathcal{R}, \Sigma, I\}$ and a set of queries $\mathcal{Q}$ on it and produces a DVM schema $G = \{\mathcal{A}, \mathcal{S}\}$ as defined earlier.*

The first set of rules that produce the DVM schema are the following:

$attr(A, R) :\Leftrightarrow \{A \mid A \in attr(R)\}$.

$pk(A_i, A_{i+1}, \ldots A_j, R) :\Leftrightarrow \{(A_i, A_{i+1}, \ldots, A_j) \mid \forall n = i, \ldots, j \ A_n \in R \wedge \exists \sigma \in \Sigma(\sigma := PK((A_i, A_{i+1}, \ldots A_j), R)\}$

$ppk(A_k, R) :\Leftrightarrow \{A_k \mid \exists pk(A_i, A_{i+1}, \ldots A_j, R) \ (A_k = A_i \vee A_k = A_{i+1} \vee \ldots A_k = A_j)\}$

$fk(A_i, A_{i+1}, \ldots A_j, R') :\Leftrightarrow \{(A_i, A_{i+1}, \ldots, A_j) \mid \exists pk(A_i, A_{i+1}, \ldots A_j, R) \wedge \exists \ \sigma \in \Sigma(\sigma := FK((A_i, A_{i+1}, \ldots A_j), R)\}$[1]

$pfk(A_k, R', R) :\Leftrightarrow \{A_k \mid \exists fk(A_i, A_{i+1}, \ldots A_j, R', R) \ (A_k = A_i \vee A_k = A_{i+1} \vee \ldots A_k = A_j)\}$

$name(R.A) :\Leftrightarrow \{'R.A' \mid \exists A_i \in R\}$

$name(R.A_i, ..R.A_j) :\Leftrightarrow \{'R.A + A_{i+1} \ldots + A'_j \mid \exists A_i, A_{i+1} \ldots, A_j \in R\}$[2]

<u>Rule 1.</u> $DVM\_NODE(name(R.A)) \leftarrow attr(A, R), \neg pfk(A, R), \neg ppk(A, R)$

<u>Rule 2.</u> $DVM\_NODE(name(R.A_i, ..R.A_j)) \leftarrow pk(A_i, A_{i+1}, \ldots A_j, R), \ attr(A_i, R), attr(A_{i+i}, R), \ldots, attr(A_j, R)$

---

[1] For simplicity, corresponding primary and foreign attributes have the same name.

[2] The naming of nodes that combine many attributes is currently implemented differently in Datamingler.

<u>Rule 3a</u>. $DVM\_EDGE(name(R.A_i, ..R.A_j), name(R.A_k), D)$ $\leftarrow$ $pk(A_i, A_{i+1}, ... A_j, R),$
DVM_NODE $(name(R.A_i, ..R.A_j)), DVM\_NODE(name(R.A_k)), D =< SQL\ expression >$
<u>Rule 3b</u>. $DVM\_EDGE(name(R.A_k), name(R.A_i, ..R.A_j), D)$ $\leftarrow$ $pk(A_i, A_{i+1}, ... A_j, R),$
DVM_NODE $(name(R.A_i, ..R.A_j)), DVM\_NODE(name(R.A_k)), D =< SQL\ expression >$
<u>Rule 4a</u>. $DVM\_EDGE(name(R.A_i, ..R.A_j), name(R'.A_n, ..R'.A_k), D)$ $\leftarrow$
$DVM\_NODE(name$
$(R.A_i, ..R.A_j)), DVM\_NODE(name(R'.A_n, ..R.A_k)), pk(A_i, A_{i+1}, ... A_j, R),$ $pk(A_n, A_{n+1},$
$... A_k,$
$R'), pfk(A_p, R', R), D =< SQL\ expression >$
<u>Rule 4b</u>. $DVM\_EDGE(name(R'.A_n, ..R'.A_k), name(R.A_i, ..R.A_j), D)$ $\leftarrow$
$DVM\_NODE(name$
$(R.A_i, ..R.A_j)), DVM\_NODE(name(R'.A_n, ..R.A_k)), pk(A_i, A_{i+1}, ... A_j, R),$ $pk(A_n, A_{n+1},$
$... A_k,$
$R'), pfk(A_p, R', R), D =< SQL\ expression >$

<u>Observations</u>: Rule (1) adds nodes for all the attributes in the database
schema that are neither parts of primary of foreign keys; rule (2) adds nodes
for (both single-attribute and composite) primary keys; rules (3a) and (3b) add
edges between nodes representing primary keys with nodes representing the rest
of the attributes of the respective relation (note that nodes for attributes that
are part of the primary key or part of some foreign key do not exist, therefore,
no edges are added for such attributes); rules (4a) and (4b) add edges between
nodes that represent primary keys for pairs relations that are connected with
primary-foreign key constraints. The creation of an edge includes the definition
of the appropriate data processing task D, which is essentially a query. For simplicity, we do not give the structure of the queries in the formal definition of
rules. As an example, in the DVM of Fig. 6 the query defining the edges between
the primary keys of Assigned and Project is: in which 'SSN, Code' is the key
and 'Code' the list for one direction, and the opposite pair is the key, list for the
other direction.

4.2   Step 2: Mapping Construction Based on User Queries

Additional rules augment $M((\mathcal{R}, \Sigma, I), \mathcal{Q})$ with attributes and edges referenced
based on queries in $\mathcal{Q}$. The mapping creates DVM nodes for attributes that are
in the query result (therefore all attributes appearing in the respective $\mathcal{K}$ and
$\mathcal{L}$). The mapping of attributes that are already present in the DVM schema as
corresponding nodes does not result in new nodes, but the mapping of attributes
that are in a query result and do not belong to the database schema, are added
as new nodes. These can be attributes representing aggregations or, in general,
functions on attributes of the database schema. Next, the edges are added: for
each $Q \in \mathcal{Q}, Q = \{\mathcal{K}, \mathcal{L}\}$ the node that corresponds to $\mathcal{K}$ is connected with the
node that corresponds to $\mathcal{L}$ and the opposite. The key-list structure of such an
edge is built according to the DVM definition, based on a projection of the query
result on the attribute sets $\mathcal{K}$ and $\mathcal{L}$. The additional rules are the following:

Definition 7. *The additional rules are defined as follows:*

$Q.attr(A, R) :\Leftrightarrow \{A \mid A \in (Q.\mathcal{K} \cup Q.\mathcal{L})\}.$
$Q.\mathcal{K}(A_i, A_{i+1}, ... A_j) :\Leftrightarrow \{(A_i, A_{i+1}, ..., A_j) \mid \forall n = i, ..., j\ A_n \in Q.\mathcal{K})\}$
$Q.\mathcal{L}(A_i, A_{i+1}, ... A_j) :\Leftrightarrow \{(A_i, A_{i+1}, ..., A_j) \mid \forall n = i, ..., j\ A_n \in Q.\mathcal{L})\}$

<u>Rule 5</u>. $DVM\_NODE(name(A_i, .., A_j)) \leftarrow Q.K(A_i, A_{i+1}, \ldots A_j)$
<u>Rule 6</u>. $DVM\_NODE(name(A_i, .., A_j)) \leftarrow Q.L(A_i, A_{i+1}, \ldots A_j)$
<u>Rule 7a</u>. $DVM\_EDGE(name(A_i, .., A_j), name(A_n, .., A_k), D) \leftarrow Q.K(A_i, A_{i+1}, \ldots A_j),$
$Q.L(\overline{A_n, A_{n+1}, \ldots A_k}), D = < SQL\ expression >$
<u>Rule 7b</u>. $DVM\_EDGE(name(A_n, .., A_k), name(A_i, .., A_j), D) \leftarrow Q.K(A_i, A_{i+1}, \ldots A_j),$
$Q.L(\overline{A_n, A_{n+1}, \ldots A_k}), D = < SQL\ expression >$

<u>Observations</u>: Rules (5) and (6) adds the nodes for the attributes in the result of the query, such that there is one node that represents the attributes that belong to the key and one node that represents the attributes that belong to the list, of the query. In practice, the key of the query (but also the list) may be a single attribute; in this case the respective node is already in the DVM from the application of the rules in Step 1. Also, in practice, the list of the query may be the output of aggregation function or an arithmetic expression on the database schema attributes. These correspond to new nodes that are added in the DVM schema. Rules (7a) and (7b) add the pair of edges that corresponds to the query, between the nodes that represent the key and the list of the query.

4.3   Discussion

There are several issues relevant to any schema and/or database instance translation. As discussed in [20], two important issues are the preservation of information as well as of queries. Furthermore, it is important to clarify if the translation is also monotonic, as well as if it preserves semantics. In the following we discuss these issues, concerning the translation of a relational database to a DVM schema, in the context of the motivation for the creation of the DVM model.

Information preservation and query preservation For the translation of a relational database schema to a DVM schema, as defined above, it is not mandatory to have a non-empty database instance I. In case there is a non-empty instance, queries over it, do add information in the DVM schema. In the classical sense, 'information preservation' denotes that there is an inverse mapping from the translated schema back to the original one, which, if computed, will populate the original schema resulting in the original instance I. For the case of RM to DVM translation, we would need some type of mapping M' that takes as input the DVM schema and the mapping $M((\mathcal{R}, \Sigma, I), \mathcal{Q})$ and produces I. Of course if only part of the RM schema is translated or $\mathcal{Q}$ is not empty, i.e. the DVM schema is augmented with query results, then it is not possible to compute M'. Nevertheless, let as consider the case that $\mathcal{Q}$ is empty and all elements of the relational schema are translated to the DVM. Then the later is produced based on the rules of step 1. These rules do not translate relational elements to DVM elements in a 1-1 manner. However, with a close look one could observe that the data processing tasks that create the edges in this case would be queries the structure of which can reveal a 1-1 connection of the output key-list structure and the relational attributes from which it is derived (even if these attributes do not correspond to DVM nodes). Therefore, based on the DVM structure and the structure of the dataprocessing tasks, it would be possible to construct a set of inverse rules that construct $(\mathcal{R}, \Sigma, I)$. Yet, the structure of the data processing

Q1:
```
Select Location AS L, sum(Hours) AS S
From Assigned A, Project P
Where A.Code = P.Code
Groupby Location
```

Q2:
```
Select DeptName AS DN, count (*) AS C
From Department D, Employee E
Where D.DeptCode = E.DeptCode
Groupby DeptName
```

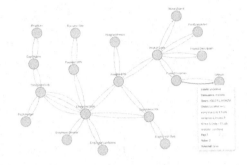

Fig. 8. Queries to be mapped in the DVM schema

Fig. 9. The DVM schema augmented with additional attributes from queries and an excel file

tasks is not part of the DVM definition, and, in general, information preservation is not in the goals of the DVM model. In fact, it opposes part of the basic motivation for its proposal: the DVM model aims to translate only interesting information; therefore, (a) not all the information in a database instance I may be of interest, and, thus, should be translated to the DVM schema, and (b) there may be information that is not explicit in I that is of interest, and, thus, this should be translated in the DVM schema (e.g. aggregations, filters etc. expressed by specific queries), and (c) it is not in the scope of the DVM to provide means for the population or the alteration of the original data sources.

Along the same lines, the notion of query preservation is related to guaranteeing that a query Q expressed on one schema can be translated as Q' and answered on the other schema, producing a result that contains exactly the same information. Continuing the discussion above, it is not the role of the DVM to provide means for query translation, but rather for query answering on the DVM schema, employing the key-list structures of the data processing tasks.

Monotonicity A common property of a mapping that translates one schema S and instance I to another schema S^t and instance I^t, is that the translation is monotonic, i.e. for another instance I' adhering to the same initial schema S, if $I \subseteq I'$ then for the translation $I^{t'}$ it holds that $I^t \subseteq I^{t'}$. The goal of the DVM is not to translate whole instances, but translate schemas and possibly part of an instance. Therefore, the previous definition of monotonicity is not applicable. However, if we consider monotonicty only in terms of schema translation, ignoring instance translation, the mapping of a relational database schema to a DVM presented above is monotonic. This is a direct derivation from the structure of the rules of Step 1 and Step 2: The rules in Step 2 do not contain negation, therefore, they are monotonic; rules (1) and (2) in Step 1 contain negation but only for subgoals that represent static information (information concerning attribute participation in primary and foreign keys); therefore they are also monotonic.

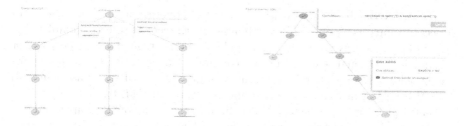

Fig. 10. A first DVM query   **Fig. 11.** A second DVM query

5 Use Case Study

In Sect. 4 we showed how a relational database can be mapped to a DVM. In a real use-case scenario we would expect to employ the DVM model to create a respective schema that coalesces information from multiple sources. Here we extend the DVM of Fig. 7 with additional nodes using data stored in (a) the relational database of Fig. 6, and (b) an excel spreadsheet. We als demonstrate the simplicity in query formulation using DVMs versus standard SQL.

Assume we want to use the database to add the attribute LHours to (Project: Location), that contains the sum of hours per project's location. We have to use SQL query "Q1" in Fig. 8 and specify that L is matched to the existing node Project:Location and S is matched to the new node LHours. This can be easily, visually specified with DataMingler's manager tool. Similarly, one can define the new node EmpNum that contains the total number of employees per department's name, associated with Dept:Name node, via the SQL query "Q2" in Fig. 8. These new attributes are shown in Fig. 9, which also shows the data processing task of the edge Project:Location → LHours. Note that the nodes Project:Location and Project:Code are colored as "entities" (blue) in the DVM after the addition of information from queries.

As it is common for most enterprises to use spreadsheets for analyzing and storing data, we assume that the company uses an excel file to monitor the project progress, keeping KPIs per project. The first excel column contains the project's code and columns m (MoneySpent) and p (PrctCompleted), denote money already spent in the project and the completion percentage, respectively. We want to add these columns as attributes of Project:Code. We could use a library program (available with DataMingler) that reads an excel file and outputs a pair of columns of this. For MoneySpent we would choose columns 1 and m as output, specifying that 1 corresponds to Project:Code (which already exists in the DVM) and m corresponds to MoneySpent (which does not exist and will be created in the DVM.) PrctCompleted is similarly defined, using columns 1 and p. Edges between the nodes are labeled with the appropriate invocation of the excel reader. The new attributes are shown in Fig. 9. Using this DVM, let us consider two query examples that are difficult or impossible to express in SQL.

Example 2. Show the SSN of employees that share at least a project, completed more than 50%, with their managers.

The above can be expressed with nested SQL queries, intersecting the set of projects (completed > 50%) of an employee and the set of projects (completed > 50%) of her manager. However, the completion percentage of a project does not exist in the relational schema but is a column in an excel file. In some systems (e.g. Oracle), one can define a spreadsheet as an external table and use it in SQL statements. Alternatively, it can be imported using an ETL task.

While query formulation in DVM is beyond the scope of this paper and can be found in [4], the query is shown in Fig. 10. The path $X000 \rightarrow X001 \rightarrow X002$ associates an employee's SSN with a list of her project codes given by the edge $X000 \rightarrow X001$. The path $X000 \rightarrow X003 \rightarrow X004 \rightarrow X005 \rightarrow X006$ associates an employee's SSN with a list of her manager's project codes, constrained by $X007$ (completion percentage) to be greater than 50. This corresponds to edge $X000 \rightarrow X003$. The selection condition at $X000$ checks the intersection of these lists. Constructing this tree is intuitive in DataMingler, navigating the DVM. □

Example 3. For each SSN, we want the list of supervisee's SSNs, the count of this list, and the total hours that these supervisees work into projects (in 4 columns). The result of this query is not relational, since the second column is a list - it is a typical dataframe query one would express in Python or R. However, even if we pivot the result to make it relational, the query expression would require the combination of multiple non-trivial SQL queries. The corresponding DVM query is shown in Fig. 11 and should be straightforward: each path from the root corresponds to the computation required for each output column.   □

6   Conclusions

The DVM is a novel graph-based conceptual model, which enables the determination of a schema by defining data processing tasks over existing data, rather than defining a-priori a schema and "fit" existing data onto it. The DVM can represent existing data (persistent, transient, derived) of structured and unstructured sources. Structured and semi-structured sources can be formally translated to a DVM, whereas unstructured sources, such as files, wrappers can be easily developed by the engineers. We show the formal translation of the relational model to the DVM and we discuss a use case that involves a relational database and other data sources, with the purpose to exhibit the easiness and intuitiveness of creating the DVM schema, and further, expressing queries on it.

References

1. Calvanese, D., De Giacomo, G., Lembo, D., Lenzerini, M., Rosati, R.: Dl-lite: Tractable description logics for ontologies. vol. 2, pp. 602–607 (01 2005)
2. Calvanese, D., Gal, A., Lanti, D., Montali, M., Mosca, A., Shraga, R.: Mapping patterns for virtual knowledge graphs (12 2020)
3. Chatziantoniou, D., Kantere, V.: Data Virtual Machines: Data-Driven Conceptual Modeling of Big Data Infrastructures. In: Workshops of EDBT 2020 (2020)

4. Chatziantoniou, D., Kantere, V.: Datamingler: A novel approach to data virtualization. In: ACM SIGMOD, pp. 2681–2685 (2021)
5. Chatziantoniou, D., Kantere, V.: Just-in-time modeling with datamingler. In: Lukyanenko, R., Samuel, B.M., Sturm, A. (eds.) Proceedings of the ER Demos and Posters 2021 co-located with 40th International Conference on Conceptual Modeling (ER 2021), St. John's, NL, Canada, October 18–21, 2021. CEUR Workshop Proceedings, vol. 2958, pp. 43–48. CEUR-WS.org (2021). http://ceur-ws.org/Vol-2958/paper8.pdf
6. Chatziantoniou, D., Kantere, V., Antoniou, N., Gantzia, A.: Data virtual machines: Simplifying data sharing, exploration & querying in big data environments. In: IEEE International Conference on Big Data, Big Data 2022, Osaka, Japan, December 17–20, 2022, pp. 373–380. IEEE (2022)
7. Denodo: Data Virtualization: The Modern Data Integration Solution (2019)
8. Doan, A., Halevy, A.Y., Ives, Z.G.: Principles of Data Integration. Morgan Kaufmann (2012)
9. Fact++: https://fact-project.org/FACT++
10. Gartner: Market Guide for Data Virtualization. https://www.gartner.com/en/documents/3893219/market-guide-for-data-virtualization (2018)
11. Halevy, A.Y.: Answering queries using views: a survey. VLDB J. 10(4), 270–294 (2001)
12. de Hert, P., Papakonstantinou, V., Malgieri, G., Beslay, L., Sanchez, I.: The right to data portability in the GDPR: towards user-centric interoperability of digital services. Comput. Law Secur. Rev. 34(2), 193–203 (2018)
13. Kolaitis, P.G., Panttaja, J., Tan, W.C.: The complexity of data exchange. In: PODS. ACM (2006)
14. Menolli, A., Pinto, H., Reinehr, S., Malucelli, A.: An incremental and iterative process for ontology building. In: ONTOBRAS (2013)
15. Mountantonakis, M., Tzitzikas, Y.: Large-scale semantic integration of linked data: A survey. ACM Comput. Surv. 52(5) (Sep 2019). https://doi.org/10.1145/3345551, https://doi.org/10.1145/3345551
16. Nsl, D.I.: Virtual knowledge graphs: An overview of systems and use cases 1, 201–223 (11 2019)
17. Oracle Corp.: Oracle Data Service Integrator (2020)
18. OWL 2.0: https://www.w3.org/TR/owl2-overview
19. Protégé: https://protege.stanford.edu
20. Sequeda, J.F., Arenas, M., Miranker, D.P.: On directly mapping relational databases to rdf and owl. In: In the Intern, Conf. WWW, pp. 649–658 (2012)
21. Virtualization, W.D.: (2020). https://en.wikipedia.org/wiki/Data_virtualization

FLOWER: Viewing Data Flow in ER Diagrams

Elijah Mitchell[1]([⊠]) [iD], Nabila Berkani[2], Ladjel Bellatreche[3],
and Carlos Ordonez[1]

[1] University of Houston, Houston, USA
emitchell5@uh.edu
[2] Ecole nationale Supérieure d'Informatique, Algiers, Algeria
[3] ISAE-ENSMA, Poitiers, France

Abstract. In data science, data pre-processing and data exploration
require various convoluted steps such as creating variables, merging data
sets, filtering records, value transformation, value replacement and normalization. By analyzing the source code behind analytic pipelines, it
is possible to infer the nature of how data objects are used and related
to each other. To the best of our knowledge, there is scarce research on
analyzing data science source code to provide a data-centric view. On the
other hand, two important diagrams have proven to be essential to manage database and software development projects: (1) Entity-Relationship
(ER) diagrams (to understand data structure and data interrelation-
ships) and (2) flow diagrams (to capture main processing steps). These
two diagrams have historically been used separately, complementing each
other. In this work, we defend the idea that these two diagrams should be
combined in a unified view of data pre-processing and data exploration.
Heeding such motivation, we propose a hybrid diagram called FLOWER
(FLOW+ER) that combines modern UML notation with data flow symbols, in order to understand complex data pipelines embedded in source
code (most commonly Python). The goal of FLOWER is to assist data
scientists by providing a reverse-engineered analytic view, with a data-
centric angle. We present a preliminary demonstration of the concept of
FLOWER, where it is incorporated into a prototype that traces a representative data pipeline and automatically builds a diagram capturing
data relationships and data flow.

1 Introduction

Data pre-processing is a time-consuming and difficult step in Big Data Analytics
(BDA); However, it remains essential in cleaning and normalizing data in order
to make analysis possible. The difficulties of this step arise from both the data
and the processes transforming them. It is important to mention that in BDA,
data often comes from diverse sources with differing structures and formats.
The accuracy of the final analysis strongly depends on the processes applied to
this data, which include data cleaning, transformation, normalization, and so on.

R. Wrembel et al. (Eds.): DaWaK 2023, LNCS 14148, pp. 356–371, 2023.
https://doi.org/10.1007/978-3-031-39831-5_32

These processes conducted by the data analyst typically generate many intermediate files and tables in an often disorganized manner. By deeply analyzing these processes from an explainability point of view, we realize that easy understanding is limited to whatever documentation exists as well as the memory of the analyst authoring them. This represents a real obstacle to deobfuscating BDA systems.

In the context of BDA, data entities and transformations must be considered together to facilitate the understanding of the whole pre-processing pipeline. Several closely related works on ER diagrams [1,2], Flow diagrams [3], and data transformation [4,5] have been done by other researchers. In this work, we propose a hybrid diagram called FLOWER (FLOW + ER) which combines data structure and the flows of a given BDA project. The goal is to assist big data analysts in data pre-processing by examining existing pipelines for data flow relationships. We illustrate its utility by demonstrating real tools that can automatically generate FLOWER diagrams from existing pipelines. Engineers and analysts leveraging these automated methods will be capable of describing and presenting complex code bases and data pipelines with minimal human input.

2 Our Proposed Hybrid Diagram

Before detailing FLOWER, we give fundamental concepts, definitions, and hypotheses related to the ER model, databases, and data sets.

Let E be a set of entities (objects), linked by relationships (references). We follow modern UML notation, where entities are represented by rectangles and relationships are shown by lines, with crowfeet on the "many" side. There exists an identifying set of attributes for each entity (i.e. a primary key or PK, an object identifier). A small (non-disruptive), but powerful change to ER notation is that relationships will have a direction, representing data flow.

Our diagram considers two primary kinds of entity: *Source* (stateful) entities containing raw data, such as SQL tables, files or other data sources, and *transformation* (flowing) entities representing intermediate steps in the data pipeline, such as scripts or partially merged tables. We assume that the input entities can be a bag, without a PK. However, when interpreting non-tabular information, it is common to vectorize or otherwise transform it into a tabular representation. Therefore, a line number or feature index for a text file may act as the primary key. In the case of images, the image file name is commonly a PK as well.

Our Hypothesis: Not all files used for data analysis have PKs or attributes, as in traditional databases. However, we presuppose that data pre-processing programs do produce entity names, keys, attributes, and relationships that can be captured from source code (i.e. file name, feature variables, columns, and other names). We extend the term "primary key" or PK from the relational model to refer more generally to object identifiers, the set of attributes uniquely identifying particular instances of an entity in some data pipeline. Considering these and provided source code, inferred attributes (like the sentiment features of text data, or the formation components extracted from a decomposed image) for FLOWER can be reverse-engineered through automated analysis.

3 Capturing Data Flow with Generated ER Diagrams

In this section, we discuss the specific notation of FLOWER along with considerations to be made when modeling real-world data systems.

3.1 Extending ER Diagram Notation

We allow the relationship lines of ER to include an arrow indicating data flow direction. This direction can also be interpreted as input and output, going from input entities to output entities. The arrow is shown only for relating transformation entities to others. That is, in the case of "raw" source data sets that can be represented as normalized tables, there is no arrow between entities. The arrow represents (1) a processing dependence between two entities, and (2) data flow direction that indicates one entity is used as input.

This is a minor yet powerful change that enables navigating all data elements in the system under consideration (such as a Data Lake), providing a flow-aware view of big data processing. We emphasize that the entities remain linked by "keys", or identifying attributes for records. A FLOWER diagram retains traditional keys, and also considers keys derived from attributes used to link entities under transformation in order to unify relational and non-relational data entities.

3.2 Entities Beyond Databases

An entity in FLOWER is broad in the sense that it can represent any object used in the context of BDA in various formats. An entity may be a file, matrix, relational table, dataframe or more depending on what the analyst chooses to consider. Entities are broadly classified as:

- Source (raw, stateful) entities, representing raw data, loaded into the Data Lake or other system.
- Transformation (flowing, data pre-processing) entities being the output of some system (e.g., Hadoop), tool (statistics or machine learning) or some programming languages used in data science (e.g., Python, R, SQL).

We focus on representing data transformations for BDA, including machine learning, graphs, and text documents. Our diagram does not yet include the "analytic output" such as the parameters of the ML model, IR metrics like precision/recall, and graph metrics, which would be the subject of further work. FLOWER considers three major categories of data transformations:

1. *Merge*, which splices ("joins" in database terms) multiple entities by one or more attributes, which is a weakly typed relational join operator.

2. *GroupBy*, which partitions and aggregates records based on some key. We emphasize that Data Science languages (Python, R) provide operators or functions highly similar to the SQL GROUP BY clause.
3. *Derived expressions*, which represent derived attributes coming from a combination of functions and value-level operators (e.g. equations, string manipulation, arithmetic expression, nested function calls). These can be grouped together, or be separately categorized as same-type and different-type operations depending on the types of the inputs and outputs.

A FLOWER diagram encompassing these transformations may contain far more than simple table data sources. Many heterogeneous data types used in a pipeline may be considered as entity candidates for the purposes of FLOWER. Information on these entities can be retrieved either by the analyst or through the use of automated tools. We provide an (inexhaustive) list of potential sources:

– Importing ER diagrams available from existing transactional or analytical databases. We assume ER diagrams are available for a relational database DDL or exported from an ER diagram tool as CSV files.
– Automatic entity and attribute identification from metadata embedded in the file itself. We assume CSV files are the typical file format for spreadsheet data, logs and mathematical software. Other popular formats such as JSON can capture non-tabular but nonetheless interesting data.
– For text files like documents or source code, there may be statistics on strings for words, numbers, symbols, and so on, as well as text features computed from NLP techniques. In this case, we assume an IR library or tool preprocesses the file and converts the useful data into tables, matrices or data frames. We propose that the "keys" and attributes of these files be identified by their real-world usage, and so can be discovered by examining how they are used by the pipeline.
– Automatic data set and attribute name identification for data sets built by Python, R, or SQL code, generalizing a previous approach with SQL queries [6]. In our section on validation, we explore applications of this approach with a prototypical diagramming tool.

In the process of diagram generation, we may also define the transformation type and create new transformed entities either manually or as informed by automated observation. Data scientists may perform several transformations discussed above, generating temporary entities which we may also wish to capture. In the case of a "Merge," the entity structure may change, but the attribute values remain the same. A "Group by" may use one or more grouping attributes along with or without aggregations (sum, count, avg). In general, aggregations return numbers, but using only "Group by" will return the attribute values as their types. Mathematical transformations will mostly return derived attributes, though there are some analogs. We discuss in Sect. 3.3 a number of transformations that may be considered analogous to these operations.

Following these considerations, the new transformed entities are linked to and from the original entities using an arrow. Keys can be preserved or intuited based on the kind of operation performed. This FLOWER diagram may have applications leading to new insights for analysts in navigating source code, reusing functions, and avoiding creation of redundant data sets.

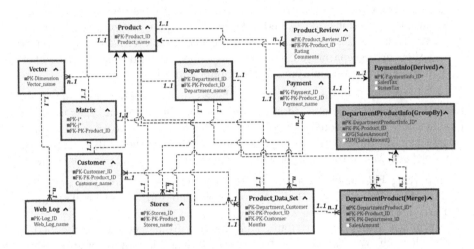

Fig. 1. Example FLOWER diagram: ER diagram enriched with UML notation.

We show in Fig. 1 a FLOWER diagram example enriched with UML notation provided from a hypothetical store data processing pipeline. Arrows show direction of flow, and are compatible with other ER extensions such as those describing attributes and cardinality. We can see the flow of the transformed entities as from the source entities by following the direction of the arrow. Following a data-oriented approach, the model underlying a FLOWER diagram can be stored as JSON files or in metadata management systems for later updates and analysis.

3.3   Equivalent Operations

The exact relationship between software entities and the data pipeline will be application dependent; However, tools created to support FLOWER can be designed with the most common design practices in mind, and be supplied with extensions to support alternative workflows. The concept of *equivalent operations* greatly simplifies this task, as we can map the behavior of program code to transformations analogous to relational databases in certain contexts.

Because FLOWER is an extension to the concept of ER diagrams, we start from ER entities (SQL tables, generally) and extend the concept of "Entity" to describe any of the source entity types we choose to consider. In this way, we can map many kinds of data-bearing objects into source entities, together with SQL-equivalent operations (Table 1).

The usefulness of FLOWER is somewhat complicated by a well-known obstacle to static analysis: Many popular data analysis languages, in particular Python

Table 1. Examples of entity candidates and their SQL-equivalent operations

Entity Type	Constructed From	Transforms	Merges	Group Bys	PK
SQL Table	SQL	SQL operations	MERGE, JOIN	GROUP BY	PK
pandas DataFrame	from_csv, from_sql	vectorized operations	merge(), join()	groupby()	index or column
numpy array/matrix	np.array()	np.sin()	@, np.matmul()	projection to lower dimensional space	row or column index
image	open	color shifting	pasting	aggregate channel statistics	row or column index
text file	text feature extraction	feature biasing	data set combination	word, character, token summarization	label

and R, are dynamically or "weakly" typed, meaning that many operations and attributes do not have known behavior or values until runtime. Static analysis of weakly typed languages is therefore limited to the attributes we can infer from program semantics. By leveraging the notion of equivalent operations, we argue that programs made to analyze data pipelines for FLOWER therefore can assume certain behaviors for a subset of program statements such as documented APIs (pandas, numpy), and adopt general policies for statements with unknown (or inaccessible, at the time of parsing) behavior.

Equivalent operations present a novel perspective in understanding data relationships by broadening of the scope of our understanding of a system as we evaluate more objects as entities. Because entity operations can be roughly grouped together into equivalences, we can make more consistent inferences while developing our FLOWER diagram, and more easily account for these functions when describing patterns for automated tools to search for.

3.4   Data Flow Analysis

In this section, we introduce a method suitable for data flow analysis of source code. The primary languages for data analysis today tend to be imperative and object-oriented, therefore we focus on these paradigms in describing how implementation of an analysis tool may be achieved.

Conceptually, this method transforms imperative statements into stateful operations that may be parsed into directed provenance graphs. These graphs can then be easily analyzed and transformed for presentation, which in our case leads to creation of a FLOWER diagram. We define *Flow* and *State* objects, from which relationships may be observed:

Given: *States*, where a State is an object containing data representing the conceptual state of an entity in the ER model at a specific phase within the pipeline.

$$State := \{name, operation, parents, writes, reads, attributes\}$$

where *name* is the text of the code statement producing it, *operation* is the name of the operation (such as a function call) that produces it, *parents* is the set of the parent states the State is derived from, $reads := \{R_1, R_2...\}$ is the set of resources (such as file name strings) read in its construction, $writes := \{W_1, W_2...\}$ is the set of resources that it may be written to, and $attributes := \{A_1, A_2, ...\}$ is a list of inferred attributes. A State can only be derived from a read operation or another State, so there will not exist any state that does not have an external resource in its ancestry.

Given: A *Flow*, containing a set of *statements* $:= \{T_1, T_2...\}$ in a given imperative language, a set of *initial* $:= \{S_1, S_2...\}$ States, and initially empty sets of reads and writes

$$Flow := \{initial, statements, reads := \{\}, writes := \{\}\}$$

where a Flow can be used to describe anywhere a sequence of statements may exist with an optional set of initial states. In Python or R, scopes like modules and function calls are Flows, where a function's arguments form its initial states. For object methods, the owning object itself may be an initial state, as might be (separately) its attributes. The following operations are available on a State in a Flow:

- Read: A State may be constructed from one or more read operations in a single statement, which are put in its list of reads. This is the only type of State which might have no parent States. The operation for constructing this state will be a *read*.
- Update: States cannot be modified directly by operations within a Flow; Instead, each modification to a state results in a new State whose parents are the modified state along with any other states included in the modification operation. The operation for this state will be the operation used in the update. The new State replaces the old State within the controlling Flow.
- Assign: A State may be Assigned to another State label while retaining its own, without modification. Anywhere the original or assigned label is updated, the state referred to by both is updated and replaced according to the update operation above. This includes States belonging to other Flows: If a State in a Flow's initial state is assigned or updated, the State is updated and replaced there as well as in the current Flow. For languages with copy operations, copying is treated as an update, not an assign.
- Write: A State may be written out to one or more resources. These resources are added to the *write* list in the State.

We also define *opaque* and *transparent* operations. An opaque operation is one that does not have behavior known to the program. A transparent operation is one that does have known behavior, such as behaviors referenced from available source code or otherwise making the operation known to the program ahead of time. Reads and writes are necessarily transparent, as they must be known to the program to be defined as read or write.

A Flow must first have its statements normalized into a list of *assignment expressions* of the form

$$Expr := \{\{S_1, S_2...\}, operation, \{E_1, E_2...\}\}$$

where S is one or more labels of a state to be updated (or created, if the state is not already present in the Flow), *operation* is the name of the operation used in this statement, and E is one or more expressions that may be parent States or non-State values. This process requires flattening each statement into an atomic assignment operation and recursively tracing the execution of nested flows to determine transformation relationships. We can extract information pertaining to data relationships between entities and flow states as follows:

1. Define and collect information on data sources and data transformations (imperative/OO language source code, in this case) under consideration.
2. Define a Flow for each data transformation with the relevant initial States.
3. For each Flow, pre-process its statements into basic assignment expressions.
4. Sequentially analyze each Flow's expressions to form a directed graph of State nodes. Nested Flows like transparent function calls are processed recursively and may modify the containing Flow's States. calls are ignored.
5. For each Flow, either: Treat the Flow as a transformation entity under FLOWER (considered as a single transformation) with its inputs and outputs based on all reads and writes; Or, collapse the State graph into a number of transformation nodes based on ancestry, preserving attributes and PKs. In both cases, reads and in the Flow with no linked write in the graph and vice versa are dropped as extraneous.

Because we only care about general data flow and cannot necessarily determine runtime behavior, control flow structures such as conditionals and loops are treated as inline statements. Assuming code correctness, all branches of the code are anticipated to be reached at some point, so we assume that State transformations in a branch should be captured regardless of whether conditions would cause it to be so. Recursive function calls are also treated as opaque operations. This approach sacrifices some accuracy, but allows for linear time complexity $O(n)$ and lets us avoid the halting problem.

3.5   FLOWER Diagram

We formalize the construction of the FLOWER diagram as follows:

Entities := $\{E_1, E_2, ..., E_n\}$, a list of ER model entity objects $E_i := \{identifier, attributes, entity\_type\}$, where

identifier is a unique identifier such as a string specifying the entity name.
attributes := $\{A_1, A_2, ..., A_m\}$ is a list of entity attributes. If (through observation or prior knowledge) it is determined that an attribute (or set of attributes) uniquely identifies a given instance of an entity, it is marked as a

primary key (PK). More importantly, attributes which make references across different data structures are labeled as foreign keys (FK) and are similarly marked.

entity_type is a type specifier for whether this is a source or transformation/derived entity.

Relationships $:= \{R_1, R_2, ..., R_k\}$, a list of relationship description objects. The relationships among entities are represented by foreign keys. Cardinalities of entities in each relationship are defined: 1:1 relationships can be merged into one entity because they share the same primary key. Relations having cardinalities 1:N or N:1 (1 to many, many to 1) exist between distinct entities. M:N relationships (many to many) connect both entities, taking their respective foreign keys as its primary key. Relations are defined in the form $\{E_i, E_j, Rel_{type}, cardinality\}$, where

E_i **and** E_j are identifiers referencing entities in the ER model.

rel_type is a type specifier as to whether this is a normal or arrow relation. If this is an arrow relation, E_i must be the source and E_j must be the entity pointed to.

cardinality is the specifier for the cardinality of the relation from E_i to E_j: one-to-one (1:1), many-to-one (M:1), one-to-many (1:M), or many-to-many (M:N).

In order to derive the FLOWER diagram, we use the output from the source code parsing/analysis as input for a diagram program reading two complementary files describing entities and relationships. In our example, we output these files as JSON files, from which we generate the model. The following excerpt explains the structure of these files:

    entities.json

```
[    // Source entities
    { key: "Product",
      items: [
        { name: "Product_ID",
          iskey: true,
          figure: "Decision",
          color: "red" }, ...],
      colorate: "#fff9ff" },

    // Transformation entities
    { key: "DepartmentProductInfo(GroupBy)",
      items: [
        { name: "DepartmentProductInfo_ID*",
          iskey :true, ... },
        { name: "FK-Product_ID",
          iskey: true, figure: "", ... }, ...],
      colorate: "#82E0AA" }, ...
]
```

relationships.json

```
[   // Source - Source (line)
    { from: "Sale", to: ''Product",
      color: "black", dash: [3, 2], arr:''Standard",
      width: "1", text: ''1..1", toText: "},

    // Transformation - Any (arrow)
    { from: "Product_Data_Set",
      to: ''DepartmentProduct(Merge)",
      color: "black", dash: [3, 2], arr:"LineFork",
      width: "1", text: "1..1", toText: "n..1" }, ...
]
```

For example, the Product entity given above may have its attributes inferred from corresponding source code like this Python snippet:

```python
import pandas as pd
...
Product = pd.read_csv("products.csv")
Sale = pd.read_csv("sales.csv")

# Analysis observes pattern of use suggesting primary key:
ProductSales = Product.merge(Sales, how="left",
    left_on= "Product_ID", right_on= "Product_ID_fk")
...
```

Given two JSON files following this arrangement, a FLOWER model is derived portraying an ER representation of the observed pipeline. For each JSON object from entities file, an entity $E_i \in Entities$ is created with its attributes described in the items elements. The primary keys are defined. Likewise, for each JSON object from the relations JSON file, a relationship such as $relation = < (E_i, Card_i, Flow_i), .., (E_j, Card_j, Flow_j) >$ is created. Foreign keys are defined from source to target relation. Cardinalities are associated with each entity-relation. The flow direction is also defined, if applicable.

3.6 Strengths and Limitations of Our Approach

FLOWER's greatest strength lies in its compatibility with the ER diagram. By adding a single symbol, an arrow, FLOWER unlocks an entire paradigm of data inter-connectivity where previously only static relationships were considered. Importantly, the arrow shows dependence and data flow while preserving ER structure. The simplicity of the arrow does not correspond to its utility: by introducing the concept of flow to ER, the worlds of Data Warehousing and Data Pre-processing can be reconciled with minimal modification. Not only are we able to see where data goes and how it is transformed, we are also able to combine it with many other extensions to ER with which it is meant to be compatible. In particular, we are able to treat UML diagrams as a specialized form of ER diagram in this manner, as we show in our prototype. While the depiction of flow in a data pipeline is far from a novel concept, we believe the notion of extending ER for the purpose of looking backwards at existing or proposed pipelines as opposed to a domain-specific or ad-hoc representation is. Other approaches also exist for management of existing metadata, but do not yet specify the nature of actual data used ([7]: structured, unstructured, programs, etc.), or analyze the way the information is actually structured and extracted [8].

FLOWER excels particularly as the output of automated reverse engineering. Potentially complex or undocumented data pipelines can be observed and explained in part or whole in moments. In practice, days to weeks of analyst time might be saved when approaching problems such as documenting legacy systems. While ER is traditionally used in describing a relational schema before implementation, it can also be leveraged through use of FLOWER to help explain existing schemas using the same kind of diagram that would be used in their creation. There is an additional advantage in FLOWER being descriptive instead of prescriptive, as it can also enable the user to connect additional attributes to the ER diagram that traditionally do not (and in non-diagnostic cases, should not) be present, such as specific PF/FK relations to entities inside and outside the database.

Implementation of automated FLOWER diagramming will not be without its challenges. Real world software supporting FLOWER must include some or all of the properties of modularity, nested analysis, some measure of attribute and type inference (as with linters and IDEs), and auxiliary source inspection. [9] gives methods of observing and describing data exchange in heterogenous schemas; We propose extending this knowledge in further work to include observed flowing data entities and describing these schemas with FLOWER. Our prototype only considers a subset of Python and known library transformations, which could be improved with the outlined capabilities to be robust enough for real world use.

4   Validation

FLOWER is only useful if it is possible to procure data sufficient to generate an accurate diagram. As discussed above, weak typing and other obstacles to static analysis make finding the relationships between nodes in a real-world data pipeline challenging. In this section, we demonstrate the possibility of a sufficient analysis tool by presenting the results of an actual prototype implementation. This prototype performs a subset of the automated reverse engineering techniques described to inspect and explain real world code.

4.1   Hardware and Software

The specifications of the computer used to implement and test our prototype are as follows: , Intel(R) Core(TM) i7-8700K CPU @ 3.70 GHz; Memory: 48.0 GB; Operating System: Linux (Ubuntu 18.04.4 LTS). Our prototype runs on pure CPython 3.10.10, with no other requirements. The primary modules from

Python 3.10 used were *ast* for traversing python's abstract syntax tree, and *json* for output. Other modules utilized were *typing, dataclasses, functools, uuid, glob* and *argparse*. Our diagram generator was developed using HTML, Bootstrap, CSS, JavaScript (JS), and the GoJS library. Git was used for versioning this project.[1]

4.2   Input Data

We validated the efficacy of our approach by building a prototype implementation for analyzing Python data pipelines. Our prototype focuses on the read and write patterns of pandas' read_* and to_* functions. All other functions are considered opaque, so no nested flow analysis is performed. The prototype also only inspects string literals when collecting resource information. We listed methods of succeeding these limitations in future work in Sect. 3.6.

We provided the software a path pointing to a collection of scripts used in prior research. These scripts used hard coded resource path strings, which is what our prototype looks for in defining reads and writes. We consider the source code as fragments, absent their context or files they interact with, to simulate a complicated or disparate system of legacy code.

4.3   Parsed and Inferred Metadata

The default output of our prototype parser is a JSON file describing each Flow node together with its inputs and outputs, along with "interesting" internal State nodes forming the graph inside the given Flow. An interesting node is a State with a read operation, a write operation, or having more than one ancestors or descendant. It includes a list of all operations on non-interesting ancestors leading up to and including the current node. This approach summarizes States such that each line of non-interesting States followed by an interesting State form a single transformation entity within FLOWER.

For the diagram itself, a command flag causes the software to output entity and relationship JSON files suited to our FLOWER diagram generating program. This instructs the program to output the Flows and any files they connect to as Entities, and their connections as relationships. Optionally, it may also output more detailed files containing the interesting node summaries as entities.

4.4   Results: Diagram Output

We developed a basic graph visualization tool in JavaScript for the purpose of demonstrating automated FLOWER diagram generation. The tool reads the formatted data, in the form of JSON files for entities and relations, and generates a visualization of the complete system in FLOWER format. The end result of

---

[1]   Full code is available on GitHub at https://github.com/Big-Data-Systems/FLOWERPrototype.

this process is a diagram very similar to one that might have been used on the reverse side to plan such a system.

In Figure 1 we displayed a FLOWER diagram generated from source and entity files describing a hypothetical data set. This data set would be a wide range of files and processes dealing in statistics about stores, customers, products, and sales, which would be extracted from external and internal sources relating history sales data, customer information and buyers' opinions. The green entities represent transformation objects such as data scripts, AWS Lambda, etc. that would be detected or declared by the analyst and added to the model. This fully-featured graph illustrates FLOWER's capabilities.

The GUI presented to the user allows them to view the FLOWER model displayed as a dynamic graph. The user can navigate between and interact with the entities to examine their relationships. Further, the PK and FK columns are labeled and the cardinalities are displayed above the arrow link. The transformation entities are shown in color green and ER entities in white. The arrows represent the data flow. An analyst working in combination with these tools can fine tune the parameters of the intermediate files and inputs to create a more accurate diagram.

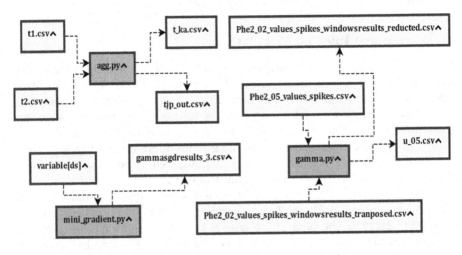

Fig. 2. Results diagram for real BDA source code

We provided the results of our analysis prototype to our visualizer, producing Fig. 2. As expected, the pipelines do not connect between unrelated data sources, only their related fragments. The Python scripts are expressed as green boxes, showing them as intermediate steps in the process of data transformation.

4.5 Efficiency Considerations

Because we statically analyze source code (text data), processing time is negligible. All our source code analyses and diagram generation ran in a fraction of a second. Data sets are not loaded into memory: the system only needs to scan source code. The version of Python analyzed, 3.10, uses a PEG (parsing expression grammar) in parsing the tree. This grammar has an exponential worst-case parsing time complexity, which python handles by loading the entire program into memory and allowing for arbitrary backtracking and caching. Our analysis leverages Python's existing parser (the *ast* module) to parse code into machine readable syntax trees, then runs over each statement linearly to inspect behavior. The parsing step therefore is no less efficient than the parsing done when the program is actually run, and our own inspection algorithm runs in O(n) time thereafter.

5   Related Work

In closely related research, [10] proposed an extension of ER diagram by adding "data transformation" entities in order to visualize tasks of data mining projects developed in the context of relational database warehouses. Two types of transformations were considered: denormalization and aggregation. Lanasri et al. [6] proposed a tool, ER4ML, that assists data scientists to visualize and understand the different data transformations in pre-processing data for a Machine Learning system. In a similar manner to [10], the authors focused only on transformations (denormalization and aggregations) that can be handled by SQL queries. This advanced work is state of the art, but does not elaborate on understanding transformations performed by common data science languages like Python, and it does not capture flows beyond databases. In an alternative line of work, [5] solved data normalization with minimal human interaction. A layout algorithm for automatic drawing of the data flow diagram was presented in [3]. This layout algorithm receives an abstract graph specifying connectivity relations between the elements as input and produces a corresponding diagram as output. The authors of [1] presented a method for entity resolution that infers relationships between observed entities, and uses those relationships to aid in mapping identities to underlying entities.

In similar context and in order to prevent data lakes from being invisible and inaccessible to users, there exist various solutions for data lake management [11], including data lake modeling, metadata management, and data lake governance. Other works focused only on generating a graph-based model to describe intra-object, inter-object, and global metadata [12] or assessing the Data Vault model's suitability for modeling a zoned data lake [13].

There is work on the detection of relationships among different data sets in data lakes [12,13]. However, a data lake contains diverse sources, therefore the description of the process of extracting information (including images, documents, programs; and so on) and identifying relationships between them is required. In addition, these works propose their own models for the representation of metadata. This stands in contrast to our approach, which is based on a common standard for data lake metadata representation, the ER model which enriched with modern UML notation. Until now, no tools currently exist to explore data pre-processing in data lakes. Our vision for FLOWER seeks to unify data warehousing, database design and data pre-processing.

6   Conclusions

Data science needs innovative techniques to capture data structure and inter-relationships, following and extending proven database design techniques with respect to ER diagrams and relational databases. We believe this diagram proposal is a step in the right direction.

Evidently, there is a lot of work to be done. Our prototype examines top-level code, not processing nested flows. Additionally, it only inspects relationships between objects, not considering inherited attributes or keys. Extending to a more mathematically secure and functionally capable tool that can encompass Python as well as other languages, as well as incorporating previous work in SQL, will be the next big task in developing FLOWER. This includes executing portions of the code on input files and tracking values in order to understand data types, dependencies and provenance. We also must study how to represent data filters, similar to the relational selection operator, in order to define a complete algebra and capture the whole data flow accurately. A complete platform would furthermore include parsing database query scripts [10] (e.g. SQL, SPARQL, MongoDB) and combining such outputs with existing UML diagram information to get a full data picture. Collecting opinions and observations from real data scientists will also be necessary to evaluate usefulness, ease of use and flexibility.

References

1. Mugan, J., et al.: Entity resolution using inferred relationships and behavior. In: IEEE International Conference on Big Data. IEEE Computer Society, 2014, pp. 555–560 (2014)
2. Guo, G.: An active workflow method for entity-oriented data collection. In: Woo, C., Lu, J., Li, Z., Ling, T.W., Li, G., Lee, M.L. (eds.) ER 2018. LNCS, vol. 11158, pp. 76–81. Springer, Cham (2018). https://doi.org/10.1007/978-3-030-01391-2_15
3. Batini, C., Nardelli, E., Tamassia, R.: A layout algorithm for data flow diagrams. IEEE Trans. Software Eng. **12**(4), 538–546 (1986)
4. Sebrechts, M., et al.: Model-driven deployment and management of workflows on analytics frameworks. In: IEEE International Conference on Big Data, 2016, pp. 2819–2826 (2016)
5. Pham, M., Knoblock, C.A., Pujara, J.: Learning data transformations with minimal user effort. In: IEEE International Conference on Big Data (BigData), 2019, pp. 657–664 (2019)
6. Lanasri, D., Ordonez, C., Bellatreche, L., Khouri, S.: ER4ML: an ER modeling tool to represent data transformations in data science. Proc. ER Forum Poster Demos Session **2469**, 123–127 (2019)
7. Eichler, R., Giebler, C., Gröger, C., Schwarz, H., Mitschang, B.: HANDLE - a generic metadata model for data lakes. In: Song, M., Song, I.-Y., Kotsis, G., Tjoa, A.M., Khalil, I. (eds.) DaWaK 2020. LNCS, vol. 12393, pp. 73–88. Springer, Cham (2020). https://doi.org/10.1007/978-3-030-59065-9_7

8. Quix, C., Hai, R., Vatov, I.: Metadata extraction and management in data lakes with gemms. Complex Syst. Inform. Model Quart. **9**, 67–83 (2016)

9. Fagin, R., Haas, L.M., Hernández, M., Miller, R.J., Popa, L., Velegrakis, Y.: Clio: schema mapping creation and data exchange. In: Borgida, A.T., Chaudhri, V.K., Giorgini, P., Yu, E.S. (eds.) Conceptual Modeling: Foundations and Applications. LNCS, vol. 5600, pp. 198–236. Springer, Heidelberg (2009). https://doi.org/10.1007/978-3-642-02463-4_12

10. Ordonez, C., Maabout, S., Matusevich, D.S., Cabrera, W.: Extending ER models to capture database transformations to build data sets for data mining. Data Knowl. Eng. **89**, 38–54 (2013)

11. Giebler, C., Gröger, C., Hoos, E., Schwarz, H., Mitschang, B.: Leveraging the data lake: current state and challenges. In: Ordonez, C., Song, I.-Y., Anderst-Kotsis, G., Tjoa, A.M., Khalil, I. (eds.) DaWaK 2019. LNCS, vol. 11708, pp. 179–188. Springer, Cham (2019). https://doi.org/10.1007/978-3-030-27520-4_13

12. Scholly, E., et al.: Coining goldmedal: A new contribution to data lake generic metadata modeling. In: DOLAP, vol. 2840, 2021, pp. 31–40 (2021)

13. Giebler, C., Gröger, C., Hoos, E., Schwarz, H., Mitschang, B.: Modeling data lakes with data vault: practical experiences, assessment, and lessons learned. In: Laender, A.H.F., Pernici, B., Lim, E.-P., de Oliveira, J.P.M. (eds.) ER 2019. LNCS, vol. 11788, pp. 63–77. Springer, Cham (2019). https://doi.org/10.1007/978-3-030-33223-5_7

Supporting Big Healthcare Data Management and Analytics: The Cloud-Based QFLS Framework

Alfredo Cuzzocrea[1,2](✉) and Selim Soufargi[1]

[1] iDEA Lab, University of Calabria, Rende, Italy
{alfredo.cuzzocrea,selim.soufargi}@unical.it
[2] Department of Computer Science, University of Paris City, Paris, France

Abstract. *QUALITOP Federated Big Data Analytics Learning System* (QFLS), a Cloud-based framework *supporting big healthcare data management and analytics for big data lakes* is herein concisely described, with an emphasis on the main data management and analytics functionalities.

Keywords: Big Healthcare Data · Big Data Management · Big Data Analytics · Cloud-Based Frameworks

1 Introduction

Healthcare domains (e.g., [12,20,24]) are one of the most relevant challenges for *big data management and analytics* research (e.g., [2–5,10]). This astounding trend is confirmed by recent proposals that have emerged in literature (e.g., [13,14,19]). Data management and analytics in healthcare domain is particularly demanding because, for instance, handled data must remain private during the analysis process, therefore *privacy issues* (e.g., [1,18,22]). In addition to this, data are disparate, i.e. not located in one single place. In the *QUALITOP* research project [21], inspired by these challenges and mentioned constraints, we design the architecture of the framework named *QUALITOP Federated Big Data Analytics Learning System* (QFLS). QFLS is capable of reaching and analyzing data from diverse sources, in an *anonymous manner*, by using an effective multidimensional analysis data model (e.g., [9]) tailored for the purpose. As a matter of fact, and within the scope of *QUALITOP*, healthcare data are restricted, so that the final goal is that of designing a *big data lake* (e.g., [8]) that is aware of the distributed nature of healthcare data (*data federation*) and of the data privacy constraint (*data anonymization*), in order to assist medical operators through providing them recommendation directives. Overall, the framework's main goal is that of enhancing the quality of life of cancer patients having undergo immunotherapy treatments. In such circumstances, QFLS processes the data locally (preserving therefore data locality) within the Cloud node at which the data reside and retrieves partial results used in a recommendation dashboard.

R. Wrembel et al. (Eds.): DaWaK 2023, LNCS 14148, pp. 372–379, 2023.
https://doi.org/10.1007/978-3-031-39831-5_33

Knowledge discovery is enabled through an effective analytical model named *Tree-Like Analytical Query* (TLAQ). Indeed, we believe that healthcare data are hierarchical by nature, so an effective way to approach these kind of data (e.g., clinical data) is to define a hierarchy of constraints from which we derive our analysis. This hierarchy of constraints is best implemented through TLAQ. Worthy of note that such hierarchical based analysis supports *precision medicine processes* by providing a personalized and accurate analysis of the data. Mathematically speaking, *TLAQ* is such that every composing node $n \in Q$ noted as Q_n, can be modeled as follows: $Q_n = \langle A_n, \Sigma_{A_n}, P_n \rangle$, where: ($i$) A_n is the target query attribute (e.g., AGE); (ii) Σ_{A_n} is a constraint over A_n (e.g., AGE must be above 60); (iii) P_n is an aggregate operator over A_n (e.g., $COUNT$). In a nutshell, QFLS is a data federation framework enabling the analysis of healthcare data in distributed (Cloud) environments in a data preserving manner. Ultimately, QFLS provides recommendation insights over the analyzed data, falling under the term *"big data analytics"*. QFLS summarizes in mainly three components: (i) a Cloud core component, named as *QFLS-Core* ; (ii) a data ingestion web-based application, named as *QUALITOP Anonymized Data Population Tool* (QADPT); (iii) a data analysis web-based application, named as *QUALITOP Anonymized Data Analytics Tool* (QADAT). Indeed, QFLS has been implemented using an on-premise Cloud architecture. Specifically, the main processing engine of QFLS is a *Hadoop cluster* composed of 21 VMs with 32GB RAM and 8-core CPU (60GB HDD). *MapReduce* was the adopted approach to analyze the high volume of data located at each of the federated nodes. More specifically, QFLS-Core represents the engine of QFLS, QADAT is responsible for providing a user-friendly interface to submit TLAQ to federated nodes, whereas QADPT enables the loading of anonymized data into each of the federated nodes.

2   QFLS at Work!

The primary requirement of QFLS is to provide a tool, with a user-friendly interface for data analysts as well as for medical experts, that leverages Cloud capabilities and enables effective analytics and prediction. In order to fulfill this requirement, we integrate a graphical editor where the user can compose a TLAQ against the selected anonymized dataset. The graphical editor is implemented using *mxGraph JavaScript* library. The user edits each of node information such as the attribute name and the target value or the predicate to use, as shown in Fig. 6. Nodes are inter-connectable to eventually form a non-cyclic graph, namely a tree structure, that embeds all of the query information to submit to the server, and to apply onto the target anonymized dataset. Figure 6 also shows the QFLS explorer through which it is possible to select a target anonymized dataset name based on its location (through the federated node logical names and their IP addresses). In the running example of Fig. 6, the *BREASTCANCER* dataset [25] is selected through the left-hand side treeview. Also, the anonymization analysis for the BREASTCANCER anonymized dataset is shown on the right-hand side anonymization analysis table. In the anonymization analysis table, the tool shows a listing of the attribute names, attribute types and their corresponding attribute anonymization percentages.

BREASTCANCER is a multi-variate type of dataset that stores data over breast cancer patients, counting 10 attributes and 286 records. Attributes are described as follows: (*i*) *class*: it describes whether or not cancer returned after a period of remission; (*ii*) *age*: it models the patient age; (*iii*) *menopause*: it describes if the patient has witnessed menopause or not; (*iv*) *tumor-size*: it models the tumor size; (*v*) *inv-nodes*: it reports the number (range [0:39]) of axillary lymph nodes that contain metastatic breast cancer visible on histological examination; (*vi*) *node-caps*: it describes if the cancer does metastasise to a lymph node; (*vii*) *deg-malig*: it models the degree of malignancy of the cancer; (*viii*) *breast*: it reports the concerned breast (left or right) interested by the cancer; (*ix*) *breast-quad*: it describes the concerned zone of the breast interested by the cancer; (*x*) *irradiat*: it models whether the cancer has been irradiated or not.

On top of the BREASTCANCER anonymized dataset, to give an example, a possible TLAQ is shown in Fig. 1.

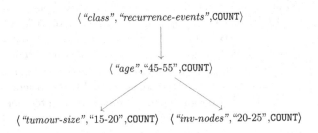

Fig. 1. Example TLAQ

The example TLAQ of Fig. 1 hierarchically investigates the target BREAST-CANCER anonymized dataset as to provide in-depth analytics of the dataset. In Fig. 1, at the root node, the query node ⟨ *"class"*,"recurrence-events",COUNT⟩ applies the filtering on the attribute *class* by means of the attribute value *recurrence-events*, and then the COUNT aggregation is retrieved, the associated Spark code is shown in Fig 2. Then, on the so-computed dataset view, the query node ⟨ *"age"*,"45–55",COUNT⟩ hierarchically applies first the filtering on the attribute *age* by means of the attribute values between 45 and 55, and then the COUNT aggregation is retrieved again, the associated Spark code is shown in Fig 3. Last, hierarchically on the so-computer dataset view, the other two query nodes with constraints: ⟨ *"tumour-size"*,"15–20",COUNT⟩ and ⟨ *"inv-nodes"*,"20–25",COUNT⟩ are applied, respectively, with analogous semantics. The associated Spark code is shown in Fig 4 and Fig 5 respectively.

```
// (class,'recurrence-events',COUNT)
dataset.filter(dataset.col('class')
.equalTo('recurrence-events'))
.agg(count(dataset.col('class')));
```

Fig. 2. Spark Code at Root Node Query of the Example TLAQ Shown in Fig. 1

```
// (age,'45-55',COUNT)
dataset.filter(dataset.col('class')
.equalTo('recurrence-events'))
.filter(dataset.col('age')
.equalTo('45-55'))
.agg(count(dataset.col('age')));
```

Fig. 3. Spark Code at Child of Root Node Query of the Example TLAQ Shown in Fig. 1

```
// (tumour-size,'15-20',COUNT)
dataset.filter(dataset.col('class')
.equalTo('recurrence-events'))
.filter(dataset.col('age').equalTo('45-55'))
.filter(dataset.col('tumour-size')
.equalTo('15-20'))
.agg(count(dataset.col('tumour-size')));
```

Fig. 4. Spark Code at Left Grandchild of Root Node Query of the Example TLAQ Shown in Fig. 1

```
//(inv-nodes,'20-25',COUNT)
dataset.filter(dataset.col('class')
.equalTo('recurrence-events'))
.filter(dataset.col('age').equalTo('45-55'))
.filter(dataset.col('inv-nodes')
.equalTo('20-25'))
.agg(count(dataset.col('inv-nodes')));
```

Fig. 5. Spark Code at Right Grandchild of Root Node Query of the Example TLAQ Shown in Fig. 1

It should be noted here that TLAQ introduce a kind of user-defined queries and they are not subject to strict containment of the columns nor are limited or restricted to "natural hierarchies", like *OLAP hierarchies* [15], since its definition is driven by the precision medicine procedure which makes it perfect to investigate healthcare data because of their inherent nature is compatible with the specific nature of so-called *lazy hierarchies*.

After the evaluation of the example TLAQ, the obtained analytics is shown in Fig. 6. Here, each node of the retrieved structure stores the aggregate result due to evaluating the corresponding node query. In addition to this, for each node, it is possible to display the distribution associated with a visualization attribute that is selected beforehand evaluating the TLAQ. Specifically, in Fig. 6, the distribution of the visualization attribute *age* at the root node of the example TLAQ analytics is displayed. In addition, the tool allows the user to download the analytics in various forms of PDF reports, which can be used for further medical investigations.

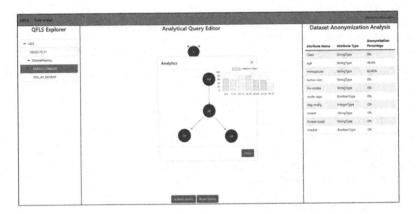

Fig. 6. TLAQ Analytics Retrieved from the Example TLAQ Shown in Fig. 1

3   Experimental Assessment and Analysis

In this Section, we provide our comprehensive experimental evaluation and analysis of QFLS by stressing its privacy-preserving data publishing capabilities via the *DIVersity Anonymization* (DIVA) algorithm [17], a state-of-the-art algorithm that provide *data diversification* while ensuring the data preservation requirement, against a synthetic healthcare dataset. The derived experimental results allow us to test the effectiveness of our algorithm which is demanded to run within the internal layer of the QUALITOP big data lake.

In our experimental campaign, we considered the synthetic healthcare dataset $SBHD_U$, which stores healthcare data distributed according to a *Uniform distribution* [7].

In order to generate our synthetic dataset, we applied simple yet effective *sampling-based techniques* (e.g., [23]) over a (small) real-life healthcare dataset, and, from these data, we generated our synthetic data. This allows us to capture the complexity of the investigated domain by adopting the *sampling generative function* $\Gamma(f)$ that varies according to the *Uniform* distribution. Therefore, we obtained the synthetic healthcare dataset $SBHD_U$ having $1,200,000$ tuples.

It should be noted that, for each synthetic dataset, we achieved a number of tuples that we retained adequate for the scope of big data processing in Cloud computing environments (e.g., [26]). This also allows us to generate datasets that are, finally, totally convergent to the effective goal of our research.

In our experimental campaign, we performed two kinds of experiments, by combining metrics and experimental parameters described in the previous Sections. For each kind of experiment, when we selected a metrics m as output and an experimental parameter p as ranging input parameter, we fixed all the other experimental parameters to a certain value. In the first kind of experiment, we measure the *anonymization accuracy* AA with respect to the ranging of the number of *diversity constraints* $|\Sigma|$ [17], which describe the constraints over the small groups in the dataset, as to preserve their diversity with respect to the bigger groups. It is almost obvious that, by increasing the number of diversity constraints, the anonymization accuracy is stressed more. In the second kind of experiment, we measure the time t necessary to evaluate the collection of synthetic TLAQ of structure shown in Fig. 1, still with respect to the number of diversity constraints $|\Sigma|$. In this case, the more is the number of diversity constraints, the more is the time needed to evaluate input TLAQ against the target big healthcare dataset, due to the intrinsic modus operandi of the core DIVA algorithm .

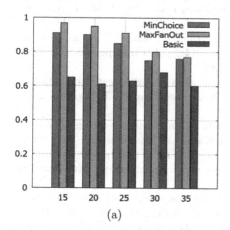
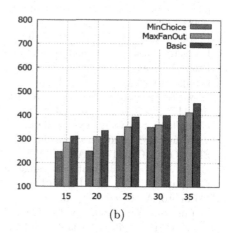

(a)   (b)

Fig. 7. Anonymization Accuracy AA vs Number of Diversity Constraints $|\Sigma|$ for the $SBHD_U$ Dataset (a) - Time t vs Number of Diversity Constraints $|\Sigma|$ for the $SBHD_U$ Dataset (b)

Figure 7(a) shows the variation of the anonymization accuracy AA with respect to the ranging of the number of diversity constraints $|\Sigma|$, for all the three variants of the core DIVA algorithm (i.e., *MinChoice*, *MaxFanOut*, *Basic* - [17]), when the synthetic dataset $SBHD_U$ is considered. Figure 7(b) shows instead the variation of the time t with respect to the ranging of the number of diversity constraints $|\Sigma|$, still for the dataset $SBHD_U$ and for all the three variants of the core DIVA algorithm.

4   Conclusions and Future Work

This paper has described in details definitions, requirements, models and techniques of QFLS, a Cloud-based framework for *supporting big healthcare data management and analytics from big data lakes.* QFLS anatomy and main functionalities have been described, along with the main software solutions proposed with the framework. Concepts and guidelines deriving from the proposed framework also open the door to emerging research challenges for the future. Among those, an interesting research line consists in extending our framework as to deal with advanced machine learning techniques, such as those experimented in other research efforts (e.g., [6,11,16]).

Acknowledgements. This research has been funded by the EU H2020 QUALITOP research project - Call Reference: H2020 - SC1-DTH-01-2019; Project Number: 875171.

References

1. Abbasi, A., Mohammadi, B.: A clustering-based anonymization approach for privacy-preserving in the healthcare cloud. Concurr. Comput. Pract. Exp. **34**(1), e6487 (2022)
2. Balbin, P.P.F., Barker, J.C.R., Leung, C.K., Tran, M., Wall, R.P., Cuzzocrea, A.: Predictive analytics on open big data for supporting smart transportation services. Procedia Comput. Sci. **176**, 3009–3018 (2020)
3. Campan, A., Cuzzocrea, A., Truta, T.M.: Fighting fake news spread in online social networks: actual trends and future research directions. In: IEEE BigData, 2017, pp. 4453–4457. IEEE Computer Society (2017)
4. Chaudhuri, S.: What next?: a half-dozen data management research goals for big data and the cloud. In: Benedikt, M., Krötzsch, M., Lenzerini, M. (eds.) ACM SIGMOD-SIGACT-SIGART PODS, 2012, pp. 1–4. ACM (2012)
5. Chen, J., et al.: Big data challenge: a data management perspective. Frontiers Comput. Sci. **7**(2), 157–164 (2013)
6. Coronato, A., Cuzzocrea, A.: An innovative risk assessment methodology for medical information systems. IEEE Trans. Knowl. Data Eng. **34**(7), 3095–3110 (2022)
7. Cramér, H.: Random Variables and Probability Distributions. No. 36, Cambridge University Press, Cambridge (2004)
8. Cuzzocrea, A.: Big data lakes: models, frameworks, and techniques. In: IEEE International Conference on Big Data and Smart Computing (BigComp), pp. 1–4. IEEE (2021)

9. Cuzzocrea, A., Furfaro, F., Mazzeo, G.M., Saccà, D.: A grid framework for approximate aggregate query answering on summarized sensor network readings. In: Meersman, R., Tari, Z., Corsaro, A. (eds.) OTM 2004. LNCS, vol. 3292, pp. 144–153. Springer, Heidelberg (2004). https://doi.org/10.1007/978-3-540-30470-8_32

10. Cuzzocrea, A., Leung, C.K., MacKinnon, R.K.: Mining constrained frequent itemsets from distributed uncertain data. Future Gener. Comput. Syst. **37**, 117–126 (2014)

11. Cuzzocrea, A., Martinelli, F., Mercaldo, F., Vercelli, G.V.: Tor traffic analysis and detection via machine learning techniques. In: IEEE BigData, 2017, pp. 4474–4480. IEEE Computer Society (2017)

12. Dash, S., Shakyawar, S.K., Sharma, M., Kaushik, S.: Big data in healthcare: management, analysis and future prospects. J. Big Data **6**(1), 1–25 (2019). https://doi.org/10.1186/s40537-019-0217-0

13. Ding, J., Errapotu, S.M., Guo, Y., Zhang, H., Yuan, D., Pan, M.: Private empirical risk minimization with analytic gaussian mechanism for healthcare system. IEEE Trans. Big Data **8**(4), 1107–1117 (2022)

14. Ghayvat, H., et al.: CP-BDHCA: blockchain-based confidentiality-privacy preserving big data scheme for healthcare clouds and applications. IEEE J. Biomed. Health Inform. **26**(5), 1937–1948 (2022)

15. Gray, J., et al.: Data Cube: A relational aggregation operator generalizing group-by, cross-tab, and sub totals. Data Min. Knowl. Discov. **1**(1), 29–53 (1997)

16. Leung, C.K., Cuzzocrea, A., Mai, J.J., Deng, D., Jiang, F.: Personalized deepinf: enhanced social influence prediction with deep learning and transfer learning. In: IEEE BigData, 2019, pp. 2871–2880. IEEE (2019)

17. Milani, M., Huang, Y., Chiang, F.: Preserving diversity in anonymized data. In: Proceedings of the 24th International Conference on Extending Database Technology, EDBT 2021, Nicosia, Cyprus, 23–26 March 2021, pp. 511–516 (2021)

18. Onesimu, J.A., Karthikeyan, J., Eunice, J., Pomplun, M., Dang, H.: Privacy preserving attribute-focused anonymization scheme for healthcare data publishing. IEEE Access **10**, 86979–86997 (2022)

19. Parimanam, K., Lakshmanan, L., Palaniswamy, T.: Hybrid optimization based learning technique for multi-disease analytics from healthcare big data using optimal pre-processing, clustering and classifier. Concurr. Comput. Pract. Exp. **34**(17), e6986 (2022)

20. Patil, H.K., Seshadri, R.: Big data security and privacy issues in healthcare. In: IEEE Congress on Big Data, 2014, pp. 762–765. IEEE Computer Society (2014)

21. QUALITOP: The QUALITOP project (2023). https://h2020qualitop.liris.cnrs.fr/wordpress/index.php/project/

22. Singh, S., Rathore, S., Alfarraj, O., Tolba, A., Yoon, B.: A framework for privacy-preservation of IoT healthcare data using federated learning and blockchain technology. Future Gener. Comput. Syst. **129**, 380–388 (2022)

23. Som, R.K.: Practical sampling techniques. CRC Press, Boca Raton (1995)

24. Sun, J., Reddy, C.K.: Big data analytics for healthcare. In: ACM SIGKDD KDD, 2013, p. 1525. ACM (2013)

25. UCI: Breast cancer data set (2023). https://archive.ics.uci.edu/ml/datasets/breast+cancer

26. Wang, L., Ma, Y., Yan, J., Chang, V., Zomaya, A.Y.: pipsCloud: high performance cloud computing for remote sensing big data management and processing. Future Gener. Comput. Syst. **78**, 353–368 (2018)

Beyond Traditional Flare Forecasting: A Data-driven Labeling Approach for High-fidelity Predictions

Jinsu Hong[(✉)][ID], Anli Ji[ID], Chetraj Pandey[ID], and Berkay Aydin[ID]

Georgia State University, Atlanta, GA 30303, USA
{jhong36,aji1,cpandey1,baydin2}@gsu.edu

Abstract. Solar flare prediction is a central problem in space weather forecasting. Existing solar flare prediction tools are mainly dependent on the GOES classification system, and models commonly use a proxy of maximum (peak) X-ray flux measurement over a particular prediction window to label instances. However, the background X-ray flux dramatically fluctuates over a solar cycle and often misleads both flare detection and flare prediction models during solar minimum, leading to an increase in false alarms. We aim to enhance the accuracy of flare prediction methods by introducing novel labeling regimes that integrate relative increases and cumulative measurements over prediction windows. Our results show that the data-driven labels can offer more precise prediction capabilities and complement the existing efforts.

Keywords: Solar Flares · Metadata · Space Weather Analytics

1 Introduction

Solar flares are one of the most pivotal space weather events that can have a significant influence on Earth and the near-Earth environment when accompanied by other eruptive phenomena such as coronal mass ejections. Some effects include but are not limited to power grid outages, disruption of navigation and positioning satellites, and increased radiation levels at high altitudes or space missions [5]. Thus, a robust prediction of solar flares and the accompanying events is essential to alert and prevent catastrophic impacts on Earth.

In flare forecasting literature, there are two main approaches: active region (AR) based models and full-disk models. The AR-based models are often formulated as image or time series classification [8] and each flare is associated with one active region. Full-disk solar flare prediction models [9] are usually framed as image classification problems and make use of all flares regardless of the AR associations. Solar flares, commonly detected using X-ray flux data from Geostationary Operational Environmental Satellites (GOES), are logarithmically categorized into five major classes (X, M, C, B, or A) by their peak X-ray flux measured in the 1–8Å passbands [4]. In traditional solar flare prediction tasks (be it active region-based or full-disk), time series data or solar full-disk images are typically labeled based on the maximum intensity of the flares. However,

R. Wrembel et al. (Eds.): DaWaK 2023, LNCS 14148, pp. 380–385, 2023.
https://doi.org/10.1007/978-3-031-39831-5_34

these labels have limited ability to reflect the solar cycle, despite some existing extensions and alternatives [11,12].

The flare prediction models attempt to predict the occurrence of a pre-defined 'strong' flaring event within a prediction window (usually 12, 24, or 48h). The flare with maximum intensity (i.e., X-ray flux) in the prediction window is used for labeling the instances. However, there are three different limitations in the flare detection and labeling methods. First, quantifying the magnitude of X-ray flux for each AR is not feasible because X-ray flux measurements are global [4]. These global X-ray flux values can be misleading as they do not accurately represent the emitted radiation from individual ARs. Second, relying solely on the maximum intensity of a flare in the prediction window, however, disregards the background X-ray flux, fluence (integrated flux), or remaining flares' information. Lastly, the use of arbitrary thresholds for binary labels, such as >M1.0 or >C5.0, can further reduce the generalization capabilities of prediction models [10]. In this paper, we propose a more comprehensive approach that considers background conditions and cumulative indices to enhance solar flare prediction.

The rest of the paper is organized as follows: In Sect. 2, we describe the generation of new solar flare labels. In Sect. 3, we present a case study that reveals the feasibility of using these new labels in prediction tasks. Finally, in Sect. 4, we present our conclusion with future work.

2 Methodology

2.1 Relative Increase of Background X-ray Flux

To address the limitations of the existing label based on GOES classification, we propose a new methodology for generating labels based on the relative increase of background X-ray flux and complementary cumulative indices. We utilize 1-minute averaged GOES X-ray flux data to generate the labels for solar flares based on relative X-ray flux increase (referred to as $rxfi$). To ensure accuracy in determining $rxfi$, we first establish the background X-ray flux for a specific flare. We consider the 24-hour period prior to the start of the flare and exclude certain intervals from the background X-ray flux calculation. These exclusions include: (1) periods between the start and end times of other flares, (2) X-ray flux measurements that exceed the initial X-ray flux value at the start time of the flare of interest, and (3) measurements identified as low-quality by the instrument. After applying these filtering steps, we calculate the background X-ray flux by averaging the valid X-ray flux values within the specified period. Finally, the $rxfi$ value is obtained by dividing the peak X-ray flux by the background X-ray flux.

We provide a practical example of how we calculate the $rxfi$ for the M1.5 flare on 2015-09-20 in Fig. 1, which shows the X-ray flux from 2015-09-19T04:55 to 2015-09-21T04:55. Measurements from 2015-09-19T04:55 to 2015-09-20T04:55 are used for calculating the background X-ray flux for the M1.5 flare. The orange dotted line displays the filtered X-ray flux obtained using Cases 1 and 2. The blue line shows the cleaned/calibrated background X-ray flux. In this example, we obtain a background X-ray flux of $3.74 \times 10^{-7} W m^{-2}$ for the M1.5 flare.

Since the peak X-ray flux of the M1.5 flare is $1.5 \times 10^{-5} Wm^{-2}$, we find $rxfi = \frac{1.5 \times 10^{-5}}{3.74 \times 10^{-7}} = 40.11$.

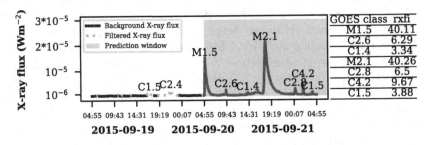

Fig. 1. An illustration of GOES 1–8 Å Solar X-ray flux observation.

2.2 Data-driven Labeling for Solar Flares

We utilize aggregated indices for our data-driven labels, which are as follows: (1) GC^{Max}: the GOES class of the maximum intensity flare in a given prediction window (24h in this work), (2) $rxfi^{Max}$: the flare with the maximum $rxfi$ value in the prediction window, (3) GC^{Σ}: the weighted sum of the GOES subclass values in the prediction window, calculated as $\sum C_i + 10 \times \sum M_j + 100 \times \sum X_k$, and (4) $rxfi^{\Sigma}$: the sum of the $rxfi$ values in the prediction window.

The new indices are obtained for both individual ARs and full-disk. In the AR-based approach, for a given time point (t_i), the procedure checks whether there exists a set of flares in the prediction window (in $[t_i, t_i + 24h]$). Based on the given set of flares, each time point (which is associated with an instance) is labeled as maximum or cumulative indices. Note that for the time series-based flare classifiers we trained for our case study (presented in Sect. 3), we use an observation window of 12h (meaning, for a time point t_i, multivariate time series instances are obtained from $[t_i - 12h, t_i]$).

To better describe our new labels, we present an example of a prediction window in Fig. 1, which shows seven flares from 2015-09-20T04:55 to 2015-09-21T04:55. We create three new labels in this example: In the case of $rxfi^{Max}$, the value of 40.26 (from M2.1 flare) will be the label, as it is the highest. For the GC^{Σ}, there are two M class flares and five C-class flares; therefore, the index for the GC^{Σ} is 48.5 ($(1.5 + 2.1) \times 10 + (2.6 + 1.4 + 2.8 + 4.2 + 1.5) = 48.5$). Lastly, the $rxfi^{\Sigma}$ over this time period is calculated as 110.05 (40.11 + 6.29 + 3.34 + 40.26 + 6.5 + 9.67 + 3.88). These three different labels are used for a time series of the 12-hour observations or a solar full-disk image at a specific time point. The sliced time series dataset and full-disk labels can be found in the data repository [6].

3 Case Study: Flare Prediction with Data-Drive Labels

3.1 Data Collection

Our baseline is active region-based and implemented by using the SWAN-SF [2], which is a multivariate time series dataset comprising 24 magnetic field parameters, covering from 2010 to 2018. For the purpose of solar flare prediction in our case study, we utilized six magnetic field parameters studied in [7]. We partitioned our data using the tri-monthly partitioning introduced in [10] and further split the SWAN-SF into four partitions where each one covers three months of data over the entire given dataset; i.e., Partition 1 covers instances from January to March, Partition 2 from April to June, Partition 3 from July to September, and Partition 4 from October to December. In this work, we used Partition 4 for testing and the rest of the three partitions for training.

3.2 Classification Method and Evaluation

In our case of analyzing multivariate time series from active region patches, we utilized a time series forest (TSF) with column ensemble technique to work with multiple parameters given [3]. The outputs of classifiers from each separate parameter are then aggregated to form a final prediction using equal voting. The other notable hyperparameters are as follows: number of estimators is set to 50 and maximum tree depth is set to 3. More details on our implementation can be found in the project repository [1].

Utilizing data-driven labels: We utilized four different AR-based indices to label our instances: GC^{Max}, $rxfi^{Max}$, GC^{Σ}, and $rxfi^{\Sigma}$. For instances labeled with a GC^{Max} of M- or X-class flares, we designated them as flaring instances, while instances labeled with B- or C-class flares or flare quiet regions were designated as non-flaring instances. To label instances with the $rxfi^{Max}$, we assigned each instance a tentative numeric label and discretized them using predefined thresholds, where instances with a $rxfi$ greater than the threshold were considered flaring, and those below the threshold were considered non-flaring. We also set thresholds for the GC^{Σ} and $rxfi^{\Sigma}$. This resulted in the creation of 19 thresholds (from 10 to 100 with a step size of 5) for binary classification and four different prediction tasks.

Model Evaluation: To evaluate the models, we utilized a binary confusion matrix and used popular forecast skill scores in solar flare prediction: these are True Skill Statistics (TSS) and Heidke Skill Score (HSS) [8]. We set nine different class weights (i.e., 1:10, 1:15, ..., and 1:50) for each label type to counter the class imbalance issue and report our results.

As depicted in Fig. 2, we present a comparison of models with four distinct labels: GC^{Max} with a threshold of M1.0, $rxfi^{Max}$ with a threshold of 45, GC^{Σ} with a threshold of 20, and $rxfi^{\Sigma}$ with a threshold of 45. Each of the three new labels has 19 models trained with 19 different thresholds ranging from 10 to 100 with a step size of 5. However, for brevity, we report only one result for each. Additional results from the models, along with saved candidate models, can be found in our project repository [1]. We also note that the TSF models are treated as black-box ensembles in the scope of this paper, and individual results from univariate time series classifiers are not independently reported.

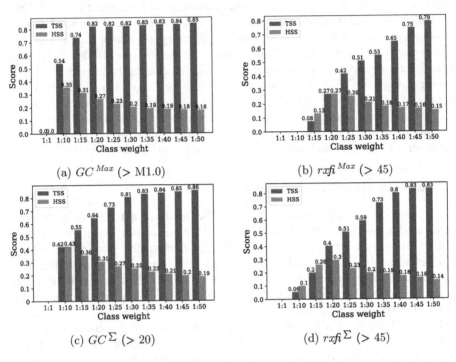

(a) GC^{Max} ($>$ M1.0)   (b) $rxfi^{Max}$ ($>$ 45)

(c) GC^{Σ} ($>$ 20)   (d) $rxfi^{\Sigma}$ ($>$ 45)

Fig. 2. The forecast skill scores for AR-based flare prediction models trained using instances labeled with data-driven labels and four distinct thresholds

Remarks: As the class weights increase, the trends of TSS and HSS scores vary. This is because the number of true positives and false positives increases. The models trained with new data-driven labels have similar results to the model trained with the GC^{Max}. Importantly, direct comparison between the labels may not be appropriate due to their inherent differences, such as varying imbalance ratios, background X-ray flux, and thresholds. However, it is anticipated that the $rxfi$ would reflect the solar cycle, which influences the background X-ray flux. Models utilizing $rxfi^{Max}$ are intrinsically less prone to false alarms during solar minima. Therefore, Max $rxfi$ can be served as an alternative label due to the high degree of flexibility in selecting thresholds and superior performance during solar minima. The evaluation metrics used in this study indicate that the proposed labels can enhance the performance of flare forecasting models beyond random guessing. Moreover, with the optimal threshold and class weights, the models trained with the new labels have the potential to outperform the existing labeling techniques.

4   Conclusion and Future Work

In this work, we have generated a new collection of flare labels (i.e., $rxfi^{Max}$, GC^{Σ}, $rxfi^{\Sigma}$) that can be used to complement the existing labeling techniques in space weather forecasting, specifically in solar flare prediction, for both active region-based and full-disk classification. We have presented a preliminary study evaluating the feasibility of using these new labels for active region-based prediction. Our results indicate that the proposed labels can serve as valuable additions to the existing labeling techniques, and their combination can improve the capability of solar flare prediction. We plan to extend this work to test different discretization thresholds and apply them to full-disk flare prediction models.

Acknowledgements. This work is supported in part under two grants from NSF (Award #2104004) and NASA (SWR2O2R Grant #80NSSC22K0272).

References

1. Source code. https://bitbucket.org/gsudmlab/data_driven_labels/src/main/
2. Angryk, R.A., et al.: Multivariate time series dataset for space weather data analytics. Sci. Data **7**(1), 227 (2020)
3. Deng, H., Runger, G., Tuv, E., Vladimir, M.: A time series forest for classification and feature extraction. Inf. Sci. **239**, 142–153 (2013)
4. Fletcher, L., et al.: An observational overview of solar flares. Space Sci. Rev. **159**(1–4), 19–106 (2011)
5. Fry, E.K.: The risks and impacts of space weather: policy recommendations and initiatives. Space Policy **28**(3), 180–184 (2012)
6. Hong, J., Ji, A., Pandey, C., Aydin, B.: A data-driven Labels for solar flare predictions (2023). https://doi.org/10.7910/DVN/1U2Q3C
7. Ji, A., Arya, A., Kempton, D., Angryk, R., Georgoulis, M.K., Aydin, B.: A modular approach to building solar energetic particle event forecasting systems. In: 2021 IEEE Third International Conference on Cognitive Machine Intelligence (CogMI), pp. 106–115. IEEE (2021)
8. Ji, A., Aydin, B., Georgoulis, M.K., Angryk, R.: All-clear flare prediction using interval-based time series classifiers. In: 2020 IEEE International Conference on Big Data (Big Data), pp. 4218–4225. IEEE (2020)
9. Pandey, C., Angryk, R.A., Aydin, B.: Solar flare forecasting with deep neural networks using compressed full-disk HMI magnetograms. In: 2021 IEEE International Conference on Big Data (Big Data), pp. 1725–1730. IEEE (2021)
10. Pandey, C., Angryk, R.A., Aydin, B.: Deep neural networks based solar flare prediction using compressed full-disk line-of-sight magnetograms. In: Lossio-Ventura, J.A., et al. (eds.) SIMBig 2021. CCIS, vol. 1577, pp. 380–396. Springer, Cham (2022). https://doi.org/10.1007/978-3-031-04447-2_26
11. Ursi, A., et al.: The first agile solar flare catalog (2023)
12. Zhang, H., et al.: Solar flare index prediction using SDO/HMI vector magnetic data products with statistical and machine-learning methods. The ApJ Suppl. Ser. **263**(2), 28 (2022)

HKS: Efficient Data Partitioning
for Stateful Streaming

Adeel Aslam$^{(\boxtimes)}$ ⓘ, Giovanni Simonini ⓘ, Luca Gagliardelli ⓘ,
Angelo Mozzillo ⓘ, and Sonia Bergamaschi ⓘ

Università degli Studi di Modena e Reggio Emilia, Modena, Italy
{adeel.aslam,giovanni.simonini,luca.gagliardelli,angelo.mozzillo,
sonia.bergamaschi}@unimore.it

Abstract. Data partitioning among processing instances of *distributed stream processing systems* (DSPSs) plays a significant role in the performance of overall stream processing. Several data partitioning schemes, including round-robin and hash-based key-splitting strategies, are employed in this context. However, stateful operations introduce challenges such as data aggregation overhead and load imbalance among processing instances due to the skewed nature of real data. In this paper, we propose a partitioning strategy (HKS) that considers the popularity of the tuples on the fly and partitions them according to their frequency: higher frequent tuples are routed by employing power-of-the-two-choices, whereas low ones by using a single hash function. We perform a comprehensive experimental evaluation on synthetic and real-world data sets on well-known Apache Storm DSPS. Results demonstrate the superior performance of the HKS against state-of-the-art data partitioning schemes in terms of load imbalance and aggregation cost.

Keywords: Data partitioning · Stream processing · Load imbalance

1 Introduction

The distributed and volatile characteristics of continuous data pose significant challenges when attempting to process it efficiently using a single machine, particularly in terms of achieving high throughput and low latency [1]. For this reason, *Distributed Stream Processing Systems* (DSPSs) are widely employed such as Storm, Flink, Spark Streaming, and Heron that employ clusters of commodity hardware to process the incoming stream data in a parallel and distributed fashion.

Typically, an application is submitted to a DSPS in the form of a Directed Acyclic Graph (DAG), in which each node represents a processing logic executed in parallel on processing elements, whereas edges depict the data flow among the nodes [2]. The most challenging task in this scenario is performing data partitioning among the computational instances for different data operations. Ideally,

Department of Information and Communication Technology, DBGROUP.

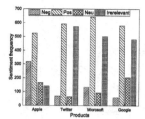 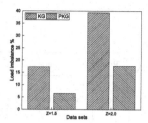 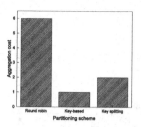

Fig. 1. Keys frequency   **Fig. 2.** Load imbalance   **Fig. 3.** Aggregation cost

we would like to distribute the computation evenly among all the processing elements, however, stateful computation needs aggregation of partial results that create a huge overhead of tuples accumulation from sub-processing instances and require another level of reducer for completion of computation. Similarly, the hash-based approaches pose challenges of load imbalance under the presence of skewed data distribution [4].

We perform a series of empirical observations on real-world and synthetic data. Initially, we analyze the open-source Twitter sentiment corpus[1]. Figure 1 depicts the uneven distribution of data with some sentiments having significantly higher frequencies than alternatives and exhibits skewness that raises the challenge of load imbalance for key-based data partitioning schemes as depicted by Fig. 2 [5,6]. Similarly, Fig. 3 shows the aggregation cost difference for partitioning schemes. We conduct this experiment against synthetic skewed zipf (Z=1.5) data where the result depicts that each tuple need 6 out of 10 processing tasks to aggregate the results for non-key based scheme.

In this study, we propose a novel data partitioning strategy that considers the frequency of the keys on the fly and routes the hotkeys to two processing elements. Similarly, all cold keys use a hash grouping strategy. In this way, the load imbalance generated due to the hotkeys is reduced by routing them to two processing tasks, and extra aggregation overhead is reduced for stateful data processing. We named it HKS[2] due to the augmentation of single and two hash-based (a hybrid of key and partial key grouping) approaches [1,6].

2   Related Work and Preliminaries

Toshniwal et al. [1] discuss the state-of-the-art data partitioning scheme. It states that KG is designed for stateful data processing, where key-grouping routes the tuples to a single processing element depending on the hash computation. Many studies propose key splitting approaches that use more than one worker process to downstream the tuples for minimizing the load generated due to highly frequent keys [4–6,8,9]. However, these approaches also raise aggregation overhead and require another level of reducer for the accumulation of partial

---

[1] https://github.com/zfz/twitter_corpus.

[2] https://github.com/AdeelAslamUnimore/StreamInequalityJoinSTA.

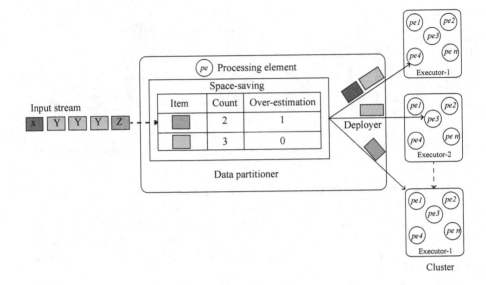

Fig. 4. HKS: Data partitioner for DSPS

results. According to these studies, it can be argued that the use of a small weight key predictive algorithm helps us to remove the load imbalance generated due to frequent keys.

A stream processing $S = \{k_1, k_2, k_3....\}$ is a continuous flow of data that requires run-time analysis where k_i represents a single tuple. An application to the DSPS composes of DAG $G = (V, E)$, where nodes (V) are data operators and edges (E) are the data flow among processing elements. A node (V) consists of set of processing elements $PE = \{pe_1, pe_2, pe_3....pe_n\}$ that perform the actual business logic. The data flow from upstream instances of V_i to the downstream instances of V_j uses a data partitioning strategy.

3   Frequency-aware Partitioner

The proposed data partitioner consists of two modules: predictive key analyzer (space-saving algorithm) [7] and data partitioner (deployer). The key predictor forecasts the frequency of tuples on the fly and routes the tuples to the downstream processing instances depending on the prediction results as depicted by Fig. 4. Moreover, the other module uses the prediction result for tuple partitioning depending on key frequencies. We name this predictive data partitioning strategy a hybrid key strategy (HKS).

3.1   Predictive Key Analyzer

The predictive key analyzer module employed the *space-saving* (SS) algorithm to identify the most frequent tuples from the streaming data [7]. This space-efficient *space-saving* (SS) algorithm lies after the upstream instances that capture and classify the hot and cold items on the fly as shown in Fig. 4. Each time a new item arrives in the stream, the SS algorithm updates it structure by incrementing the frequency count of the corresponding item, or by replacing a less frequent item with the new item if the size of the data structure is already full. These classified tuples are then routed to downstream instances with an additional flag indicating their frequency count, allowing downstream processing tasks to treat differently on identified tuples.

3.2 Data Partitioner

The second component of our proposed HKS is a data partitioner that takes the classified tuples from the predictive analysis component for data distribution. All hotkeys in a stream are forwarded to the two downstream instances using a popular power-of-two-choices strategy, where input hotkeys are equally assigned to the processing elements ensuring the balanced distribution of workload for highly frequent items. Similarly, cold items use a single hash function for the selection of downstream instances as depicted by Fig. 4.

4 Evaluation

To evaluate the proposed hybrid data partitioner for DSPS, we deployed it over the well-known Apache Storm [3]. The Storm cluster contains 10 machines with heterogeneous resources with services such as Kafka stream (2 nodes), zookeeper (2 nodes; 1 for storm cluster, 1 for Kafka cluster), and storm cluster with nimbus (master), and supervisor (5 slaves). Moreover, we compare the proposed HKS against round-robin (SG) [1], hash-based (KG) [1], power-of-two-choices (PKG) [6], and direct choices (DC) [5]. We use two distinct datasets BLOND[3] and synthetic data with 50 and 100 processing elements.

4.1 Performance Metrics

For evaluation, we consider the load imbalance among downstream processing tasks, aggregation overhead of tuples, throughput, and latency as performance metrics. Moreover, we only compare the proposed HKS scheme with the most relevant state-of-the-art data partitioning scheme. For example, considering load imbalance the round-robin grouping distributes the load among all processing elements equally and generates no imbalance, however, hash-based grouping cause imbalance due to the presence of skew in data distribution.

4.2 Results

Figure 5 depicts that for 50 PEs the hotkey tuple aggregation for SG is 48, and 40 for DC, while a proposed HKS has only 3 and has less aggregation overhead.

---

[3] https://www.nature.com/articles/sdata201848.

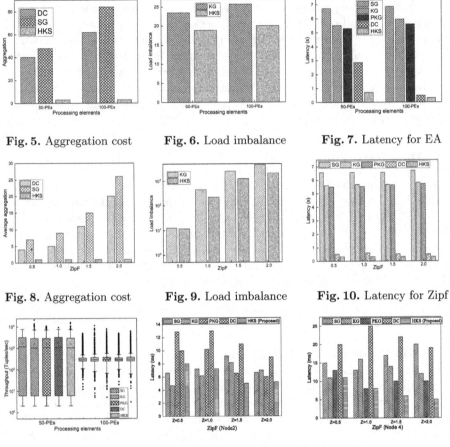

Fig. 5. Aggregation cost   **Fig. 6.** Load imbalance   **Fig. 7.** Latency for EA

Fig. 8. Aggregation cost   **Fig. 9.** Load imbalance   **Fig. 10.** Latency for Zipf

Fig. 11. Throughput for EA   **Fig. 12.** Latency with Node=2   **Fig. 13.** Latency with Node=4

Similar performance is noticed for 100 PEs (65 and 83). Figure 6 depicts 1.3x and 1.4x has a lower load imbalance of HKS than hash-based KG. Moreover, Fig. 7 depicts the latency comparison for partitioning strategies. Results depict that the proposed HKS has performed 6x, 4.4x, 4.1x, and 2.3x better than SG, KG, PKG, and DC for stateful computation.

Figures 8 and 9 show the result of average aggregation overhead and load imbalance for proposed HKS against state-of-the-art approaches for the synthetic zip$f$ data set against 100 PEs. The result depicts the superior performance of HKS against other state-of-the-art schemes. Figure 10 shows the latency measurement for varying zip$f$ datasets. Results of the figure depict that the proposed HKS has lower average processing latency than alternatives for every zip$f$ case. Figure 11 describes the tuple throughput for electric appliances; where it is depicted that the proposed HKS has slightly better performance than alternatives. Similarly, Fig. 12 and Fig. 13 shows the superior performance of latency for proposed HKS than alternatives with different cluster node sizes. HKS considers the hotness of the tuple where highly frequent tuples are routed to two PEs that decreases the load imbalance, moreover, non-hotkeys use a single hash to avoid aggregation overhead.

5   Conclusion

Data partitioning plays an important role in the performance of overall stream processing. Numerous challenges exist during the partitioning of data for stateful data computation, such as load imbalance and tuple aggregation overhead during the presence of skew and more distinct tuples. In this study, we propose an HKS stream-aware solution that monitors the frequency and routes the hot tuple using PKG and non-hot keys using KG. Experimental results prove that the proposed approach helps to reduce imbalance and less aggregation overhead that ultimately leads to better performance.

References

1. Toshniwal, A., et al.: Storm@twitter. In: Proceedings of the International Conference on Management of Data (SIGMOD), pp. 147–156, ACM, Snowbird, Utah, USA (2014)
2. Liu, X., Buyya, R.: Resource management and scheduling in distributed stream processing systems: a taxonomy, review, and future directions. ACM Comput. Surv. (CSUR) **53**(3), 1–41 (2020)
3. Apache Storm. https://storm.apache.org/. Accessed 4 Jan 2023
4. Zapridou, E., Mytilinis, I., Ailamaki, A.: Dalton: learned Partitioning for distributed data streams. Proc. VLDB Endowment **16**(3), 491–504 (2022)
5. Nasir, M.A.U., Garg, S., Agrawal, A., Balazinska, M., Howe, B.: When two choices are not enough: balancing at scale in distributed stream processing. In: Proceedings of the International Conference on Data Engineering (ICDE), pp. 589–600, Helsinki, Finland (2016)
6. Nasir, M.A.U., Garg, S., Agrawal, A., Balazinska, M., Howe, B.: The power of both choices: practical load balancing for distributed stream processing engines. In: Proceedings of the International Conference on Data Engineering (ICDE), pp. 137–148, Seoul, South Korea (2015)
7. Metwally, A., Agrawal, D., Abbadi, A.E.: An integrated efficient solution for computing frequent and top-k elements in data streams. ACM Trans. Database Syst. (TODS) **31**(3), 1095–1133 (2006)
8. Gedik, B.: Partitioning functions for stateful data parallelism in stream processing. VLDB J. **23**(4), 517–539 (2014)
9. Abdelhamid, A.S., Aref, W.G.: PartLy: learning data partitioning for distributed data stream processing. In: Proceedings of the Third International Workshop on Exploiting Artificial Intelligence Techniques for Data Management, pp. 1–4 (2020)

A Fine-Grained Structural Partitioning Approach to Graph Compression

François Pitois[1,2]([✉]) [ID], Hamida Seba[1] [ID], and Mohammed Haddad[1] [ID]

[1] Université Lyon, UCBL, CNRS, INSA Lyon, LIRIS, UMR 5205,
69622 Villeurbanne, France
{Francois.pitois,hamida.seba,mohammed.haddad}@univ-lyon1.fr
[2] LIB, Université de Bourgogne, Dijon, EA 7534, France

Abstract. To compress a graph, some methods rely on finding highly compressible structures, such as very dense subgraphs, and encode a graph by listing these structures compressed. However, structures can overlap, leading to encoding the same information multiple times. The method we propose deals with this issue, by identifying overlaps and encoding them only once. We have tested our method on various real-world graphs. The obtained results show that our approach is efficient and outperforms state of the art methods. The source code of our algorithms, together with some sample input instances, are available at https://gitlab.liris.cnrs.fr/fpitois/fgsp.git.

Keywords: Graph compression · graph summarising · graph partitioning · graph mining

1 Introduction

Graphs are ubiquitous. Also called networks, they are found in several real-world applications and services: social networks, road networks, biological structures, etc. In most of these applications, the considered graphs are voluminous or are growing significantly or even exponentially. This motivated several works that aim to compress or simplify graph data in order to simply save storage space or to make the study and analysis of these graphs easier.

One of the most used methods in graph compression consists to look for structures in the graph and to encode them. However, existing methods consider these structures at a high level ignoring overlaps [7,12,14]. This leads to encoding multiple times the overlapping parts of the considered structures, which is redundant. We propose to deal with this issue. We first detect some structures in the graph, and we deduce a partition of the vertices that coincides with them. Then, for each pair of parts, we will encode the edges present between these two parts. Such an object is called a *block*. With this encoding, we ensure that each edge is encoded only once.

The remainder of the paper is organised as follows: Sect. 2 describes existing methods for graph compression and analyses them. Then, in Sect. 3, we present

R. Wrembel et al. (Eds.): DaWaK 2023, LNCS 14148, pp. 392–397, 2023.
https://doi.org/10.1007/978-3-031-39831-5_36

our main contribution. Finally, in Sect. 4, we presents the experiments we undertook to evaluate our approach.

2   Related Work

There are many graph compression algorithms in the literature and many surveys are devoted to them [1,9]. Most of graph compression methods are based on finding regularities in the graph [6] like particular structures that are easy to compress, such as triangle [8], cliques or quasi-cliques. The approaches differ on how these structures are found in the graph and how they are encoded within the computed summary. In [11], the authors propose a summarizing approach that iteratively aggregates similar nodes, i.e., those that share the largest number of common neighbours. At each iteration, the selected nodes to be aggregated are validated by an objective function that represents the cost of the compressed output using Minimum Description Length (MDL). The graph is encoded with a two-part representation containing a summary S and a set of correcting edges C. In [7], the authors consider a predefined set of structures, mostly found in real-world graphs, namely stars, bipartite graphs, cliques, and paths, and propose to construct a succinct description of a graph in these "vocabulary" terms. They first mine candidate subgraphs using graph partitioning and community detection algorithms such as Metis [5] and Louvain [3], and then, they identify the optimal summarization using MDL principle. In [14], the authors use only one structure, the clique, and allow a small fraction of them to overlap. In [12], the authors use a clustering algorithm to partition the original set of vertices into a number of clusters, which will be super-nodes connected by super-edges to form a complete weighted graph. In [2], the authors use MDL to measure motif relevance based on motif capacity to compress a graph. However, contrarily to the work of [7], the motifs are not predefined structures. In [10], the authors extend the approach proposed in [7] by dealing with overlapping structures. This is achieved by considering a matrix of overlaps as a heuristic to choose the retained structures and divide the graph G into two graphs G_S and G_E, then encode these two graphs. One of the strengths of this approach is that it can select overlapping structures. However, whenever two structures overlap, this method encodes the overlapping part twice, which is redundant.

3   Proposed Approach

3.1   Preliminaries

A graph G is defined as a two-set structure $G = (V, E)$ where V is a set of vertices, also called nodes, and E is a set of edges. Let $G = (V, E)$ be an undirected graph without self-loops. Let $G = (V, E)$ be a graph, and let $S_1, S_2 \subset V$ be two subset of vertices of G. Then the induced subgraph $G[S_1, S_2]$ is the graph whose vertex set is $S_1 \cup S_2$ and whose edge set consists of all of the edges in E that have one endpoint in S_1 and the other in S_2. A *structure* is a subgraph of G of the form $G[S_1, S_2]$, with S_1 and S_2 either equal or disjoint. A lot of algorithms are dedicated to find structures that have some general properties or shaped, such as quasi-clique, star or dense subgraph detection algorithm.

3.2   Principle

Let $\mathcal{P}$ be a partitioning of the vertex set of G. Let p_i, p_j be two parts of $\mathcal{P}$. We call the subgraph $G[p_i, p_j]$ a *block* of G. A decomposition of a graph into blocks is depicted in Fig. 1. The idea of our compression scheme is to find a vertex partition of G that matches structures found in G. The main problem with structures of a graph is the fact that they can overlap. Hence, the goal is to break structures into pieces so that overlapping parts merge into one new

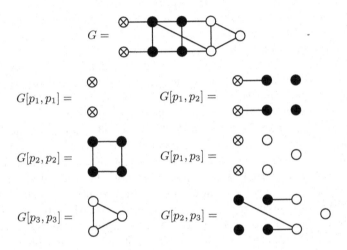

Fig. 1. Example of a partition of a graph as well as associated blocks. Vertices of graph G are partitioned into parts p_1 (drawn with crossed circles), p_2 (drawn with filled circles) and p_3 (drawn with empty circles).

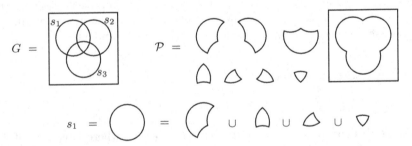

Fig. 2. Example of a partition induced by three structures. The graph G is represented by a square. There are three structures s_1, s_2 and s_3 that overlaps, represented by circles. It induces a partition $\mathcal{P}$ made of eight parts. Each structure can be expressed as a disjoint union of parts; the case of s_1 is shown.

part. Given a collection of structures of G, namely $\{s_1, \ldots, s_k\}$, one can derive a partition $\mathcal{P}$ of the vertex set of G as follows: two vertices u and v of G are in the same part if and only if they belong exactly to the same structures. An overview of this idea is depicted in Fig. 2. Let us denote this partition $\mathcal{P}(s_1, \ldots, s_k)$. To compress a graph using a partition, it suffices to encode each block induced by the partition. This encodes successfully every edge of G.

3.3 Algorithm

Our algorithm is divided in three parts. These parts are independently implemented, meaning that one can refine one part without impacting the others.

- Part 1: we run a number of algorithms that find structures in a graph. We decided to use SlashBurn [4], BigClam [15], Louvain [3], and Metis [5]. Any set of algorithm can be used here. These algorithms are mainly graph partitioning algorithms that find communities and more generally dense subgraphs in a given graph. This produces a collection of structures $s_1, \ldots, s_k$.
- Part 2: we use a greedy heuristic to find a partition $\mathcal{P}$. We start with a structure set $S = \emptyset$ and with the coarsest partition $\mathcal{P} = \{V(G)\}$. At each step, we choose greedily a structure s_i and we add it to S, such that an estimation of the encoding $\texttt{estimate}(G, \mathcal{P}(S))$ is minimal. We stop as soon as we cannot find a new structure s_i that makes $\texttt{estimate}(G, \mathcal{P}(S))$ decrease. Structures not chosen during this step are discarded and not used at all in the compressed representation.
- Part 3: we explicitly encode the graph G using the found partition $\mathcal{P}$.

Our code is available at https://gitlab.liris.cnrs.fr/fpitois/fgsp.git. The three parts are kept separated.

4 Experimental Results

We evaluated our approach, denoted FGSP (for Fine-Grained Structural Partitioning), on several real-world graphs taken from the Network Repository [13]. We compared our algorithm with GraphZIP [14]. Our algorithm was executed on a Intel Core laptop with i5 CPU, running using one core at 1.70 GHz. Our results are summarised in Table 1. In general, our approach raises results that are 5 to 10 times better, allowing for a better compression. The downside is time, as our algorithm can spend several hours compressing a graphs, whereas GraphZIP is always faster than a minute. This is not a problem, as the main focus of our algorithm is to provide a nice compression for very large graphs, despite the time needed to achieve it. Figure 3 shows adjacency matrices of two graphs at the end of the greedy heuristic (Part 2 of our algorithm).

Table 1. Comparison between **GraphZIP** and our algorithm **FGSP** (in Bytes). **Raw encoding** corresponds to the graph given as a list of edge.

Graph	Nodes	Edges	Raw encoding	GraphZip	FGSP
hamming8-2	256	31,616	226,252	110,838	713
ia-infect-dublin	410	2,765	20,597	13,287	1,420
web-google	1,299	2,773	22,977	12,216	1,597
ia-email-univ	1,133	5,451	41,802	32,405	4,949
web-BerkStan	12,305	19,500	199,261	135,813	15,694
ca-HepPh	11,204	117,619	1,178,601	498,661	73,958
ca-AstroPh	17,903	196,972	2,121,144	1,423,607	198,996
web-sk-2005	121,422	334,419	4,083,711	2,040,783	256,992
ca-MathSciNet	332,689	820,644	10,482,316	8,370,757	1,254,030

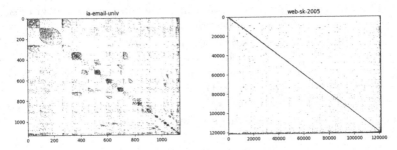

Fig. 3. Adjacency matrices of two graphs after executing FGSP.

Conclusion

In this paper, we proposed a new approach to enhance graph compression by structural partitioning. Our algorithm allows compressing graphs efficiently by obtaining a fairly satisfactory bit rate per edge. This work can be extended on several axes: First of all, the first part, which consists of launching structure mining algorithms, can surely be optimized and accelerated by better choosing the algorithms that generate these structures. We use currently 4 algorithms, i.e., SlashBurn [4], BigClam [15], Louvain [3], and Metis [5], but a more in-depth study need to be undertaken to choose the algorithm that provides the best structures or to design a structure mining algorithm specific to compression. Then, it will be very interesting to find other heuristics that could replace or enhance the greedy heuristic used to find the set of structures.

Acknowledgements. This work is funded by the French National Research Agency under grant ANR-20-CE23-0002.

References

1. Besta, M., Hoefler, T.: Survey and taxonomy of lossless graph compression and space-efficient graph representations. CoRR abs/1806.01799 (2018). http://arxiv.org/abs/1806.01799
2. Bloem, P., de Rooij, S.: Large-scale network motif analysis using compression. Data Min. Knowl. Discov. **34**(5), 1421–1453 (Sep 2020). https://doi.org/10.1007/s10618-020-00691-y
3. Blondel, V.D., Guillaume, J.L., Lambiotte, R., Lefebvre, E.: Fast unfolding of communities in large networks. J. Stat. Mech. Theor. Exper. **2008**(10), P10008 (2008)
4. Kang, U., Faloutsos, C.: Beyond 'caveman communities': hubs and spokes for graph compression and mining. In: Proceedings - IEEE International Conference on Data Mining, ICDM, pp. 300–309 (12 2011). https://doi.org/10.1109/ICDM.2011.26
5. Karypis, G., Aggarwal, R., Kumar, V., Shekhar, S.: Multilevel hypergraph partitioning: applications in VLSI domain. IEEE Transactions on Very Large Scale Integration (VLSI) Systems **7**(1), 69–79 (1999)
6. Kiouche, A.E., Baste, J., Haddad, M., Seba, H.: A neighborhood-preserving graph summarization. CoRR abs/2101.11559 (2021). https://arxiv.org/abs/2101.11559
7. Koutra, D., Kang, U., Vreeken, J., Faloutsos, C.: Summarizing and understanding large graphs. Stat. Anal. Data Min. ASA Data Sci. J. **8**(3), 183–202 (2015). https://doi.org/10.1002/sam.11267
8. Lagraa, S., Seba, H.: An efficient exact algorithm for triangle listing in large graphs. Data Min. Knowl. Discov. **30**(5), 1350–1369 (2016). https://doi.org/10.1007/s10618-016-0451-4
9. Liu, Y., Safavi, T., Dighe, A., Koutra, D.: Graph summarization methods and applications: a survey. ACM Comput. Surv. **51**(3) (2018). https://doi.org/10.1145/3186727
10. Liu, Y., Safavi, T., Shah, N., Koutra, D.: Reducing large graphs to small supergraphs: a unified approach. Soc. Netw. Anal. Min. **8**(1), 1–18 (2018). https://doi.org/10.1007/s13278-018-0491-4
11. Navlakha, S., Rastogi, R., Shrivastava, N.: Graph summarization with bounded error. In: Proceedings of the 2008 ACM SIGMOD International Conference on Management of Data, pp. 419–432. SIGMOD '08, Association for Computing Machinery, New York, NY, USA (2008). https://doi.org/10.1145/1376616.1376661
12. Riondato, M., García-Soriano, D., Bonchi, F.: Graph summarization with quality guarantees. Data Min. Knowl. Discov. **31**(2), 314–349 (2017). https://doi.org/10.1007/s10618-016-0468-8
13. Rossi, R.A., Ahmed, N.K.: The network data repository with interactive graph analytics and visualization. In: Proceedings of the Twenty-Ninth AAAI Conference on Artificial Intelligence (2015). http://networkrepository.com
14. Rossi, R.A., Zhou, R.: GraphZIP: a clique-based sparse graph compression method. J. Big Data **5**(1), 1–14 (2018)
15. Yang, J., Leskovec, J.: Overlapping community detection at scale: a nonnegative matrix factorization approach. In: Proceedings of the Sixth ACM International Conference on Web Search and Data Mining, pp. 587–596. ACM (2013)

Author Index

Printed in the United States
by Baker & Taylor Publisher Services

Printed in the United States
by Baker & Taylor Publisher Services